S0-BQM-175

Copper and Lymphomas

Authors

Martin J. Hrgovcic, M.D., D.Sc. (Med.), F.A.C.N.

Consultant in Internal Medicine and Oncology
Senior Partner, Diagnostic Clinic of Houston
Clinical Associate Internist (V)
Department of Internal Medicine
The University of Texas System Cancer Center
M. D. Anderson Hospital and Tumor Institute
and Clinical Associate
Department of Internal Medicine
The University of Texas Medical School
Houston, Texas

C. C. Shullenberger, M.D.

Internist and Emeritus Professor of Medicine
Department of Internal Medicine
The University of Texas System Cancer Center
M. D. Anderson Hospital and Tumor Institute
and Professor of Medicine
University of Texas Medical School at Houston
and Medical Director
Gulf Coast Regional Blood Center
Houston, Texas

CRC Press, Inc.
Boca Raton, Florida

Library of Congress Cataloging in Publication Data
Hrgovcic, Martin J., 1926-
Copper and lymphomas.

Bibliography: p.
Includes index.
1. Lymphoma--Diagnosis. 2. Copper--Diagnostic use.
3. Patient monitoring.I. Shullenberger, C. C.,
1914- . II. Title.
RC280.L9H73 1984 616.99'40756 83-14335
ISBN 0-8493-6331-4

This book represents information obtained from authentic and highly regarded sources. Reprinted material is quoted with permission, and sources are indicated. A wide variety of references are listed. Every reasonable effort has been made to give reliable data and information, but the author and the publisher cannot assume responsibility for the validity of all materials or for the consequences of their use.

Direct all inquiries to CRC Press, Inc., 2000 Corporate Blvd., N.W., Boca Raton, Florida, 33431.

International Standard Book Number 0-8493-6331-4

Library of Congress Card Number 83-14335
Printed in the United States

THE AUTHORS

Martin J. Hrgovcic, M.D., D.Sc. (Med.), F.A.C.N., is a Consultant in Internal Medicine and Oncology, Senior Partner of the Diagnostic Clinic of Houston, Clinical Associate Internist (V) in the Department of Internal Medicine, The University of Texas System Cancer Center, M. D. Anderson Hospital and Tumor Institute at Houston, and a Clinical Associate, Department of Internal Medicine, The University of Texas Medical School, Houston, Texas.

Dr. Hrgovcic obtained his medical training in the University of Zagreb Medical School, Zagreb, Croatia, Yugoslavia, received his M. D. degree in 1953 and became a Diplomate of the Board of Internal Medicine in 1961. He received his degree of Doctor of Medical Sciences from Zagreb University in 1968 for his dissertation entitled ''Clinical Significance of Serum Copper Levels in Hemoblastoses''. He served as an Instructor in Medicine in the University of Zagreb Medical School from 1961 to 1965, as a Clinical Fellow in the Department of Developmental Therapeutics, The University of Texas Cancer System Center, M. D. Anderson Hospital and Tumor Institute at Houston from 1965 to 1967, as Chairman of Chemotherapy in the Institute for Malignant Diseases, Zagreb, Croatia, Yugoslavia from 1967 to 1969. From 1970 through 1973, he was a Research and Faculty Associate in the Department of Medicine, Hematology Section, The University of Texas Cancer System Center, M. D. Anderson Hospital and Tumor Institute at Houston. In 1973, he went into private practice, joining the Diagnostic Clinic of Houston as a consultant in the fields of Oncology/Hematology and Internal Medicine.

Dr. Hrgovcic is a member of the American Society of Internal Medicine, the American Society of Clinical Oncology, the American Society of Hematology, the Texas Society of Internal Medicine, the Houston Society of Internal Medicine, the American Medical Association, the Texas Medical Association, and the Harris County Medical Society. He is a Fellow of the American College of Nutrition. He has received the Mike Hogg Scientific Paper Award for his research in ''Serum Copper Levels in Hematologic Malignant Diseases'' from the Division of Continuing Education, the University of Texas Graduate School of Biomedical Sciences at Houston in 1966.

Dr. Hrgovcic is the author or co-author of more than 30 publications and three chapters in medical text books. His major research interests have been in trace elements in malignant diseases, with particular attention to copper.

C. C. Shullenberger, A.B., M.D., is Professor of Medicine Emeritus in the Department of Internal Medicine, The University of Texas System Cancer Center, M. D. Anderson Hospital and Tumor Institute at Houston, Professor of Medicine in The University of Texas Medical School at Houston, and Medical Director of the Gulf Coast Regional Blood Center, Houston, Texas.

Dr. Shullenberger received the A.B. degree from Butler University in 1935, the M.D. degree from Indiana University in 1939, and became a diplomate of the American Board of Internal Medicine in 1950. He was an intern at the Indianapolis City Hospital from 1939 to 1941, a Fellow in Internal Medicine at the Mayo Clinic from 1941 to 1944 and from 1946 to 1948. He was Lieutenant, MC, USNR from 1944 to 1946. He served as First Assistant in Internal Medicine at the Mayo Clinic from 1948 to 1949. He joined the staff of the Department of Medicine, The University of Texas M. D. Anderson Hospital and Tumor Institute in 1949 and became Chief, Section of Hematology in 1954. He was Head of the Department of Medicine in that institution from 1970 to 1977. He became Medical Director of the Gulf Coast Regional Blood Center at its inception in 1975.

Dr. Shullenberger is a member of the American Medical Association, the Texas Medical Association, the Southern Medical Association, the American Society of Hematology, and the American Association of Blood Banks. He was the recipient of The Distinguished Service

Award of The University of Texas System Cancer Center, M. D. Anderson Hospital and Tumor Institute in 1982.

Dr. Shullenberger is author or co-author of more than 100 papers. His clinical and research interests have been in clinical hematology and blook banking.

CONTRIBUTORS

Ashok M. Balsaver, M.D.
Pathologist and Senior Partner
Diagnostic Clinic of Houston
and
Clinical Assistant Professor
Department of Pathology
Baylor University College of Medicine
Houston, Texas

Martin J. Hrgovcic, M.D., D.Sc. (Med.), F.A.C.N.
Consultant in Internal Medicine and Oncology
Senior Partner, Diagnostic Clinic of Houston
Clinical Associate Internist (V)
Department of Internal Medicine
The University of Texas System Cancer Center
M. D. Anderson Hospital and Tumor Institute
and
Clinical Associate
Department of Internal Medicine
The University of Texas Medical School
Houston, Texas

C. C. Shullenberger, M.D.
Internist and Emeritus Professor of Medicine
Department of Internal Medicine
The University of Texas System Cancer Center
M. D. Anderson Hospital and Tumor Institute
and
Professor of Medicine
University of Texas Medical School at Houston
and
Medical Director
Gulf Coast Regional Blood Center
Houston, Texas

ACKNOWLEDGMENTS

The authors are particularly indepted to Mrs. Ruzica Hrgovcic, medical biochemist, for the organization, method modification, and conduct of the serum copper determinations in the pilot study. Excellent technical assistance in the preparation of this volume was provided by Mr. Hrvoje Hrgovcic. Special thanks must also be given to Dr. James J. Butler for reviewing the chapter on pathology and classification of malignant lymphoma.

TABLE OF CONTENTS

Chapter 1

INTRODUCTION AND HISTORICAL REVIEW

The element copper is distributed widely in nature and has interested physicians and other scientists over many centuries. Copper was one of the earliest metals known to man and was first used around 8000 B.C. as a substitute for stone by Neolithic man. Copper utensils were found in Egyptian tombs dating back to 5000 B.C. Around 3500 B.C. copper was alloyed with tin, heralding the onset of the Bronze Age. The Romans acquired their copper from mines located on the island of Cyprus. It was known to Romans as *aes cyprium* or "metal of Cyprus". This was shortened to cyprium and, later, to cuprum.[1-3]

Copper salts were employed in the therapy of eye diseases in Babylon, Assyria, and Egypt.[4] Hippocrates and Galen, as well as Roman physicians, used copper in their prescriptions.[5]

The presence of copper in living organisms (plants and animals) was not identified until the beginning of the 19th century. Bucholtz[6] in 1816 and Meissner[7] in 1917 discovered copper in plants, and in 1830 Sarzeau[8] found ox blood to contain copper in levels estimated at 700 μg/ℓ. This value is close to the copper content determined by modern methods. The early investigators, however, believed that the copper of plants and animals represented accidental contamination from the soil.

The first indication that copper is an essential body constituent at least in some species, came with the recognition of the copper-containing respiratory pigment, hemocyanin, in cold-blooded animals in 1878[9] and the discovery of turacin, a copper-containing red pigment in the feathers of certain birds in 1869.[10,11] Despite continued reports that most animal fluids and tissues contained small amounts of copper, the metal was not accepted as a definite physiological constituent until the 20th century.

Hart et al.[12] and Elvehjem[13] recognized copper to be an essential nutrient for rats and other animal species. The recognition in the early 1930s of naturally occurring copper deficiency disorders in a number of domestic animals was a further important step toward the establishment of copper as an essential trace element. Cattle, sheep, and pigs feeding on plants grown on copper-deficient soil were reported to be affected in North America,[14] Europe,[15,16] and in Australia.[17]

Anemia has been observed in experimental copper deprivation in rats,[12] chickens,[18] and cattle;[19,20] the most extensive studies have been made in swine.[21,22] Hypochromic microcytic anemia is found in rats, rabbits, pigs, and lambs. In cattle and in ewes, anemia is hypochromic and macrocytic, and in chicks and dogs it is normochromic and normocytic. This defect appears to be a consequence of several abnormalities in iron metabolism, including impaired iron absorption, defective transfer of iron from reticuloendothelial cells and hepatocytes to plasma, and failure to the normoblast to utilize intracellular iron for hemoglobin synthesis.[22-25] In addition to anemia, signs of copper deficiency include abnormalities in the synthesis of elastin, leading to dissecting aneurysms and intramural hemorrhages affecting major blood vessels.[26-28] Abnormalities of color and character of hair and wool have been observed, as well as bone deformities.[29-31] In second-generation copper deficiency in sheep an extensive demyelinating neurologic disease known as "swayback" or enzootic neonatal ataxia occurs.[17,32-34] If this "enzootic ataxia" is treated with copper early enough, the ataxia is relieved. If copper insufficiency is not corrected, a diffuse symmetrical demyelination develops which is most marked in the corpus callosum, the internal capsules, the white matter of the occipital and frontal lobes, and the motor pathways of the spinal cord.[32-36]

The historical development of studies of the role of copper has strongly reflected the importance of copper deficiency as a cause of disease in domesticated animals. The early appreciation of the importance of copper in normal hemopoiesis was succeeded by the conclusion that copper deficiency anemia was likely to occur only rarely in humans. The awareness that copper deficiency was a cause of disease in farm livestock stimulated an increased investigation of the role of copper in human health and disease.

Until recently, the concept of copper deficiency in humans was not widely accepted. In spite of evidence dating back to early 1930[37] it was stated that copper deficiency could not exist in man.[38,39] Its development in severely malnourished, or small premature babies reared on milk-based low-copper diets have been reported.[40-44] Patients of all age groups on prolonged total parenteral nutrition without copper supplementation are likely to develop a copper deficiency.[45-50] Prolonged zinc administration can also result in a copper deficiency.[51]

Too much copper, on the other hand, produces an equally dramatic range of lesions in experimental animals. Both excesses and deficiencies of copper can be fatal. Copper toxicity occurs in all species but is a special problem in ruminant animals.[30,52,53] The consumption of plants with environmentally high copper content, and such husbandry practices as feeding copper supplements combined with the condition of restricted elimination of absorbed copper, have caused toxic reactions in ruminant animals.[52-57] Monogastric animals, such as swine and poultry, can tolerate much higher levels of dietary copper than ruminants before exhibiting toxic symptoms, provided that the diet also includes adequate zinc, iron, and protein.[30] The dietary level of naturally occurring or synthetic chelates can also be an important factor. In sheep and cattle, as well as in humans, the existence of liver copper values above very high critical levels is usually followed by catastrophic liberation of copper into the blood with resulting hemolytic jaundice, hemoglobinuria, liver and kidney failure, and early death.[30,58-61]

Wilson's classic description in 1912 of the disease which bears his name is probably the most significant event in the medical history of copper.[62] In 1962, Menkes et al.[63] described a syndrome which was characterized by slow growth, progressive cerebral degeneration, pili torti, and an x-linked recessive inheritance caused by the impaired transport of copper out of the intestinal wall. In Wilson's disease, there is an abnormal toxic accumulation of copper in various organs and tissues, whereas in Menkes' kinky hair syndrome there is a decreased serum, liver, and brain copper concentration, but normal copper levels in red blood cells.

There are three areas of primary consideration with regard to the relation of copper to health and disease. The first area is the obvious one in which either a deficiency or an excess may precipitate a disorder. In humans, genetically determined copper deficiency and copper toxicity are considered to be the cause of Menkes' kinky hair syndrome and Wilson's disease or hepatolenticular degeneration, respectively. The second area is the involvement of copper in many phases of metabolism as an intrinsic part of many vital enzymes, in transport mechanism, and in tissue synthesis, notably collagen, elastin, and bone formation. The third area, in which we are particularly interested, is the one in which measurable changes occur in the serum copper level in patients with lymphoproliferative malignant diseases to the extent that these changes can be used as sensitive biochemical markers of disease activity and the efficacy of therapy.

In this volume we will summarize the current concepts of copper metabolism and the processes that govern the absorption, transport, storage, subsequent utilization and distribution of copper, and its alterations in physiological and pathological conditions with special emphasis on malignant lymphoma. Our knowledge of the above processes and the biological roles of copper and its significance in human health and disease is not adequate to explain the basic control mechanisms which produce a sequential series of consistent alterations in the serum copper concentrations during the course of malignant lymphoma. Equally important would be to clearly define whether the observed serum copper and/or ceruloplasmin alterations have a purposeful role as a host defense mechanism or merely represent secondary pathophysiological consequences of a disease state. Without an understanding of these mechanisms and bridging the existing gaps, the clinician will continue to find it difficult to depend more on the serum copper level as a monitor of lymphomatous disease activity, predicing its exacerbations.

Because of space limitations and the voluminous literature of recent developments in the field of copper metabolism, some important lines of information must be omitted. Several thousand papers have been published on the chemistry and biology of ceruloplasmin alone and its studies have excited the imagination of many basic science researchers, as well as clinicians.[64] The literature review has been made selective, the basis of selection being malignant diseases. References to older literature can, in general, be found in the bibliographies of the excellent articles, reviews, symposia, and books quoted.[3,30,36,65-78] No attempt has been made to refer to all recent papers in the field of copper biochemistry and metabolism. While some references to the recent work on copper metabolism known to the reviewers will be included, emphasis will be upon the significance and usefulness of serum copper determinations in the management of patients with malignant lymphomatous disease in daily oncological practice observed by the authors as well as others.

Since this volume is primarily concerned with the clinical significance of serum copper levels in malignant lymphoproliferative diseases, normal copper metabolism and its alterations in physiological and pathological conditions will be included for a better understanding and proper interpretation of serum copper variation during the course of lymphomatous disease.

In spite of the steadily increasing number of reports during the last 2 decades concerning the clinical significance and usefulness of serum copper and/or ceruloplasmin alterations in malignant lymphomatous disease, this topic has never been addressed in any symposia,[73-75,77] reviews,[65-73] or books,[30,76,78] with the exception of our studies, some of them recently reviewed.[79] Therefore, in this volume we will summarize the current knowledge of copper homeostasis and the role of copper in malignant diseases with special emphasis to malignant lymphoma. Many important gaps in our knowledge of the biological roles of copper and its significance in disease states, particularly in malignant states, need to be bridged.

The review will first deal in general terms with methodology, copper biochemistry and metabolism, and copper homeostasis in the mammalian system with preference given to the observations made in man. Thereafter, normal and pathological conditions, particularly of malignant diseases associated with an alteration of serum and tissue copper, will be briefly reviewed. Finally, the authors will discuss their extensive clinical studies regarding serum copper determinations in patients with malignant myeloproliferative and lymphoproliferative disorders, and with particular attention to Hodgkin's and non-Hodgkin's lymphoma, with special emphasis to the relationship between the clinical state of disease activity and the serum copper level. Significant variations in this relationship, including interfering substances and conditions will be discussed. The histological classification, natural history of lymphoma and staging of disease, as well as some other involved factors influencing copper homeostasis will also be included because of their relationship to copper metabolism. Related studies in experimental animals supporting clinical observations will be discussed briefly and our thoughts on the clinical significance of serum copper and/or ceruloplasmin alterations in patients with malignant lymphoproliferative diseases will be outlined.

REFERENCES

1. **Sigerist, H. E.,** Primitive and archaic medicine, in *A History of Medicine,* Vol. 1, Oxford University Press, New York, 1951, 564.
2. **Sayers, R. R.,** *Copper and Health,* 3rd ed., Copper and Brass Association, New York, 1951.
3. **Sass-Kortsak, A.,** Copper metabolism, in *Advances in Clinical Chemistry,* Vol. 8, Sobotka, H. and Stewart, C. P., Eds., Academic Press, New York, 1965, 1-167.
4. **Adelstein, S. J. and Vallee, B. L.,** Copper metabolism in man, *N. Engl. J. Med.,* 265, 892, 1961; 265, 941, 1961.
5. **Karcioglu, Z. A. and Sarper, R. M.,** Preface, in *Zinc and Copper in Medicine,* Karcioglu, Z. A. and Sarper, R. M., Eds., Charles C Thomas, Springfield, Ill., 1980, 13.
6. **Bucholtz, C. F.,** Chemische Untersuchung der Vanillenschoten *(Silizua vanillae), Repert. Pharm.,* 2, 253, 1816.
7. **Meissner, W.,** Sur la presence du cuivre dans les cendres vegetaux, *Ann. Chim. Phys.,* 4, 106, 1817.
8. **Sarzeau, A.,** Sur la presence du cuivre dans les vegetaux et dans le sang, *J. Pharm. Sci. Accesoires,* 16, 505, 1830.
9. **Frederick, L.,** Recherches sur la physiologie du pouple commun *(Octopus vulgaris), Arch. Zool. Exp. Gen.,* 7, 535, 1878.
10. **Church, A. H.,** Researches on turacin, an animal pigment containing copper, *Philos. Trans. R. Soc.,* 159, 627, 1869.
11. **Church, A. H.,** Turacin, a remarkable animal pigment containing copper, *R. Inst. Proc.,* 14, 44, 1892.
12. **Hart, E. B., Steenbock, H., Waddel, J., and Elvehjem, C. A.,** Iron in nutrition. VII. Copper as a supplement to iron for hemoglobin building in the rat, *J. Biol. Chem.,* 77, 797, 1928.
13. **Elvehjem, C. A.,** The biological significance of copper and its relation to iron metabolism, *Physiol. Rev.,* 15, 471, 1935.
14. **Neal, W. M., Becker, R. B., and Shealy, A. L.,** A natural copper deficiency in cattle rations, *Science,* 74, 418, 1931.
15. **Sjolemma, B.,** Kupfermangel als Ursache von Krankheiten bei Phlanzen and Tieren, *Biochem. Z.,* 267, 151, 1930.
16. **Sjolemma, B.,** Kupfermangel als Ursache von Tierkrankheiten, *Biochem. Z.,* 295, 372, 1938.
17. **Bennetts, H. W. and Chapman, F. E.,** Copper deficiency in sheep in Western Australia: a preliminary account of the etiology of enzootic ataxia in lambs and an anemia of ewes, *Aust. Vet. J.,* 13, 138, 1937.
18. **Hill, C. H. and Matrone, G.,** Studies on copper and iron deficiency in growing chickens, *J. Nutr.,* 73, 425, 1961.
19. **Mills, C. F., Dalgarno, A. C., and Wenham, G.,** Biochemical and pathological changes in tissues of friesian cattle during experimental induction of copper deficiency, *J. Nutr.,* 35, 309, 1976.
20. **Suttle, N. F. and Angus, K. V.,** Experimental copper deficiency in the calf, *J. Comp. Pathol.,* 86, 595, 1976.
21. **Lahey, M. E., Gubler, C. J., Chase, M. S., Cartwright, G. E., and Wintrobe, M. M.,** Studies on copper metabolism. II. Hematologic manifestations of copper deficiency in swine, *Blood,* 7, 1053, 1952.
22. **Lee, G. R., Nacht, S., Lukens, J. N., and Cartwright, G. E.,** Iron metabolism in copper deficient swine, *J. Clin. Invest.,* 47, 2058, 1968.
23. **Gubler, C. J., Lahey, M. E., Chase, M. S., Cartwright, G. E., and Wintrobe, M. M.,** Studies on copper metabolism. III. The metabolism of iron in copper deficient swine, *Blood,* 7, 1075, 1952.
24. **Lee, G. R., Williams, D. M., and Cartwright, G. E.,** Role of copper in iron metabolism and heme synthesis, in *Trace Elements in Human Health and Disease,* Vol. 1, Zinc and Copper, Prasad, A. S., Ed., Academic Press, New York, 1976, chap. 23.
25. **Kiel, H. L. and Nelson, V. E.,** The role of copper in hemoglobin regeneration and in reproduction, *J. Biol. Chem.,* 93, 49, 1931.
26. **Harris, E. D., Rayton, J. K., Balthrop, R. A., DiSilvestro, R. A., and Garcia-De-Quevedo, M.,** Copper and the synthesis of elastin and collagen, in *Biological Roles of Copper (Ciba Found. Symp. 79),* Excerpta Medica, Amsterdam, 1980, 163-182.
27. **Shields, G. S., Coulson, W. F., Kimbal, D. A., Carnes, W. H., Cartwright, G. E., and Wintrobe, M. M.,** Studies on copper metabolism. XXXII. Cardiovascular lesions in copper deficient swine, *Am. J. Pathol.,* 41, 603, 1962.
28. **O'Dell, B. L., Hardwick, B. C., Reynolds, G., and Savage, J. E.,** Connective tissue defect resulting from copper deficiency, *Proc. Soc. Exp. Biol. Med.,* 108, 402, 1961.
29. **Marston, H. D. and Lee, H. J.,** Nutritional factors in wool production by Merino sheep, *Aust. J. Sci. Res.,* 1, 376, 1948.
30. **Underwood, E. J.,** Trace elements, in *Human and Animal Nutrition,* 4th ed., Academic Press, New York, 1977, 56-108.

31. **Baxter, J. H. and Van Wyk, J. J.,** A bone disorder associated with copper deficiency, *Bull. Johns Hopkins Hosp.*, 93, 1, 25, and 41, 1953.
32. **Bennetts, H. W. and Beck, A. B.,** Enzootic Ataxia and Copper Deficiency of Sheep in Western Australia, Bull. 147, Counc. Sci. Ind. Res., Commonwealth of Australia, Melbourne, 1942, 1—52.
33. **Howell, J. McC.,** The pathology of swayback, in *Trace Element Metabolism in Animals,* Mills, C. F., Ed., E & S. Livingstone, Edinburgh, 1970, 103.
34. **Suttle, N. F. and Field, A. C.,** Production of swayback by experimental copper deficiency, in *Trace Element Metabolism in Animals,* Mills, C. F., Ed., E & S. Livingstone, Edinburgh, 1970, 110.
35. **Hunt, C. M.,** Copper and neurological function, in *Biological Roles of Copper (Ciba Found. Symp. 79),* Excerpta Medica, Amsterdam, 1980, 247-266.
36. **Wintrobe, M. M., Cartwright, G. E., and Gubler, C. J.,** Studies on the function and metabolism of copper, *J. Nutr.*, 30, 395, 1953.
37. **Josephs, H. W.,** Treatment of anemia in infants with iron and copper, *Bull. Johns Hopkins Hosp.*, 49, 246, 1931.
38. **Wintrobe, M. M.,** *Clinical Hematology,* 5th ed., Lea & Febiger, Philadelphia, 1961, 150.
39. **Adelstein, S. J. and Valee, B. L.,** Copper, in *Mineral Metabolism,* Comar, C. L. and Bronner, F., Eds., Academic Press, New York, 1962, chap. 32.
40. **Cordano, A., Baertl, J. M., and Graham, G. G.,** Copper deficiency in infancy, *Pediatrics,* 34, 324, 1964.
41. **Graham, G. G. and Cordano, A.,** Copper depletion and deficiency in the malnourished infant, *Johns Hopkins Med. J.*, 124, 139, 1969.
42. **Holtzman, N. A., Chrarache, P., Cordano, A., and Graham, G. G.,** Distribution of serum copper deficiency, *Johns Hopkins Med. J.*, 126, 34, 1970.
43. **Al-Rashid, R. A. and Spangler, J.,** Neonatal copper deficiency, *N. Engl. J. Med.*, 285, 841, 1971.
44. **Askenazi, A., Levin, S., Djaldetti, M., Fishel, E., and Benvenisti, D.,** The syndrome of neonatal copper deficiency, *Pediatrics,* 52, 525, 1973.
45. **Karpel, J. T. and Peden, V. H.,** Copper deficiency in long term parenteral nutrition, *J. Pediatr.*, 80, 32, 1972.
46. **Heller, R. M., Kirchner, S. G., O'Neill, J. A., Hough, A. J., Howard, L., Kramer, S. S., and Green, H. L.,** Skeletal changes of copper deficiency in infants receiving prolonged total parenteral alimentation, *J. Pediatr.*, 92, 947, 1978.
47. **McCarthy, D. M., May, R. J., Mather, M., and Brennan, M. P.,** Trace metals and essential fatty acid deficiency during total parenteral nutrition, *Am. J. Dig. Dis.*, 23, 1009, 1978.
48. **Dunlap, W. M., James, G. W., and Hume, D. M.,** Anemia and neutropenia caused by copper deficiency, *Ann. Intern. Med.*, 80, 470, 1974.
49. **Vilter, R. W., Bozian, R. C., Hess, E. V., Zellner, D. C., and Petering, H. G.,** Manifestations of copper deficiency in a patient with systemic sclerosis on intravenous alimentation, *N. Engl. J. Med.*, 291, 188, 1974.
50. **Danks, D. M.,** Copper deficiency in humans, in *Biological Roles of Copper (Ciba Found. Symp. 79),* Excerpta Medica, Amsterdam, 1980, 209-225.
51. **Prasad, A. S., Brewer, G. J., Schoemaker, E. B., and Rabani, P.,** Hypocupremia induced by large doses of zinc therapy in adults, *J. Am. Med. Assoc.*, 240, 2166, 1978.
52. **Buck, W. B.,** Copper/molibdenum toxicity in animals, in *Toxicity of Heavy Metals in the Environment,* Vol. 1, Oehme, F. W., Ed., Marcel Dekker, New York, 1978, 491-515.
53. **Scheinberg, I. H.,** The effects of heredity and environment on copper metabolism, *Med. Clin. North Am.*, 60, 705, 1976.
54. **Dalgano, A. C. and Mills, C. F.,** Retention by sheep of copper from aerobic digest of pig faecal slurry, *J. Agric. Sci.*, 85, 11, 1975.
55. **Gracey, H. I., Stewart, T. A., and Woodside, J. D.,** The effect of disposing high rates of copper-rich pig slurry on grassland on the health of grazing sheep, *J. Agric. Sci.*, 87, 617, 1976.
56. **Hill, R.,** Copper toxicity, in *Copper in Farming Symposium (R. Zool. Soc. Symp.),* Copper Development Association, Potters Bar, Herdforshire, 1975, 43.
57. **Todd, J. R.,** Chronic copper toxicity of ruminants, *Proc. Nutr. Soc.*, 28, 187, 1969.
58. **Chuttani, H. K., Gupta, P. S., Gulati, S., and Gupta, D. N.,** Acute copper sulfate poisoning, *Am. J. Med.*, 39, 849, 1965.
59. **Wahal, P. K., Mittal, V. P., and Bansal, O. P.,** Renal complications in acute copper sulphate poisoning, *Ind. Pract.*, 18, 807, 1965.
60. **Davenport, S. J.,** Health hazards of metals. I. Copper, Bur. Mines, Inform. Circ. 7666, U.S. Department of Commerce, Washington, D.C., 1953, 1-114.
61. **Scheinberg, H. and Sternlieb, I.,** Copper toxicity and Wilson's disease, in *Trace Elements in Human Health and Disease,* Vol. 1, Zinc and Copper, Prasad, A. S. and Oberlen, D., Eds., Academic Press, New York, 1977, chap. 25.

62. **Wilson, S. A. K.,** Progressive lenticular degeneration: a familial nervous disease associated with cirrhosis of the liver, *Brain,* 34, 295, 1912.
63. **Menkes, J. H., Alter, M., Steigleder, G. K., Weakley, D. R., and Sung, J. H.,** A sex-linked recessive disorder with retardation of growth, peculiar hair and focal cerebral and cerebellar degeneration, *Pediatrics,* 29, 764, 1962.
64. **Frieden, E.,** Ceruloplasmin: a multi-functional metalloprotein of vertebrate plasma, in *Biological Roles of Copper (Ciba Found. Symp. 79),* Excerpta Medica, 1980, 93-124.
65. **Frieden, E.,** Ceruloplasmin: the serum copper transport protein with oxidase activity, in *Copper in the Environment, Part II,* Nriagu, J. O., Ed., John Wiley & Sons, New York, 1979, 241-284.
66. **Scheinberg, I. H.,** Ceruloplasmin: a review, in *Biochemistry of Copper,* Peisach, J., Aisen, P., and Blumberg, W., Eds., Academic Press, New York, 1966, 513-524.
67. **Cartwright, G. E. and Wintrobe, M. M.,** Copper metabolism in normal subjects, *Am. J. Clin. Nutr.,* 14, 224, 1964; 15, 94, 1964.
68. **Peisach, J., Aisen, P., and Blumberg, W. E., Eds.,** *The Biochemistry of Copper,* Academic Press, New York, 1966, 475-513.
69. **Widdowson, E.M.,** Trace elements in human development, in *Mineral Metabolism in Pediatrics,* Barltrop, S. and Burland, W. L., Eds., Blackwell Scientific, Oxford 1969, chap. 6.
70. **Dowdy, R. P.,** Copper metabolism, *Am. J. Clin. Nutr.,* 22, 887, 1969.
71. **Mills, C. F., Ed.,** *Trace Element Metabolism for Animals,* E. & S. Livingstone, Edinburgh, 1970, 92, 231, 264, 277, 354, and 441.
72. **Evans, G. W.,** Copper homeostasis in mammalian system, *Physiol. Rev.,* 53, 535, 1973.
73. **Evans, G. W.,** New aspects of biochemistry and metabolism of copper, in *Zinc and Copper in Clinical Medicine,* Hambridge, K. M. and Nichols, B. L., Eds., S.P. Medical and Scientific Books, New York, 1976, 113-118.
74. **Burch, R. E. and Sullivan, J. F., Eds.,** *Symposium on Trace Elements (Med. Clin. North Am.),* W. B. Saunders, Philadelphia, 60(4), 655-849, 1976.
75. **Prasad, A. S., Ed.,** *Trace Elements in Human Health and Disease,* Vol. 1, Zinc and Copper, Academic Press, New York, 1976, chap. 22 to 25.
76. **Prasad, A. S.,** Copper, in *Trace Elements and Iron in Human Metabolsim,* Prasad, A. S., Ed., Plenum Press, New York, 1978, chap. 2.
77. *Biological roles of Copper (Ciba Found. Symp. 79),* Excerpta Medica, Amsterdam, 1980, 1-343.
78. **Karcioglu, Z. A. and Sarper, R. M., Eds.,** *Zinc and Copper in Medicine,* Charles C Thomas, Springfield, Ill., 1980, chap. 1 to 18.
79. **Hrgovcic, M.,** Copper in myeloproliferative and lymphoproliferative disorders, in *Zinc and Copper in Medicine,* Karcioglu, Z. A. and Sarper, R. M., Eds., Charles C Thomas, Springfield, Ill., 1980, 481-520.

Chapter 2

METHODS OF COPPER DETERMINATION

I. INTRODUCTION

In recent years, many elements occurring only in trace amounts have been found to play important roles in plant and animal life.[1,2] As a result, much effort has gone into lowering the detectability limit and increasing the precision with which trace elements are measured in biological substances.[3-12] From a theoretical standpoint, one technique may be superior to others for detecting trace elements in a given sample. For use as a practical analytic tool, however, other factors must be considered, i.e., the amount of sample needed, sample preparation and analysis time, simplicity of equipment and operation, cost per sample, and the amount of information obtained and problems with interference.[3-5,12-14]

Because of the ubiquitous presence of copper in nature and especially in the modern environment, contamination of samples before or during analysis can seriously affect the results of trace element determinations depending upon the methodology used.[15-16] For example, ordinary distilled water from copper or tinned copper stills contains about 10 to 200 μg/ℓ of copper.[17] Since only trace amounts of copper are present in biological material (the usual range being 0.1 to 10 μg/g or 0.1 to 10 ppm), the use of distilled water for the rinsing of glassware or as a diluent of samples and reagents should be avoided. Distilled deionized water must be used and its quality frequently checked. Glassware must be of high-quality glass, free from excessive etching, meticulously cleaned, and washed in dilute acid followed by extensive rinsing in copper-free distilled water. Filter paper, dialyzing membranes, etc. may contain copper and may also retain copper from the samples.[17] The reagents must be of the highest analytical purity and be carefully screened for copper contamination. Solvents and concentrated acids may have to be redistilled. As is customary in trace element analysis, reagent blanks and standards must be included with practically each batch of determinations.[14,17,18]

A variety of surgical instruments such as needles for venipuncture, liver biopsy blades for sampling tissue specimens, etc. are used to collect surgical specimens. Special steel needles and plastic syringes were found to be satisfactory for the collection of blood by venipuncture. Platinum Plus® surgical blades or plastic knives are suitable for tissue copper determination. Polyethylene vials and test tubes should be used since polyethylene has minimum absorption properties for most elements except lead and chromium. Prior to use, these vials and tubes should be rinsed with deionized water, soaked in reagent-grade nitric acid, and finally rinsed with deionized water. Only Teflon® or polyethylene stoppers should be used.[18] Strict precautions should be taken to prevent contamination during the collection, storage, and analysis. The importance of these precautionary measures to avoid contamination cannot be overemphasized in work with trace element analysis. Contamination is probably the chief reason for the wide variations of trace elements reported in biologic material.[14-18]

The following are the most widely used methods for copper determination:

Colorimetric method — Based on the color reaction (read spectrophotometrically) between serum copper and appropriate color reagents.[19-24] Since 95% of serum copper is tightly bound to protein (ceruloplasmin), special care must be taken to free the copper from the copper-protein complex. This is accomplished by utilizing acid hydrolysis, protein precipitation, or chelation.[25,26] This method has the great advantage of not requiring expensive apparatus. The sensitivity of this method is adequate for clinical use. Specificity and accuracy are also satisfactory. Interference by drugs causing color production in body fluids may give high (false-positive) copper results.[27] Another disadvantage is that the colorimetric method is time consuming and requires a minimum of 1 mℓ of serum for each copper determination. Special strict precautions must be undertaken to avoid contamination from various sources during sample preparation, storage, and analysis.

Neutron activation analysis — Has been used for analysis of biological samples for copper and zinc determination along with many other elements.[9,12,28-35] This technique is highly sensitive but requires long irradiation time. It lends itself to direct simultaneous determinations of copper, zinc, and many other elements, as well as copper-zinc ratio. The biological material may be in liquid or solid form with sample size from less than 0.1 mg to several grams. Radiochemical separation requires the removal of 24-Natrium which is always present in biological samples and this makes neutron activation analysis inconvenient for copper determination when compared with other techniques. Sample preparations require encapsulation which can be performed within minutes, and hours are needed for irradiation.

Atomic absorption spectroscopy technique — Described by Walsh[36] in 1955, it is the most convenient and commonly used method for single trace element (copper) or simultaneous copper and zinc determinations from the same sample, either in liquid or small solid form.[7,10-12] Sample ashing, acid digestion, and/or extractions are generally required. The preferred sample form is liquid and the tissue samples must be solubilized before analysis, with the exception of hair and nails.[37-39] Atomic absorption analysis is the method of choice for copper determinations in biological fluids, particularly serum or plasma. This method requires only 0.2 mℓ of serum which is diluted with 0.8 mℓ of deionized water. Direct copper measurement takes only 30 sec. The sensitivity of this procedure is adequate and few interferences are encountered. Serum protein precipitation with trichloroacetic acid, followed by chelation and extraction and analysis against extracted standards, will improve the accuracy and precision of the analysis. Disadvantages of the atomic absorption technique include the initial cost of equipment, occasional problems with interference, and, for researchers interested in multitrace-element analysis, its limitation to one or two elements at a time (copper and zinc). There are many articles, reviews, and books describing the underlying theory and applications of this technique to analytical problems.[36,40-51] An important innovation is the recent development of nonflame atomization technique (emission spectrometry), which might be even more sensitive than atomic absorption for measurement of certain trace elements.[40,41,52-55] Emission spectrometry, however, is more susceptible to interference and temperature variations than atomic absorption and the latter method is generally preferred. The relative advantages and disadvantages of flame and nonflame atomization for atomic absorption spectrometry have been discussed by several investigators.[40,41,56-60]

X-ray fluorescent spectroscopy — Is a fundamentally different technique which offers a clear advantage under certain circumstances for the determination of copper and zinc simultaneously with a number of other trace elements in biological samples and for the direct measurement of the zinc-copper ratio, as well as the absolute concentration of both metals. Several good reviews and articles covering the basic physics of the energy-dispersive photon-excited X-ray fluorescent technique have appeared in the recent literature.[6,12,34,61-71] X-ray fluorescent samples are made from 2 mℓ of blood serum, predoped with 0.2 mℓ of nickel atomic absorption standard (20 ppm), steam dried to form 12 mm disks, flattened, dried, and then dry-ashed to remove the organic matrix material. X-rays from a Mo X-ray tube with an Se secondary fluorescence radiator are used to produce characteristic X-rays from elements in a disc.[4,13] These X-rays are counted as a function of energy using a Si (Li) detector and a multichannel analyzer. Typical running times are 20 min per sample. An on-line minicomputer is programed to sum the number of characteristic X-ray counts for each element in a X-ray spectrum. These data are corrected to account for a small number of background counts in the spectrum produced by Compton scattering of the Se X-rays in the sample. Premeasured efficiencies of the total system for Fe, Cu, and Zn relative to Ni and the known concentration of Ni in each sample are used to evaluate the data for absolute concentrations of Fe, Cu, and Zn. The advantages and drawbacks are discussed later in comparison of our analytical techniques.[4,13]

Proton-induced X-ray emission (PIXE) analysis — Was introduced in 1970 in Sweden and since then, many reviews and books have appeared in literature.[72-80] Valkovic et al.[4,75] used a 100-ppm spike of yttrium as an internal calibration standard to determine 2 ppm Zn and 1.4 ppm Cu in human serum by PIXE analysis, together with Cu, Zn, V, Cr, Mn, Fe, Co, Ni, Se, and I in tissues. Proton excitation samples are prepared by predoping 0.5 mℓ of serum with 0.1 mℓ of yttrium atomic absorption standard (500 ppm). A 0.04 mℓ drop of the resultant mixture was allowed to dry on a 300 $\mu g/cm^2$ sandwich backing of high purity aluminum and Formvar®. The samples are exposed to 100 nA of 3-MeV protons from a tandem Van de Graaff accelerator for 20 min.[3,14] Along with characteristic X-rays of elements in the sample, protons also produce ''knock-on'' electrons which deaccelerate and produce low energy *bremsstrahlung*. To eliminate most of this unwanted background, which would have overloaded the electronics, a 0.1 cm polystyrene filter is used to completely attenuate all radiation below 2.5 keV. The remaining radiation is counted as a function of energy using an Si (Li) detector and a multichannel analyzer system interfaced to an IBM® 1800 computer.[81] In a typical sample spectrum, characteristic X-rays from K, Ca, Fe, Ni, Cu, Zn, Br, and the dopant, Y, are present. The number of counts in X-ray peaks are evaluated by programing the computer to ''fit'' each peak to a Gaussian plus quadratic background function.[15] Premeasured efficiencies of the total system for all elements relative to Y and the known concentration of Y in each sample are used to evaluate the computed areas for absolute concentrations of K, Ca Fe, Cu, Zn, and Br.[3,14]

The efficiency, precision, advantages, and pitfalls of each technique and direct comparison of various analytical procedures based on our experience will be discussed in the section covering our methods.

II. HISTOCHEMICAL METHODS FOR THE DETERMINATION OF PRESENCE OF COPPER

Several histochemical methods are available for the demonstration of copper in tissue sections. These are of value mainly in evaluation of cholestatic liver diseases and have been reviewed in many reports.[82-86] The following methods are widely used:

Rubeanic acid (dithioxamide[82]) — This method yields a green-black reaction with copper. It is relatively insensitive and false-negatives are not uncommon.

Timm's silver sulfide method[87] — This method is more sensitive and less specific than the above.

Rhodanine (5-*p*-dimethylaminobenzylidine rhodanine)[88] — This method gives an orange-red color with copper. A negative rhodanine stain does not exclude presence of copper in tissues.

Shikata's orcein method — Yields a black reaction product with copper.[83,88] This method is used in demonstrating tissue copper in primary biliary cirrhosis and Indian childhood cirrhosis. This stain demonstrates presence of lysosomal copper, probably as a copper-associated protein.

X-ray fluorescence — This can be used in fine sections but has not been used much in human pathologic conditions.[89-91]

Details of these and other methods are available in several reports.[82-91] Lindquist's[85] report summarizes six procedures and discusses limitations and pitfalls of these methods. The six staining methods reviewed in his report are rhodanine, alizarin blues, rubeanic acid, diphenylcarbazone, diethyldithiocarbamate, and thiocarbohydrazide-osmium black technique.

III. METHODS OF CERULOPLASMIN DETERMINATION

Ceruloplasmin can be measured in serum as well as other body fluids. Several methods are available.[19,92-100] Enzymatic methods measure the color development of the oxidized substrate. This assay is simple, rapid, and frequently used.[94-97] Since ceruloplasmin has immunologic specificity, this property can be used in measuring ceruloplasmin concentration in agar diffusion techniques (RID).[93,98] Colorimetric method consists of measuring disappearance of the blue color of ceruloplasmin after reduction. The difference in optical density before and after reduction depends on ceruloplasmin concentration.[99,100]

Since ceruloplasmin contains over 90% of total serum copper and in view of the high degree of correlation between serum copper and ceruloplasmin concentrations found in various physiological and pathological conditions, measurement of the serum copper level gives an idea of the ceruloplasmin concentration.[20,22,101] The question whether both copper and ceruloplasmin should be used in clinical studies or if one of the two would suffice has been raised. Ray et al.,[92] based on the findings of a linear relationship between ceruloplasmin and copper values in disease states, particularly lymphoma, feel that there is no reason to perform both estimations. They have chosen to eliminate the ceruloplasmin determinations since the copper assay by atomic absorption is inherently more accurate than the test for ceruloplasmin. On the other hand, Linder et al.,[93] based on their data, believe that at least for lung cancer, the enzymatic and immunological assays of ceruloplasmin are superior to assays of total serum copper in terms of the degree and consistency of changes associated with the tumor.

REFERENCES

1. **Underwood, E. I.,** *Trace Elements in Human and Animal Nutrition,* 4th ed., Academic Press, New York, 1977.
2. **Mills, F. C., Ed.,** *Trace Element Metabolism in Animals,* Proc. WAAP/IBP Int. Symp., Aberdeen, 1969, E. & S. Livingstone, Edinburgh, 1970.
3. **Valkovic, V., Liebert, R. B., Zabel, T., Larson, H. T., Miljanic, D., Wheeler, R. M., and Phillips, G. C.,** Trace element analysis using PIXE spectroscopy, *Nucl. Instrum. Methods,* 114, 573, 1974.
4. **Valkovic, V.,** *Trace Element Analysis,* John Wiley & Sons, New York, 1975.
5. **Smeyers-Verbeke, J., Massart, D. L., Versieck, J., and Speecke, A.,** The determination of copper and zinc in biological materials. A comparison of atomic absorption with spectrophotometry and neutron activation, *Clin. Chim. Acta,* 44, 243, 1973.
6. **Ong, P. S., Lund, P. K., Litton, C. E., and Mitchell, B. A.,** An energy dispersive system for the analysis of trace elements in human blood serum, in *Advances in X-ray Analysis,* Vol. 16, Plenum Press, New York, 1973, 124.
7. **Sunderman, W. F., Jr.,** Atomic absorption spectrometry of trace metals in clinical pathology, *Human Pathol.,* 4, 549, 1973.
8. **Birks, L. S. and Gilfrich, J. V.,** X-ray absorption and emission, *Anal. Chem.,* 46, 360R, 1974.
9. **Adams, F.,** Instrumental and radiochemical activation analysis, in *CRC Crit. Rev. Anal. Chem.,* 1, 455, 1971.
10. **Rubeska, I. and Moldan, B.,** *Atomic Absorption Spectrophotometry,* CRC Press, Inc., Boca Raton, Fla., 1969, 1-189.
11. **Christian, G. D. and Feldman, F. J.,** Atomic absorption spectroscopy: applications in agriculture, in *Biology and Medicine,* John Wiley & Sons, New York, 1970, 1-490.
12. **Carden, J. L. and Fink, R. W.,** Methods of detection, in *Zinc and Copper in Medicine,* Karcioglu, Z. A. and Sarper, R. M., Eds., Charles C Thomas, Springfield, Ill., 1980, chap. 1.
13. **Scheer, J.,** Comparison of sensitivities in trace element analysis by X-ray tube, X-ray fluorescence and by PIXE, *Nucl. Instrum. Methods,* 142, 333, 1977.

14. **Wheeler, R. M., Liebert, R. B., Zabel, T., Chatuvedi, R. P., Valkovic, V., Philips, G. C., Ong, P. S., Cheng, E. L., and Hrgovcic, M.,** Techniques for trace element analysis: X-ray fluorescence, X-ray excitation with protons, and flame atomic absorption, *Med. Phys.*, 1, 68, 1974.
15. **Thiers, R. E.,** Contamination in trace element analysis and its control, *Methods Biochem. Anal.*, 5, 273, 1957.
16. **Robertson, D. E.,** Contamination problems in trace element analysis and ultrapurification, in *Ultrapurity*, Zief, M. and Speights, R., Eds., Marcel Dekker, New York, 1972, 207-251.
17. **Sass-Kortsak, A.,** Copper metabolism, in *Advances in Clinical Chemistry*, Vol. 8, Sobotka, H. and Stewart, C. P., Eds., Academic Press, New York, 1965, 1-67.
18. **Sarper, R., Karcioglu, A. Z., Carden, L. J., and Fink, R. W.,** Sample preparation, in *Zinc and Copper in Medicine*, Karcioglu, A. Z. and Sarper, R., Eds., Charles C Thomas, Springfield, Ill., 1980, chap. 2.
19. **Gubler, C. J., Lahey, M. E., Ashenbrucker, H., Cartwright, G. E., and Wintrobe, M. M.,** Studies on copper metabolism. I. Method for the determination of copper in whole blood, red blood cells, and plasma, *J. Biol. Chem.*, 196, 209, 1952.
20. **Jensen, K. B., Thorling, E. B., and Anderson, C. J.,** Serum copper in Hodgkin's Disease, *Scand. J. Haematol.*, 1, 63, 1964.
21. **Hrgovcic, M., Tessmer, C. F., Minckler, T. M., Mosier, B., and Taylor, G.,** Serum copper levels in lymphoma and leukemia: special reference of Hodgkin's disease, *Cancer*, 21, 743, 1968.
22. **Rice, E. W.,** Spectrophotometric determination of serum copper with oxalydihidrazice, *J. Lab. Clin. Med.*, 55, 325, 1960.
23. **Zack, B. and Ressler, N.,** Serum copper and iron on a single sample, *Clin. Chem.*, 4, 43, 1958.
24. **Stoner, R. I. and Dassler, W.,** Spectrophotometric determination of copper following extraction with 1,5-diphenylcarbohydrazide in benzene, *Anal. Chem.*, 32, 1207, 1960.
25. **Smith, G. F. and McCurdy, W. H.,** 2,9-Dimethyl-1,10-phenanthroline. New specific in spectrophotometric determination of copper, *Anal. Chem.*, 24, 371, 1952.
26. **Smith, G. F. and Wilkins, D. H.,** A new colorimetric reagent specific for copper, *Ann. Chem.*, 25, 510, 1953.
27. **Hrgovcic, M., Tessmer, C. F., Mumford, D. M., Ong, P. S., Gamble, J. F., and Shullenberger, C. C.,** Interpreting serum copper levels in Hodgkin's disease, *Tex. Med.*, 71, 53, 1975.
28. **Danielson, A. and Steinnes, E.,** A study of some selected trace elements in normal and cancerous tissue by thermal neutron activation analysis, *J. Nucl. Med.*, 11, 260, 1970.
29. **DeSoete, D., Gijbels, R., and Hoste, J.,** *Neutron Activation Analysis*, John Wiley & Sons, New York, 1972.
30. **Leddicotte, G. W.,** Activation analysis of biological trace elements, *Methods Biochem. Anal.*, 19, 345, 1971.
31. **Nadkarni, R. A., Flieder, E. E., and Ehmann, W. D.,** Instrumental neutron activation analysis of biological materials, *Radiochim. Acta*, 11, 97, 1969.
32. **Nadkarni, R. A. and Morrison, G. H.,** Multielement analysis of human blood serum by neutron activation analysis, *Radiochem. Radioanal. Lett.*, 24, 103, 1976.
33. **Schwartz, A. E., Friedman, E. W., Leddicotte, G. W., and Fink, R. W.,** Trace elements in normal and malignant human breast tissue, *Surgery*, 76, 325, 1974.
34. **MacDonald, G. L.,** Recent developments in X-ray spectrometry, *CRC Crit. Rev. Anal. Chem.*, 4, 281, 1975.
35. **Strain, W. H., Rob, C. G., Pories, W. J., Childers, R. C., Thompson, M. F., Jr., Hennessen, J. A., and Graber, F. M.,** in Modern Trends in Activation Analysis, Vol. 1, Devoe, J. R. and LaFleur, P. D., Eds., Spec. Publ. 312, National Bureau of Standards, U.S. Government Printing Office, Washington, D.C., 1969, 98-207.
36. **Walsh, A.,** The application of atomic absorption spectra to chemical analysis, *Spectrochim. Acta*, 7, 108, 1955.
37. **Jacob, R. A., Kleray, L. M., and Logan, G. M.,** Hair as a biopsy material. V. Hair metal as an index of hepatic metal in rats: copper and zinc, *Am. J. Clin. Nutr.*, 31, 477, 1978.
38. **Hambridge, V. M.,** Increase in hair copper concentration with increasing distance from the scalp, *Am. J. Clin. Nutr.*, 26, 1212, 1973.
39. **Harrison, W. W. and Tyree, A. B.,** The determination of trace elements in human fingernails by atomic absorption spectroscopy, *Clin. Chim. Acta*, 31, 63, 1971.
40. **L'vov, B. V.,** The analytical use of atomic absorption spectra, *Spectrochim. Acta*, 17, 761, 1961.
41. **Amos, M. D., Bennett, P. A., Brodie, K. G., Lung, P. W. Y., and Matousek, J. P.,** Carbon rod atomizer in atomic absorption and fluorescence spectroscopy and its clinical application, *Anal. Chem.*, 43, 211, 1971.
42. **Donega, H. L. and Burgess, T. E.,** Atomic absorption analysis by flameless atomization in a controlled atmosphere, *Anal. Chem.*, 42, 1521, 1970.

43. **Dean, J. A. and Rains, T. C.,** *Flame Emission and Atomic Absorption Spectrometry,* Vol. 1, 2, and 3, Marcel Dekker, New York, 1969, 1971, and 1975, respectively.
44. **Christian, G. D.,** Atomic absorption spectroscopy for the determination of elements in medical biological samples, *Fortschr. Chem. Forsch.,* 26, 77, 1972.
45. **Kirkbright, G. F. and Sargent, M.,** *Atomic Absorption and Fluorescence Spectroscopy,* Academic Press, New York, 1974.
46. **Schrenk, W. G.,** *Analytical Atomic Spectroscopy,* Plenum Press, New York, 1975, 243.
47. **Veloon, C.,** Opitcal atomic spectroscopic methods, in *Trace Analysis: Spectroscopic Methods for Elements (Chem. Anal. 46),* Windfordner, J. D., Ed., John Wiley & Sons, 1976, 152.
48. **Murphy, T. F., Nomoto, S., and Sunderman, F. W., Jr.,** Measurements of blood lead by atomic absorption spectrometry, *Ann. Clin. Lab. Sci.,* 1, 57, 1971.
49. **Elwell, W. T. and Gidley, J. A. F.,** *Atomic Absorption Spectrophotometry,* Macmillan, New York, 1962, 1-102.
50. **Robinson, J. W.,** *Atomic Absorption Spectroscopy,* Marcel Dekker, New York, 1966, 1-204.
51. **Dawson, J. B. and Heaton, F. W.,** *Spectrochemical Analysis of Clinical Material,* Charles C Thomas, Springfield, Ill., 1967, 1-118.
52. **Glenn, M. T., Savory, J., Winefordner, J. D., Hart, L., and Fraser, M.,** Use of nonflame cell atomic absorption spectrometry for measurement of copper in serum, *Clin. Chem.,* 17, 666, 1971.
53. **Glenn, M., Savory, J., Hart, L., Glenn, T., and Winefordner, J. D.,** Determination of copper in serum with a graphite rod atomizer for atomic absorption spectrophotometry, *Anal. Chim. Acta,* 57, 263, 1971.
54. **West, T. S. and Williams, X. K.,** Atomic absorption and fluorescence spectroscopy with a carbon filament atom reservoir, *Anal. Chim. Acta,* 45, 27, 1969.
55. **Kirkbright, G. F.,** The application of non-flame atom cells in atomic absorption and atomic fluorescence spectroscopy, *Analyst,* 96, 600, 1971.
56. **Massmann, H.,** Heutiger Stand der Atomabsorptionsspektrophotometrie, *Chimia,* 21, 217, 1967.
57. **Massmann, H.,** Vergleich von Atomabsorption und Atomfluoreszenz in der Graphitküvette, *Spectrochim. Acta, Part B,* 23, 215, 1968.
58. **Reeves, R. D., Patel, B. M., Molnar, C. J., and Winefordner, J. D.,** Decay of atom populations following graphite rod atomization in atomic absorption spectrometry, *Anal. Chem.,* 45, 246, 1973.
59. **Kahn, H. L., Bancroft, M., and Emmel, R. H.,** Solving precision problems in flameless AA sampling, *Res. Dev.,* 27, 30, 1976.
60. **Fuller, C. A.,** The loss of copper and nickel during pre-atomization heating periods in flameless atomic absorption determinations, *Anal. Chim. Acta,* 62, 422, 1972.
61. **Spatz, R. and Lieser, K. H.,** Critical comparison of the measuring range and detection limits in energy dispersive X-ray fluorescence analysis with tube excitation (secondary target) and with radionuclide excitation (^{109}Cd and ^{241}Am) by means of bulk powder samples on silicagel basis as an example (in German), *Z. Anal. Chem.,* 288, 267, 1977.
62. **Kaufman, L., Shosa, D., and Camp, D.,** Polarized radiation for XRF of elements with Z = 47-57, *Trans. Am. Nucl. Soc.,* 27, 205, 1977.
63. **Ryon, R. W.,** Proc. 27th Ann. Denver conf. on applications of X-ray analysis, August 1978, *Adv. X-Ray Anal.,* 20, 575, 1977.
64. **Kaufman, L. and Camp, D. C.,** Polarized radiation for X-ray fluorescence analysis, *Adv. X-Ray Anal.,* 18, 247, 1974; *IEEE Trans. Nuclear Sci.,* NS-24, 525, 1977; An automated fluorescent excitation analysis system for medical applications, *Invest. Radiol.,* 11, 210, 1976.
65. **Rhodes, J. R.,** Practical problems in chemical analysis using energy-dispersive X-ray emission spectrometry, in *Inner Shell Ionization Phenomena and Future Applications,* R. W. Fink et al., Eds. U.S. Atomic Energy Commission, Oak Ridge, Tenn., 1973.
66. **Camp, D. C.,** *An Introduction to Energy Dispersive X-Ray Fluorescence Analysis, Rep., UCRL-52489,* Lawrence Livermore Laboratory, University of California, June 1, 1978.
67. **Dzubay, T. G.,** *X-Ray Fluorescence Analysis of Environmental Samples,* Ann Arbor Science Publishers, Mich., 1977.
68. **Owers, M. J. and Shalgosky, H. L.,** Use of X-ray fluorescence for chemical analysis, *J. Phys. (London),* E7-593, 1974.
69. **Herglotz, H. K. and Birks, L. S.,** *X-Ray Spectrometry (Pract. Spectrosc. Ser.,)* Vol. 2, Marcel Dekker, New York, 1978.
70. **Müller, R. O.,** *Spectromechemical Analysis by X-Ray Fluorescence,* Transl. from German, Adam Hilger, London, 1966 and 1972.
71. **Bertin, E. P.,** *Principles and Practice of X-Ray Spectrometric Analysis,* 2nd ed., Plenum Press, New York, 1975; *Introduction to X-ray Spectrometric Analysis,* Plenum Press, New York, 1978.
72. **Johansson, T. B., Akselsson, R., and Johansson, S. A. E.,** X-ray analysis: elemental trace analysis at the 10^{-12}g level, *Nucl. Instrum. Methods,* 84, 141, 1970.

73. **Johansson, T. B., Ahlberg, M., Akselsson, R., Johanson, G., and Malquist, K.,** Analytical use of proton-induced X-ray emission, *J. Radioanal. Chem.*, 32, 207, 1976.
74. **Johnsson, S. A. E. and Johansson, T. B.,** Proton-induced X-ray emission spectroscopy in elemental trace analysis, *Nucl. Instrum. Methods,* 137, 473, 1976; 142(1/2), 1977; *Adv. X-Ray Anal.*, 15, 373, 1972.
75. **Valkovic, V.,** PIXE Applications in medicine, *Nucl. Instrum. Methods,* 142, 151, 1977.
76. **VanRinsvelt, H. A., Lear, R. D., and Adams, W. R.,** Human diseases and trace elements: PIXE, *Nucl. Instrum. Methods,* 142, 171, 1977.
77. **Deconninck, G., Demortier, G., and Bodart, F.,** Application of X-ray production by charged particles to elemental analysis, *At. Energ. Rev.*, 13, 267, 1975.
78. **Cahill, T. A.,** Ion-excited X-ray analysis of environmental samples, in *New Uses of Ion Accelerators,* Ziegler, J. R., Ed., Plenum Press, New York, 1975, chap. 1.
79. **Young, F. C., Roush, M. L., and Berman, P. G.,** Trace element analysis by PIXE, *Int. J. Appl. Radiol. Isotopes,* 24, 153, 1973.
80. **Walter, R. L., Willis, R. D., Gutknecht, W. F., and Joyce, J. M.,** Analysis of biological, clinical, and environmental samples using PIXE, *Anal. Chem.*, 146, 843, 1974.
81. **Jones, H. V. and Buchanan, J. A.,** Proc. Skytop Conf. Comput. Syst. Exp. Nucl. Phys., CONF 690301, Columbia University, New York, 1969, 266.
82. **Uzman, L. L.,** Histochemical localization of copper with rubeanic acid, *Lab. Invest.*, 5, 299, 1956.
83. **Sternlieb, I.,** Copper and the liver, *Gastroenterology,* 78, 1615, 1980.
84. **Howell, J. S.,** Histochemical demonstrations of copper in copper-fed rats and in hepatolenticular degeneration, *J. Pathol. Bacteriol.*, 77, 473, 1959.
85. **Lindquist, R. R.,** Studies on the pathogenesis of hepatolenticular degeneration; cytochemical methods for the localization of copper, *Arch. Pathol.*, 87, 370, 1969.
86. **Thompson, S. W.,** *Selective Histochemical and Histopathological Methods,* 2nd ed., Charles C Thomas, Springfield, Ill., 1966, 579.
87. **Timm, F.,** Der histochemishe Nachweis der Kupfers im Gehirn, *Histochemie,* 2, 332, 1961.
88. **Irons, R. D., Schenk, E. A., and Lee, J. C. K.,** Cytochemical methods for copper. Semiquantitative screening procedure for identification of abnormal copper levels in liver, *Arch. Pathol. Lab. Med.*, 101, 298, 1977.
89. **Twedt, D. C., Sternlieb, I., and Gilbertson, S. R.,** Clinical, morphologic and chemical studies on copper toxicosis of Bedlington terriers, *J. Am. Vet. Med. Assoc.*, 175, 269, 1979.
90. **Goldfischer, S. and Moskal, J.,** Electron probe microanalysis of liver in Wilson's disease. Simultaneous assays for copper and for lead deposited by acid phosphatase activity in lysosomes, *Am. J. Pathol.*, 48, 305, 1966.
91. **Wiesner, R. H., Barham, S. S., and Dickson, E. R.,** X-ray microanalysis: a new technique to measure hepatic copper and iron in Wilson's disease and hemochromatosis, *Gastroenterology,* 76:47, 1979.
92. **Ray, R. G., Wolf, P. H., and Kaplan, H. S.,** Value of laboratory indicators in Hodgkin's disease: preliminary results, *Natl. Cancer Inst. Monogr.*, 36, 315, 1973.
93. **Linder, M. C., Moore, J. R., and Wright, C.,** Ceruloplasmin assays in diagnosis and treatment of human lung, breast and gastrointestinal cancer, *J. Natl. Cancer Inst.*, 67, 263, 1981.
94. **Abreu, L. A.,** Determination of serum ceruloplasmin oxidative activity, *Rev. Bras. Biol.*, 21, 97, 1961.
95. **Broman, L.,** Separation and characterization of two caeruloplasmins from human serum, *Nature (London),* 182, 1655, 1958.
96. **Ravin, H. A.,** An improved colorimetric enzymatic assay of ceruloplasmin, *J. Lab. Clin. Med.*, 58, 161, 1961.
97. **Houchin, O. B.,** A raspid colorimetric method for the quantitative determination of copper oxidase activity (ceruloplasmin), *Clin. Chem.*, 4, 519, 1958.
98. **Gell, P. G. H.,** The estimation of the individual human serum proteins by an immunological method, *J. Clin. Pathol.*, 10, 67, 1957.
99. **Deutsch, H. F.,** A chromatographic-spectrophotometric method for the determination of ceruloplasmin, *Clin. Chim. Acta,* 5, 460, 1960.
100. **Scheinberg, I. H., Morell, A. G., Harris, R. S., and Berger, A.,** Concentration of ceruloplasmin in plasma of schizophrenic patients, *Science,* 126, 925, 1957.
101. **Trip, J. A. J.,** Clinical significance of ceruloplasmin, *Folia Med. Neerl.*, 12, 150, 1969.

Chapter 3

DISTRIBUTION OF COPPER

I. TISSUE COPPER

The copper content of fetal and neonatal tissues is much higher than that of adults.[1-7] This is well documented in man and many animal species.[1,7,8] The whole-body concentration of copper in newborn infants is about three times that of adults (newborn 4.7 ppm vs. adults 1.7 ppm).[1-7] This difference is largely a reflection of the increased copper content of the liver (300 to 400 μg/g/dry weight) and other organs during the perinatal period. The liver of the normal human newborn contains roughly six to eight times the copper concentration of the adult.[1,7,9,10] Within 6 months the amount of copper in the baby's liver falls to the adult value, about 33 μg/g of dry tissue, coinciding with increased capacity of the liver to synthesize ceruloplasmin, the main copper protein.[1-9] There is also a diminution in the copper concentration of the kidney, heart, spleen, and lung during the first 12 months.

The brain is the only organ in which copper concentration increases with age, approximately doubling from birth to maturity.[11] Thereafter, the amount of copper varies little throughout life. This homeostasis results from an x-linked intestinal tract transport mechanism and the capacity of the liver to regulate copper stores.[9] The latter is achieved through the synthesis of plasma and tissue copper-proteins and the excretion of copper from the body via the bile.[12-14] The highest concentrations of copper are found in the liver, brain, heart, and kidneys, in decreasing order. Lung, intestine, pancreas, spleen, skin, and endocrine glands (prostate, thymus, thyroid, and hypophysis), contain an intermediate content of this element, while muscle and bone have the lowest copper concentration.[14-16] In view of the large mass of muscle and bone, these tissues contain approximately 50% of the total body copper;[14] hepatic copper accounts for about 10 to 15% of the total.[1,14] The copper enzymes account for only a small fraction of the total body copper, the bulk being associated with proteins and other cell constituents for which the physiological functions are unclear, and probably represent intracellular storage. Intracellular copper is often found associated with amino acids, purines, pyrimidines, nucleotides, DNA, RNA, and protein.[14]

Tipton and Cook[16] have analyzed the content of 24 trace elements in 29 human tissues obtained from 150 adults from various parts of the U.S. who were free of disease and died suddenly, usually as a result of an accident. Copper was one of the elements studied. The highest copper contents were found in the liver (11.28 mg), brain (8.10 mg), heart (1.16 mg), kidney (0.86 mg), and muscle (26.70 mg). One third of total body copper was contained in the muscle. There was a higher concentration in the upper parts of the gastrointestinal tract than in the lower parts. The copper content of human hair (16.1 ± 1.19 μg/g) and nails was considerably higher than that of other organs, compared on a wet weight basis, no doubt due to the low water content of these tissues.[17,18] The mean total body copper was 75 mg with a range of 50 to 120 mg, which is lower than previously reported. According to Cartwright,[19] the adult human body contains some 100 to 150 mg of copper, of which 64 mg are found in the bones and muscle mass and 23 mg in the brain, liver, heart, spleen, kidneys, and blood. Of this total, 8 mg was in the liver and 8 mg in the brain. The copper content of the placenta is high (30 to 40 μg/g/dry weight).[3,5] Hamilton et al.[5] found the following mean copper concentrations in adult humans: liver, 14.7 ± 3.9; brain, 5.6 ± 0.2; lung, 2.2 ± 0.2; kidney, 2.1 ± 0.4; ovary, 1.2 ± 0.3; testis, 0.8 ± 0.2; lymph nodes, 0.8 ± 0.06; and muscle, 0.7 ± 0.02 μg/g wet weight.

Very high copper concentrations occur in pigmented parts of the eye, associated particularly with melanins bound to protein. The role of copper in these sites is not known.[8] The inner layer of human dental enamel contains 11.3 ppm and that of the outer 9 ppm.[8]

Smythe et al.,[20] recently reported the results of X-ray fluorescence analysis of 13 trace elements (one of which was copper) in 9 human tissues. Tissues were obtained from 120 uremic patients who had been on dialysis (105 autopsy cases and 15 kidney transplants). In addition, tissues were also obtained at the autopsy of 29 uremic patients who had not had dialysis and 64 control subjects who died from various other causes. Reduced concentrations were found in the heart tissue of uremic patients (13.8 ± 2.7 μg/g dry weight in dialyzed patients and 14.4 ± 2.4 μg/g dried tissue in nondialyzed patients) compared to control subjects (16 ± 3.3 μg/g dried tissue). Increased copper content was found in lung tissue (7.6 ± 3.5 in dialyzed and 8.8 ± 3.3 μg/g in nondialyzed) vs. controls (5.7 ± 1.8 μg/g dried tissue). No satisfactory explanation for these differences was found. Copper concentrations were also found to be reduced in kidneys (nondialyzed 9 ± 5.3 μg/g, dialyzed 6.1 ± 2.1 μg/g vs. control group 11.5 ± 5.2 μg/g dried tissue) which was not surprising because of loss of functional components in the end stage of kidney disease. Similar copper concentrations in uremic patients and controls were found in liver (mean 24 μg/g dried tissue).

There is little question that the liver plays a central role in the metabolism of copper. About 80% of the copper in the normal human liver is present in hepatic copper proteins.[9] The remainder is incorporated into specific copper proteins, such as cytochrome c oxidase and ceruloplasmin, or is taken up by lysosomes before being excreted in bile.[9,21] Ceruloplasmin synthesis is believed to be in the polyribosomes and there is mounting evidence that the lysosome is presumably the storage site for excess copper.[22] Owen[22] hypothesized three distinct hepatic copper compartments: bile, ceruloplasmin, and storage, and all three could be in two-way communication with the blood and with each other.

Whether any specific copper storage protein exists in the liver is still a matter of debate, but three different proteins have been considered in this regard, namely hepatocuprein (superoxide dismutase), metallothionein, and mitochondrocuprein.[23] It seems unlikely that superoxide dismutase acts as a storage form of copper. The proportion of hepatic copper present as metallothionein depends on hepatic copper content, age, species, and zinc status of the animal.[14,23] The function of metallothionein in the control of hepatic copper metabolism is still debatable.[23] In neonates, however, the bulk of the hepatic copper is present in mitochondria or hepatocellular lysosomes, which is a polymeric form of metallothionein, from which it is mobilized by unknown mechanisms and routes during the first few months of life.[9,14,22-24] It is worth noting that the liver retains some copper even in infants with inherited x-linked copper malabsorption and its impaired transport.[25] Porter et al.[26-29] showed that the copper in newborn liver is chiefly accounted for by mitochondrocuprein, a protein compound extraordinarily high in copper (4%), localized in the mitochondrial fraction and specific to the neonatal period and the lysosomes. The intracellular distribution of copper has also been studied.[24,26-33]

According to Evans,[14] a distinct pattern of distribution was found among the four subcellular fractions that can be separated by differential centrifugation. The microsomal fraction contains the least amount of copper (10%) within the hepatic cell.[26,30-33] It comprises fragments of the smooth endoplasmic reticulum, the rough endoplasmic reticulum, the Golgi apparatus, and other vesicular elements as well as the ribosomes. The evidence suggesting that the copper in the microsomal fraction is utilized for ceruloplasmin synthesis was presented by Whanger and Weswig.[34,35] The nuclear fraction contains approximately 20% of the total hepatic copper in the adult mammal. The exact amount of copper contained within the nucleus is difficult to assess because the nuclear fraction also contains other cellular material. However, since it is made up of nucleic acids and several basic proteins all of which bind copper, the nucleus is a potential site for temporary copper storage. Recent experiments by Derger and Eichhorn[36] indicate that a portion of nuclear copper is involved in cellular metabolism. The large granule fraction contains approximately 20% of the total hepatic copper. Mitochondria and lysosomes are contained within this fraction, the latter having a vital role in copper homeostasis. Several histochemical studies have shown localization of copper in the hepatic granules in livers from copper-fed animals[37-39] and from patients with Wilson's disease.[40-41] Sequestrations of copper within pericanalicular, acid phosphatase-rich lysosomes has been demonstrated by electron probe analysis[42] and electron microscopy.[43-47] Cytosol, the final supernatant obtained during the differential centrifugation, contains the major portion of the total hepatic copper in the adult mammal, bound to the enzyme superoxide dismutase and a low-molecular-weight protein similar to metallothionein.

Using light and electron microscopic staining techniques, Goldfischer and Sternlieb[48] indicated that in newborn infants the hepatic copper is found exclusively in the lysosomes. The decline in whole liver copper content that occurs as the mammal matures is accompanied by changes in copper distribution among the subcellular fractions.[31] In human,[32] bovine,[26] and rat liver[30,31] over 80% of the total neonatal hepatic copper is present in the mitochondrial and nuclear fractions, and less than 20% in the microsomes and soluble (supernatant) fraction, cytosol. At maturity, about 50% of the hepatic copper is within the cytosol and the remainder is distributed among nuclei, mitochondria, and the soluble fraction of the parenchymal cells in proportions that vary with age, strain, and the copper-zinc status of the animal.[30-33] The copper is either stored in these sites or released for incorporation into cytocuprein and ceruloplasmin and the various copper-containing enzymes of the cells. Porter and Hills[28] suggested that the copper-binding protein in the cytosol polymerizes, develops a high sulfur content, and then becomes the lysosomal storage component of copper.

Owen et al.[49] studied the subcellular distribution of copper in the liver of 17 normal adults, 1 newborn infant, and 35 patients with diseased liver, including 4 with primary biliary cirrhosis. Livers were homogenized and ultracentrifuged and divided into five particular fractions: nuclei and cell debris, heavy mitochondria, light mitochondria, heavy microsomes, and light microsomes as well as the supernatant fraction.

The mean total copper content from homogenized liver was 5.78 μg/g wet weight ± 2.48 SD, and supernatant copper (assessed from the supernatant after centrifugation) averaged 4.54 μg/g (SD ± 1.51) or 78.5% of the total liver copper concentrations. In one 2-day-old infant a high liver copper content (31.3 μg/g wet weight) was found and only one fourth (24.5%) was in the supernatant fraction.

As far as the subcellular copper distribution is concerned, the smallest copper concentrations were found in the lightest particles. As the concentration of copper rose in the liver in neonatal animals and in copper-laden animal or diseased human livers, in all species more copper appeared in the progressively heavier subcellular particles. These authors[49] found that the greater the total liver copper concentration, the greater the proportion in the larger subcellular particles and nuclei. The exception was patients with primary biliary cirrhosis who had high total hepatic copper, with 45 to 75% of that copper residing in the supernatant fraction.

II. COPPER IN BIOLOGICAL FLUIDS

Copper is normally found in the blood in both the red cells and plasma.[14,19,50] Approximately 60% or more of the total copper in the red cell is associated with the enzyme superoxide dismutase,[14] formerly known as erythrocuprein.[14,51,52] Erythrocuprein is probably synthesized in normoblasts in bone marrow.[53] The function of superoxide dismutase in removing the superoxide anion will be discussed later. Similar or identical proteins are found in other tissues, especially the liver (hepatocuprein) and brain (cerebrocuprein), and the general term "cytocuprein" has been proposed to apply to all such proteins.[54] Erythrocytic copper, although circulating, is not involved in transporting the metal to and from tissues.[14] The remainder of erythrocyte copper, designated the "labile pool", is more loosely bound to amino acid complexes and its function is probably to ensure an adequate supply of copper to maintain the activity of superoxide dismutase.[55,56] The total copper content of erythrocytes remains more or less constant in spite of the copper status of the animal or man under a wide range of conditions.[57-66] In both hypocupremic humans[25,61] and copper-deficient rats,[67] the copper content of the red cells remains invariable. Moreover, ingestion of excessive amounts of copper does not produce a significant increase in erythrocyte copper content.[64] The copper content of human erythrocytes is 1.2 ± 0.2 μg/mℓ.[59] The amount of copper in the individual leukocyte and platelet is approximately one quarter of the red cell content.

The copper in plasma also occurs in at least two main forms; one firmly and one loosely bound.[14,19,50] The former consists of the major protein-bound copper, ceruloplasmin, containing approximately 90 to 95% of total plasma copper or 3% of total body copper, first discovered in 1948 by Holmberg and Laurell.[68-71] Its property and function will be discussed in a subsequent section. The loosely associated copper fraction constitutes 5 to 10% of total plasma copper.[14,64,72] The plasma copper not present as ceruloplasmin is known as "direct-reacting copper" because it reacts directly with dithizone, is nondialyzable, and is reversibly bound to serum albumin.[50,64,65] The remainder is termed "indirect-reacting copper" in which ceruloplasmin-bound copper must be liberated from the protein by acids before it can be measured. This indirect-reacting copper does not transfer copper from plasma to red cells, no matter how high the plasma copper.[65] It is noteworthy that the ratio of direct- and indirect-reacting copper remains the same even in the copper-deficient state. It is believed that albumin-bound plasma copper represents copper in transit to the body tissues, especially to the liver which is the principal organ in the metabolism of this element. The copper-albumin serum pool also receives copper from the tissues. In addition it has been reported that three copper-amino acid complexes (Cu-histidine, Cu-threonine, and Cu-glutamine) have been identified in normal serum.[73] It was postulated that because these copper-amino acid complexes are of a much lower order of molecular size than the copper-albumin complex, they might mediate the transport of the copper through biological membranes. The mechanism of this mediation is suggested to be the condition of equilibrium between the concentration of copper-albumin complex and the concentration of the copper-amino acid complexes.

$$Cu^{2+} + \text{amino acid} \leftrightharpoons \text{Cu-amino acid}$$
$$Cu^{2+} + \text{albumin} \leftrightharpoons \text{Cu-albumin}$$
$$\text{Cu-albumin} + \text{amino acid} \leftrightharpoons \text{albumin} + \text{Cu-amino acid}$$

A small amount of copper also exists in the serum in an ultrafiltrable form, as free copper and copper bound to amino acids. The physiological role of the amino acid-bound copper is not known. It may serve in absorption of copper and cellular metabolism.[50,73]

The mean levels and range of serum copper level concentrations are well established in normal adults. The results obtained in several large series of healthy persons show close agreement.[74] The whole blood levels of copper are around 100 μg/100 mℓ.[24,64,74] Plasma and erythrocytes contain similar concentrations of copper.[74] Normal mean serum copper values reported by Lahey et al.[74] are 105.5 ± 16 μg/100 mℓ for men and 116 ± 16 μg/100 mℓ for women. No statistically significant difference was observed between males and females.[74,75] Explanation of sex differences will be discussed later. Slightly higher serum copper levels are found in serum (127 μg/100 mℓ) than in plasma (119 μg/100 mℓ), as a result of loss of copper and ceruloplasmin during the clotting process.[76,77] In 504 subjects studied by 9 investigators before 1950, Lahey et al.[74] reported the mean value for serum copper ranged between 84 and 126 μg/100 mℓ, with individual values ranging from 84 to 165 μg/100 mℓ.

Serum copper levels in healthy adults, both sexes, reported by other investigators from 1950 up to 1980 are presented in Table 1. As can be seen, the normal range of serum copper

Table 1
SERUM COPPER LEVELS OF NORMAL SUBJECTS

Year	Method	Number of subjects	Sex	Mean (mg/100 mℓ)	Range (mg/100 mℓ)	Ref.
1950	Colorimetric	12	M	105.5 ± 5	95—115.5	19
		11	F	114 ± 4.6	104.8—123.2	
1952	Colorimetric	40	M,F	110 ± 13	84—136	78
1953	Colorimetric	40	M	105 ± 16	68—134	74
		23	F	116 ± 16	84—143	
1956	Spectrochemical	58		98 ± 12	65—135	79
1957	Spectrophotometry	10	M	105	65—154	80
1960	Spectrophotometry	10	M	96	70—118	81
		9	F	100	87—117	
1960	Spectrochemical	109	M,F	103	48—193	82
1960	Spectrophotometry	15	M,F	118	85—163	83
1960	Spectrophotometry	120	M	110 ± 15.7	68—160	84
		85	F	120 ± 17.8	83—165	
1964	Colorimetric	95		109	69—150	50
1964	Colorimetric	24	M	102.3 ± 22.6		85
		24	F	112.2 ± 16.2		
1965	Atomic absorption	60	M,F	133	80—170	86
1966	Flame photometry	23	F	121 ± 11.4	95—140	87
1967	Atomic absorption	50		119 ± 19	70—165	88
1967	Atomic absorption	28		105 ± 16		89
1968	Atomic absorption	24		108	70—165	90
1969	Colorimetric	117	M,F	101 ± 16	69—133	91
1969	Atomic absorption	50		112	74—160	92
1969	Flame photometry	70	M,F	121.5 ± 12.1	82—148	93
1970	Atomic absorption	100	M	119 ± 18	65—165	94
1971	Emission spectroscopy	58		97 ± 32.1		95
1971	Atomic absorption	82		106	80—147	96
1973	Atomic absorption	240		104	64—184	97
1972	Atomic absorption	36	M,F	113 ± 19	70—150	98
1972	Atomic absorption	33	M	111.2 ± 3	78—150	99
		33	F	129 ± 4	85—188	
1973	Atomic absorption	75	M	101 ± 19	56—143	100
		75	F	112 ± 16	80—148	
1976	Atomic absorption	14	M	104 ± 14		101
		11	F	113 ± 12.5		
1978	Atomic absorption	40	M,F	99 ± 44	62—191	102
1980	Atomic absorption	69	F	121.41 ± 23.87	70—188	103
1980	Atomic absorption	16	M,F	112.81	88—148	104
1980	Atomic absorption	148	F	111		105

concentrations in different groups studied varied from 62 μg/100 mℓ to 191 μg/100 mℓ with the mean value between 96 and 129 μg/100 mℓ. Variations within range could be caused by many factors involved such as methods used, sensitivity of the methods, possible contamination of samples, and selection of controls in regard to age, sex, and race (some of them with possible occult underlying disease or on estrogen medications), and circadian and cyclic patterns and variations within individuals. In children a distinct age-related serum copper level pattern has been observed which will be discussed later.

Our normal range and mean of serum copper level according to age groups, including methods of determinations and comparison of various analytic trace metal techniques used, will be discussed later.

Serum ceruloplasmin-bound copper is more labile than red cell or whole blood copper and its fluctuation has been observed in a number of physiological and pathological conditions.[57-59,61,63,78,80,85,95,98-114] On the basis of RID the normal serum ceruloplasmin concentrations are 31 (25 to 37) mg/100 mℓ in man, and 36 (25 to 47) mg/100 mℓ in women.[72] The copper content of cerebrospinal fluid is low (2 to 7 μg/100 mℓ).[50,97,115]

The urinary copper is normally very low. It ranges from 10 to 60 μg of copper per 24 hr in normal adults.[11,116-119] Circadian variations in serum and urinary copper has been reported.[116,120,121] Lower urinary copper excretion is observed from 8:00 p.m. to 4:00 a.m. than at other times of the day. Alterations of serum and urinary copper concentrations in various pathological conditions will be discussed later. There is a negligible amount of copper in sweat.[122] Synovial fluid collected from patients who died with no evidence of connective tissue disease revealed the mean copper value of 27.5 ± 16.4 μg/100 g.[95] The copper content of normal human aqueous humor is 12 to 20 μg/100 mℓ, varying somewhat with the technique of analysis.[123,124] The copper content of saliva is much lower than that of the blood, mean 31.7 ± 15.1 μg/100 mℓ, and it is recirculated.[50] Bile contains highly variable amounts of copper (0.5 to 1.3 mg excreted in bile daily) which is not surprising since it is the main excretory route of copper.[50,85] The copper content of milk during lactation ranges from 0.62 to 0.89 μg/mℓ.[8,125]

REFERENCES

1. **Linder, C. M. and Munro, H. N.,** Iron and copper during development, *Enzyme,* 15, 111, 1973.
2. **Ramage, H.,** A spectrographic investigation of the metallic content of the liver in childhood, *Proc. R. Soc. B,* 113, 308, 1933.
3. **Fazekas, I. Gy., Romhanyi, I., and Rengei, B.,** Copper content of fetal organs (in Hungarian), *Kiserl. Orvostud,* 15, 230, 1963.
4. **Gerbach, W.,** Untersuchungen über den Kupfergehalt menschlincher and tierischer Organe, *Virchow's Arch. Pathol. Anat. Physiol.,* 264, 171, 1934; 295, 394, 1935.
5. **Hamilton, E. J., Minski, M. J., and Cleary, J. J.,** Problems concerning multielement assay in biological material, *Sci. Total Environ.,* 1, 341, 1972.
6. **Nusbaum, R. E., Alexander, G. J., Butt, E. M., Gilmour, T. C., and DiDio, S. L.,** Some spectrographic studies of trace element storage in human tissues, *Soil Sci.,* 85, 95, 1958.
7. **Widdowson, E. M.,** Trace elements in human development, in *Mineral Metabolism in Pediatrics,* Barltrop, D. and Burland, W. L., Eds., Blackwell Scientific, Oxford, 1969, 85-97.
8. **Underwood, E. J.,** *Trace Elements in Human and Animal Nutrition,* 4th ed., Academic Press, New York, 1977, 56-108.
9. **Sternlieb, I.,** Copper and liver, *Gastroenterology,* 78, 1615, 1980.
10. **Bloomer, L. D. and Lee, G. R.,** *Normal Hepatic Copper Metabolism in Metals and the Liver,* Powel, L. W., Ed., Marcel Dekker, New York, 1978, 179.
11. **Schroeder, H. A., Nason, A. P., Tipton, I. H., and Ballassa, J. J.,** Essential trace metals in man: copper, *J. Chronic Dis.,* 19, 1007, 1966.
12. **Mahoney, J. A., Bush, J. A., Gubler, G. J., Moretz, W. H., Cartwright, G. E., and Wintrobe, M. M.,** Studies on copper metabolism. XV. The excretion of copper by animals, *J. Lab. Clin. Med.,* 46, 702, 1955.
13. **Sternlieb, I. and Steinberg, I. H.,** Radiocopper in diagnosing liver disease, *Semin. Nucl. Med.,* 2, 176, 1972.
14. **Evans, G. W.,** Copper homeostasis in mammalian system, *Physiol. Rev.,* 53, 535, 1973.
15. **Smith, H.,** The distribution of antimony, arsenic, copper and zinc in human tissues, *Forensic Sci. Soc.,* 7, 97, 1967.
16. **Tipton, I. H. and Cook, M. J.,** Trace elements in human tissue, II. Adult Subjects from the United States, *Health Phys.,* 9, 103, 1963.
17. **Martin, G. M.,** Copper content of hair and nails of normal individuals and of patients with hepatolenticular degeneration, *Nature (London),* 202, 903, 1964.

18. **Rice, E. W. and Goldstein, N. P.,** Copper content of hair and nails in Wilson's disease (hepatolenticular degeneration), *Metab. Clin. Exp.*, 10, 1085, 1961.
19. **Cartwright, G. E.,** Copper metabolism in human subjects, in *Copper Metabolism,* McElroy, W. D. and Glass, B., Eds., Johns Hopkins, Baltimore, 1950, 274.
20. **Smythe, W. R., Allen, C. A., Craswell, P. W., Crouch, C. B., Ibels, L. S., Kubo, H., Nunnelley, L. L., and Rudolph, H.,** Trace element analysis in chronic uremia, *Ann. Intern. Med.*, 96, 302, 1982.
21. **Sternlieb, I., Morell, A. G., and Steinberg, I. H.,** The uniqueness of ceruloplasmin in the study of plasma protein synthesis, *Trans. Assoc. Am. Phys.*, 75, 228, 1962.
22. **Owen, C. A., Jr.,** Copper and hepatic function, in *Biological Roles of Copper (Ciba Found. Symp. 79),* Excerpta Medica, Amsterdam, 1980, 267-282.
23. **Bremner, I.,** Absorption, transport and distribution of copper, in *Biological Roles of Copper (Ciba Found. Symp. 79),* Excerpta Medica, Amsterdam, 1980, 23-48.
24. **Goldfisher, S. and Bernstein, J.,** Lipofuscein (aging) pigment granules of the newborn human liver, *J. Cell. Biol.*, 42, 253, 1969.
25. **Danks, D. M., Campbell, A. B., Walker-Smith, J., Stevens, B. J., Gilespie, J. M., Blomfield, J., and Turner, B.,** Menkes' kinky hair syndrome, *Lancet,* 1, 1100, 1972.
26. **Porter, H.,** Neonatal hepatic mitochondrocuprein: The nature, submitochondrial localization, and possible function of the copper accumulating physiologically in the liver of newborn animals, in *Trace Element Metabolism in Animals,* Mills, C. F., Ed., E. & S. Livingstone, Edinburgh, 1970, 237.
27. **Porter, H., Wiener, W., and Barker, M.,** The intracellular distribution of copper in immature liver, *Biochim. Biophys. Acta,* 52, 419, 1961.
28. **Porter, H. and Hills, J. R.,** The half-cystein rich copper protein of newborn liver. Probable relationship to metallothionein and subcellular localization in non-mitochondrial particles probably representing heavy lysosomes, in *Trace Element Metabolism in Animals,* Hoekstra, W. G. et al., Eds., University Park Press, Baltimore, 1974, 482.
29. **Porter, H., Sweeney, M., and Potter, E. M.,** Human hepatocuprein, isolation of a copper-protein from the subcellular soluble fraction of adult human liver, *Arch. Biochem. Biophys.*, 105, 97, 1964.
30. **Gregoriadis, G. and Sourkes, T. L.,** Intracellular distribution of copper in the liver of the rat, *Can. J. Biochem.*, 45, 1841, 1967.
31. **Evans, G. W., Myron, D. R., Cornatzer, N. F., and Cornatzer, W. E.,** Age dependent alterations in hepatic subcellular copper distribution and plasma ceruloplasmin, *Am. J. Physiol.*, 218, 298, 1970.
32. **Porter, H.,** Tissue copper proteins in Wilson's disease, *Arch. Neurol.*, 11, 341, 1964.
33. **Verity, M. A., Gambell, J. K., Reith, A. R., and Brown, W. J.,** Subcellular distribution and enzyme changes following subacute copper intoxication, *Lab. Invest.*, 16, 580, 1967.
34. **Whanger, P. D. and Weswig, P. H.,** Effects of some copper antagonists on induction of ceruloplasmin in the rat, *J. Nutr.*, 100, 341, 1970.
35. **Whanger, P. D. and Weswig, P. H.,** Effects of supplementary zinc on the intracellular distribution of hepatic copper in rats, *J. Nutr.*, 101, 1093, 1971.
36. **Berger, N. A. and Eichhorn, G. L.,** Interaction of metal ions with polynucleotides and related compounds. XV. Nuclear magnetic resonance studies of the binding of copper (II) to nucleotides and polynucleotides, *Biochemistry,* 10, 1857, 1971.
37. **Howell, J. S.,** Histochemical demonstration of copper in copper-fed rats and in hepatolenticular degeneration, *J. Pathol. Bacteriol.*, 77, 473, 1959.
38. **Wesenberg, R. L., Gwinn, J. L., and Barnes, G. R.,** Radiological findings in kinky hair syndrome, *Radiology,* 92, 500, 1969.
39. **Wolff, S. M.,** Copper deposition in the rat, *Arch. Radiol.*, 69, 217, 1960.
40. **Goldfischer, S. S.,** The localization of copper in the pericanalicular granules (lysosomes) of liver in Wilson's disease (hepatolenticular degeneration), *Am. J. Pathol.*, 46, 977, 1965.
41. **Uzman, L. L.,** The intrahepatic distribution of copper in relation to the pathogenesis of hepatolenticular degeneration, *Arch. Pathol.*, 64, 464, 1957.
42. **Goldfischer, S. and Moskal, J.,** Electroprobe microanalysis of liver in Wilson's disease. Simultaneous assay for copper and for lead deposited by acid phosphatase activity in lysosomes, *Am. J. Pathol.*, 48, 305, 1966.
43. **Barka, T., Scheuer, P., Schaffner, F., and Popper, H.,** Structural changes of liver cells in copper intoxication, *Arch. Pathol.*, 78, 331, 1964.
44. **Goldfisher, S.,** Liver cell lysosomes in Wilson's disease: acid phosphatase activity by light and electron microscopy, *Am. J. Pathol.*, 43, 511, 1963.
45. **Lindquist, R. R.,** Studies on the pathogenesis of hepatolenticular degeneration. I. Acid phosphatase activity in copper-loaded livers, *Am. J. Pathol.*, 51, 471, 1967.
46. **McNary, W. F., Jr.,** The intrahepatic and intracellular distribution of copper following chronic administration of the metal in the diet, *Anat. Rec.*, 146, 193, 1963.

47. **Schaffner, F., Sternlieb, I., Barka, T., and Popper, H.,** Hepatocellular changes in Wilson's disease. Histochemical and electron microscopic studies, *Am. J. Pathol.*, 41, 315, 1962.
48. **Goldfischer, S. and Sternlieb, I.,** Changes in the distribution of hepatic copper in relation to the progression of Wilson's disease (hepatolenticular degeneration), *Am. J. Pathol.*, 53, 883, 1968.
49. **Owen, C. A., Dickson, E. R., Goldstein, N. P., Baggenstoss, A. H., and McCall, J. T.,** Hepatic subcellular distribution of copper, *Mayo Clin. Proc.*, 52, 73, 1977.
50. **Sass-Kortsak, A.,** Copper metabolism, in *Advances in Clinical Chemistry*, Vol. 8, Sobotka, H. and Stewart, C. P., Eds., Academic Press, New York, 1965, 1-67.
51. **Mann, T. and Keilin, D.,** Hemocuprein and hepatocuprein: copper-protein compounds of blood and liver in mammals, *Proc. R. Soc. London, B*, 126, 303, 1938.
52. **Markowitz, H., Cartwright, G. E., and Wintrobe, M. M.,** Studies on copper metabolism. XVII. Isolation and properties of erythrocyte cuproprotein (erythrocuprein), *J. Biol. Chem.*, 234, 40, 1959.
53. **Lee, G. R., Williams, D. M., and Cartwright, G. E.,** Role of copper in iron metabolism and heme biosynthesis, in *Trace Elements in Human Health and Disease*, Vol. 1, Zinc and Copper, Prasad, A. S., Ed., Academic Press, New York, 1976, chap. 23.
54. **Carico, R. J. and Deutsch, H. A.,** Isolation of human erythrocuprein and cerebrocuprein, *J. Biol. Chem.*, 244, 6087, 1969.
55. **Bush, J. A., Mahoney, J. P., Gubler, C. J., Cartwright, G. E., and Wintrobe, M. M.,** Studies on copper metabolism. XXI. The transfer of radiocopper between erythrocytes and plasma, *J. Lab. Clin. Med.*, 47, 898, 1956.
56. **Neumann, P. Z. and Silverberg, M.,** Metabolic pathways of red blood cell copper in normal humans and in Wilson's disease, *Nature (London)*, 213, 775, 1967.
57. **Versieck, J., Barbier, F., Speeck, A., and Hoste, J.,** Manganese, copper and zinc concentrations in serum and packed blood cells during acute hepatitis, chronic hepatitis and post-necrotic cirrhosis, *Clin. Chem.*, 20, 1141, 1974.
58. **Bogden, J. D., Lintz, D. I., Joselow, M. M., Charles, J., and Salak, J. S.,** Effect of pulmonary tuberculosis on blood concentrations of copper and zinc, *Am. J. Clin. Pathol.*, 67, 251, 1977.
59. **Prasad, A. A., Ortega, J., Brewer, G. J., Oberleas, D., and Schoomaker, E. B.,** Trace elements in sickle cell disease, *J. Am. Med. Assoc.*, 235, 2396, 1976.
60. **Khahil, M., El Khateeb, S., Aref, K. J., and El Lozy, M.,** Plasma and red cell copper in infantile diarrhea, *Gaz. Egypt. Paediatr. Assoc.*, 22, 105, 1974.
61. **Williams, D. M., Atkins, C. L., Frens, D. B., and Bray, P. F.,** Menkes' kinky hair syndrome: studies of copper metabolism and long term copper therapy, *Pediatr. Res.*, 11, 823, 1977.
62. **Shields, G. S., Markowitz, S. H., Klassen, W. H., Cartwright, G. E., and Wintrobe, M. M.,** Studies on copper metabolism. XXXI. Erythrocytic copper, *J. Clin. Invest.*, 40, 2007, 1961.
63. **Lahey, M. E., Gubler, C. J., Cartwright, G. E., and Wintrobe, M. M.,** Studies on copper metabolism. VII. Blood copper in pregnancy and various pathological states, *J. Clin. Invest.*, 32, 329, 1953.
64. **Gubler, C. J., Lahey, G. E., Cartwright, G. E., and Wintrobe, M. M.,** Studies on copper metabolism. IX. The transportation of copper in blood, *J. Clin. Invest.*, 32, 405, 1953.
65. **Wintrobe, M. M., Cartwright, G. E., and Gubler, C. J.,** Studies on function and metabolism of copper, *J. Nutr.*, 50, 395, 1953.
66. **Fay, J., Cartwright, G. E., and Wintrobe, M. M.,** Studies on free erythrocyte protoporphyrin, serum iron, serum iron-binding capacity and plasma copper during pregnancy, *J. Clin. Invest.*, 28, 487, 1949.
67. **Dreosti, I. E. and Quicke, G. V.,** Blood copper as an indication of copper status with a note on serum proteins and leucocyte counts in copper-deficient rats, *Br. J. Nutr.*, 22, 1, 1968.
68. **Holmberg, C. G. and Laurell, C. B.,** Investigations in serum copper. I. Nature of serum copper and its relation to the iron-binding protein in human serum, *Acta Chem. Scand.*, 1, 944, 1947.
69. **Holmberg, C. G. and Laurell, C. B.,** Investigations in serum copper. II. Isolation of the copper-containing protein and the description of some of its properties, *Acta Chem. Scand.*, 2, 550, 1948.
70. **Holmberg, C. B. and Laurell, C. B.,** Investigations in serum copper. III. Ceruloplasmin as an enzyme, *Acta Chem. Scand.*, 5, 476, 1951.
71. **Holmberg, C. G. and Laurell, C. B.,** Oxidase reactions in human plasma caused by ceruloplasmin, *Scand. J. Clin. Lab. Invest.*, 3, 103, 1951.
72. **Wintrobe, M. M.,** *Clinical Hematology*, 7th ed., Lea & Febiger, Philadelphia, 1974, 150.
73. **Sarkar, B. and Kruck, A. P. T.,** Copper amino-acid complexes in human serum, in *The Biochemistry of Copper*, Peisach, J., Alsen, P., and Blumberg, V. E., Eds., Academic Press, New York, 1966, 183-196.
74. **Lahey, M. E., Gubler, C. J., Cartwright, G. E., and Wintrobe, M. M.,** Studies on copper metabolism. VI. Blood copper in normal human subjects, *J. Clin. Invest.*, 32, 322, 1953a.
75. **Heilmeyer, L., Keiderling, W., and Strove, C.,** *Kupfer and Eisen als Körpereigene Wirkstoffe und ihre Bedeutung beim Krankheitsgeschehen*, Fischer, Jena, Germany, 1941.

76. **Rosenthal, R. W. and Blackburn, A.,** Higher copper concentrations in serum than in plasma, *Clin. Chem.*, 20, 1233, 1974.
77. **McMurray, C. H.,** Copper deficiency in ruminants, in *Biological Roles of Copper (Ciba Found. Symp. 79)*, Excerpta Medica, Amsterdam, 1980, 195.
78. **Valee, B. L.,** Time course of serum copper concentrations of patients with myocardial infarctions, *Metabolism*, 1, 420, 1952.
79. **Koch, H. J., Jr., Smith, E. R., Shimp, N. F., and Connor, J.,** Analysis of trace elements in human tissues. I. Normal tissues, *Cancer*, 9, 499, 1956.
80. **Chen, P. E.,** Abnormalities of copper metabolism in Wilson's disease: a preliminary report, *Chin. Med. J.*, 75, 917, 1957.
81. **Rice, E. W.,** Spectrophotometric determination of serum copper with oxylidihydrazide, *J. Lab. Clin. Med.*, 55, 325, 1960.
82. **Herring, W. B., Leavell, B. S., Palkao, L. M., and Voe, J. H.,** Trace metals in human plasma and red blood cells. A study of magnesium, chromium, nickel, copper and zinc. I. Observations of normal subjects, *Am. J. Clin. Nutr.*, 8, 846, 1960.
83. **Welshman, S. G.,** The determination of serum copper, *Clin. Chim. Acta*, 5, 497, 1960.
84. **Cartwright, G. E., Markowitz, H., Shields, G. S., and Wintrobe, M. M.,** Studies on copper metabolism, *Am. J. Med.*, 28, 555, 1960.
85. **Jensen, K. B., Thorling, E. G., and Anderson, C. F.,** Serum copper in Hodgkin's disease, *Scand. J. Haematol.*, 1, 62, 1964.
86. **Berman, E.,** Determination of copper in biological material by atomic absorption spectrophotometry, *Perkins-Elmer At. Absorpt. News Lett.*, 4, 296, 1965.
87. **Carruthers, M. E., Hobbs, C. B., and Warren, R. L.,** Raised serum copper and ceruloplasmin levels in subjects taking oral contraceptives, *J. Clin. Pathol.*, 19, 498, 1966.
88. **Sunderman, F. W., Jr. and Roszel, N. D.,** Measurement of copper in biological materials by atomic absorption spectrophotometry, *Am. J. Clin. Pathol.*, 48, 286, 1967.
89. **Parker, M. M., Humoller, F. L., and Mahler, D. J.,** Determination of copper and zinc in biological material, *Clin. Chem.*, 13, 40, 1967.
90. **Dawson, J. B., Ellis, D. J., and Newton-John, H.,** Direct estimation of copper in serum and urine by atomic absorption spectroscopy, *Clin. Chim. Acta*, 21, 33, 1968.
91. **Hrgovcic, M., Tessmer, C. F., Minckler, T. M., Mosier, B., and Taylor, G. H.,** Serum copper levels in lymphoma and leukemia. Special reference in Hodgkin's disease, *Cancer*, 21, 743, 1968.
92. **Blomfield, J. and MacMahon, R. A.,** Micro-determination of plasma and erythrocyte copper by atomic absorption spectrophotometry, *J. Clin. Pathol.*, 22, 136, 1969.
93. **Warren, R. L., Jellifle, A. M., Watson, J. V., and Hobbs, C. W.,** Prolonged observations on variations in the serum copper in Hodgkin's disease, *Clin. Radiol.*, 20, 247, 1969.
94. **Sinha, S. N. and Gabrieli, E. R.,** Serum copper and zinc in various pathological conditions, *Am. J. Clin. Pathol.*, 54, 570, 1970.
95. **Niedermeyer, W. and Grigs, J. H.,** Trace metal composition of synovial fluid and blood serum of patients with rheumatoid arthritis, *J. Chronic Dis.*, 23, 527, 1971.
96. **Meret, S. and Henkin, R. I.,** Simultaneous direct estimation by atomic absorption spectrophotometry of copper and zinc in serum, urine and cerebrospinal fluid, *Clin. Chem.*, 17, 369, 1971.
97. **Heinemann, G.,** Eisen-Kupfer-und Zinkanalysen unter Antwendung der Atom-Absorptiones-Sperktral-photometrie, *Z. Klin. Chem. Klin. Biochem.*, 10, 467, 1972.
98. **Mortazavi, S. H., Bani-Hashemi, A., Mozafari, M., and Raffi, A.,** Value of serum copper measurement in lymphomas and several other malignancies, *Cancer*, 29, 1193, 1972.
99. **Zackheim, S. H. and Wolf, P.,** Serum copper in psoriasis and other dermatoses, *J. Invest. Derm.*, 58, 28, 1972.
100. **Ray, R. G., Wolf, H. P., and Kaplan, H. S.,** Value of laboratory indicators in Hodgkin's disease: preliminary results, *Natl. Cancer Inst. Monogr.*, 36, 315, 1973.
101. **Fisher, G. L., Byers, V. S., Shifrine, M., and Levine, A. S.,** Copper and zinc levels in serum from human patients with sarcomas, *Cancer*, 37, 356, 1976.
102. **Davidof, G. N., Votaw, M. L., Coon, W. W., Hultquist, D. E., Filter, B. J., and Wexler, S. M.,** Elevations in serum copper, erythrocytic copper, and ceruloplasmin concentrations in smokers, *Am. J. Clin. Pathol.*, 70, 790, 1978.
103. **Garofalo, J. A., Ashikari, H., Lesser, M. L., Menendez-Botet, C., Cunningham-Rundles, S., Schwartz, M. K., and Good, R. A.,** Serum zinc, copper, and the Cu/Zn ratio in patients with benign and malignant breast lesions, *Cancer*, 46, 2682, 1980.
104. **Shah-Reddy, I., Khilanani, P., and Bishop, C. R.,** Serum copper levels in non-Hodgkin's lymphoma, *Cancer*, 45, 2156, 1980.

105. **Prema, I.,** Predictive value of serum copper and zinc in normal and abnormal pregnancy, *Indian J. Med. Res.,* 71, 554, 1980.
106. **Graham, G. G.,** Human copper deficiency, *N. Engl. J. Med.,* 285, 857, 1971.
107. **Scheinberg, I. H., Cook, C. D., and Murphy, J. A.,** The concentration of copper and ceruloplasmin in maternal and infant plasma at delivery, *J. Clin. Invest.,* 33, 963, 1954.
108. **Markowitz, H., Gubler, C. J., Mahoney, J. P., Cartwright, G. E., and Wintrobe, N. M.,** Studies on copper metabolism. XIV. Copper, ceruloplasmin, and oxidase activity in sera of normal human subjects, pregnant women, and patients with infection, hepato-lenticular degeneration and nephrotic syndrome, *J. Clin. Invest.,* 34, 1498, 1955.
109. **Briggs, M., Austin, J., and Staniford, M.,** Oral contraceptives and copper metabolism, *Nature (London),* 225, 81, 1970.
110. **Cartei, G., Meani, A., and Cansarano, D.,** Rise of serum transferrin, serum iron and ceruloplasmin in men due to "esoestrolo" and "clorotriamisene", *Nutr. Rep. Int.,* 2, 343, 1970.
111. **Russ, E. M. and Raymunt, J.,** Influence of estrogens on total serum copper and ceruloplasmin, *Proc. Soc. Exp. Biol. Med.,* 92, 465, 1956.
112. **Scheinberg, I. H. and Sternlieb, I.,** Wilson's disease and the concentration of ceruloplasmin in serum, *Lancet,* 2, 1420, 1963.
113. **Andrews, G. S.,** Studies on plasma zinc, copper, ceruloplasmin and growth hormone, *J. Clin. Pathol.,* 32, 325, 1979.
114. **Linder, M. C., Moor, J. R., and Wright, K.,** Ceruloplasmin assays in diagnosis and treatment of human lung, breast and gastrointestinal cancers, *J. Natl. Cancer Inst.,* 67, 263, 1981.
115. **McCall, J.T., Goldstein, N. P., and Smith, L. H.,** Implications of trace metals in human diseases, *Fed. Proc.,* 30, 1011, 1971.
116. **Agarwal, B. N. and Agarwal, P.,** Zinc and copper in nephrology, in *Zinc and Copper in Medicine,* Karciogly, Z. A. and Sarper, R. M., Eds., Charles C Thomas, Springfield, Ill., 1980, chap. 13.
117. **Butler, E. J. and Newman, G. E.,** The urinary excretion of copper and its concentrations in the blood of normal human adults, *J. Clin. Pathol.,* 9, 157, 1956.
118. **Piper, K. G. and Higgins, G.,** Estimation of trace metals in biological material by atomic absorption spectrophotometry, *Proc. Assoc. Clin. Biochem.,* 4, 190, 1967.
119. **Spector, H., Glusman, S., Jatlow, P., and Seligson, D.,** Direct determination of copper in urine by atomic absorption spectrophotometry, *Clin. Chim. Acta,* 31, 5, 1971.
120. **Lipschitz, M. D. and Henkin, R. I.,** Circadian variation in copper and zinc in man, *J. Appl. Physiol.,* 31, 88, 1971.
121. **Munch-Peterson, S.,** The variations in serum copper in the course of 24 hours, *J. Clin. Lab. Invest.,* 2, 48, 1950.
122. **Mitchell, H. H. and Hamilton, T. S.,** The dermal excretion under controlled environmental conditions of nitrogen and minerals in human subjects, with particular reference to calcium and iron, *J. Biol. Chem.,* 178, 345, 1949.
123. **Ellis, P. Ph.,** Ocular deposition of copper in hypercupremia, *Am. J. Ophthalmol.,* 68, 423, 1969.
124. **Hemmingsen, L. and Other, A.,** Disc electrophoresis of aqueous humor, *Acta Ophthalmol.,* 45, 359, 1967.
125. **Cavell, P. A. and Widdowson, E. M.,** Intakes and excretions of iron, copper, and zinc in the neonatal period, *Arch. Dis. Child.,* 39, 496, 1964.

Chapter 4

BIOCHEMISTRY OF COPPER

Our understanding of the biological role of copper as an essential component of key metalloenzymes in a variety of catalytic reactions has escalated during the past 3 decades. Deficiencies have been described in numerous areas, and where high concentrations occurred in soils and plants, toxic conditions have been identified in ruminant animals.[1,2] From the standpoint of human health, copper roles in three physiological functions are of prime importance.[3] Copper is involved in hemopoeisis, maintenance of vascular and skeletal integrity, and structure and function of the central nervous system. This metal in the body is in a complexed form with proteins, peptides, amino acids, and probably some other organic substances. An exception may be the stomach contents where a relatively high degree of acidity may allow the solution of copper ions.[4]

Of the serum proteins, albumin and transferrin have known copper-binding properties. When, however, copper is added to serum in excess of a 1:1 molar ratio in respect to albumin, then several other globulins exhibit some copper-complexing properties. There are a few proteins of which copper is an integral part, and because of this they are termed cuproenzymes. Copper in these proteins is part of the molecular structure and not in dissociation equilibrium with ionic copper in the solution; there is a characteristic ratio of moles of protein and atoms of associated copper.[4]

Common mammalian cuproenzymes can be classified by function.[3,5-7] The largest class, the oxidases (ceruloplasmin, cytochrome c oxidase, and amine oxidase), reduce oxygen to either water or hydrogen peroxide. The hydroxylases (tyrosinase and dopamine-beta-hydroxylase) actually utilize one oxygen atom from molecular oxygen into their substrate with the resultant hydroxylation of the substrate. Superoxide dismutases occupy a unique position due to their biological and structural diversity.[8]

Ceruloplasmin was first recognized and later characterized as the major copper-containing protein in the blood by Holmberg and Laurell in a series of reports dating from 1947 to 1951.[9-12] Since then the chemical, physical, and enzymatic characteristics of the ceruloplasmin molecule have undergone extensive investigation.[13-67] Ceruloplasmin in humans is a single-chain protein found in the alpha-2-globulin fraction with a molecular weight of 132,000 and a copper content of 6 atoms per molecule (0.30 $\pm$ 0.03%), composed of three, possibly four, different types of copper.[66,67] It contains 9 sialic acids per molecule and 7 to 8% carbohydrate.[10] Because of its dark blue color, this complex is called ceruloplasmin (Latin *caeruleus* — dark blue). The blue color and the enzyme activity of ceruloplasmin disappear when copper is removed from the molecule or when the protein is treated with reducing substances, due to the valence change of the copper.[4,26] Under normal circumstances, the biological half-life of this metalloprotein is 4.2 to 5.2 days;[21,25] its daily turnover is about 0.5 to 1.0 mg.[29] Ceruloplasmin is synthesized and catabolized in the liver, the key organ of copper homeostasis.[53,54,67-69] At least two other ferroxidases have been found to exist in plasma.[43,70]

According to Frieden,[66,67] ceruloplasmin functions in copper transport, iron mobilization and oxidation, and it is the major molecular link between copper and iron metabolism. It is the primary serum antioxidant, preventing deleterious oxidation of polyenoic acids and other substrates. It can also serve as a scavenger for superoxide dismutase anion radicals, but this function is much weaker than that of the superoxide dismutase. Frieden[67] also states that ceruloplasmin serves as an acute phase reactant and endogenous modulator of the inflammatory response. Finally, ceruloplasmin may function as a regulator of epinephrine and serotonin concentrations.

Another property of ceruloplasmin is its immunosuppressive effect on lymphocyte response in vitro, probably related not to its copper content but rather to its glycoprotein nature.[71,72]

Alternative ferroxidase activities, consisting of proteins other than ceruloplasmin, substances that have strong electron-accepting groups, transforming Fe(II) to Fe(III) and compounds that strongly and preferentially chelate Fe(III) enhancing the Fe(II) to Fe(III) reaction, have been proposed.[31] Furthermore, ceruloplasmin function as an acute phase reactant of endogenous modulator in patients with estrogen imbalances, inflammations, and malignant diseases has not been unanimously accepted.[73-76] Recently, Itoh et al.[77] reported antitumor and toxohormone-neutralizing activities of ceruloplasmin against the basic protein isolated from Ehrlich carcinoma cells and sarcoma 180 cells in tumor-bearing animals. It has also been reported that ceruloplasmin serves as a powerful lipid-autooxidation inhibitor.[78]

The role of this metalloenzyme in iron metabolism has been intensively studied and it is still somewhat controversial. The recognition of anemia in the copper-deficient animal was first reported in 1928,[79] and the subject has been extensively researched by Wintrobe and associates,[13-17,20,22,23,28,36-38,61,80-91] as well as by other groups.[4,5-7,9,12,31-32,43,44,48,55-58,60,63-67,92-95] Before 1952 it was believed that copper was necessary only for iron utilization in blood formation and iron mobilization from the tissues.[96] Since then, studies by Gubler et al.[13,16] showed that copper increased radioiron absorption in rats. Furthermore, these and subsequent studies by Lee et al.[36-38,61] showed an impaired ability of copper-deficient swine to properly absorb and transfer iron to the plasma. The amounts of stainable iron in fixed sections of the duodenal mucosa of these animals was increased. Hypoferremia was observed, even though iron stores were normal or increased. Copper administration promptly increased plasma iron.

The explanation for these abnormalities was given in terms of a failure of the duodenal mucosa, the reticuloendothelial system, and the hepatic parenchymal cells to release iron to the plasma, consistent with suggestions by Osaki et al.[31] and Frieden[44] that tissue-plasma iron transfer requires the enzymatic oxidation of ferrous iron by ceruloplasmin. Accumulation of nonhemoglobin iron in the normoblasts supported the hypothesis of Lee et al.[61] that an additional intracellular defect resides in these cells which is a major cause of the anemia.

According to Lee et al.,[61] certain clinical and experimental observations do not support the Osaki and Frieden hypotheses. In patients with Wilson's disease, in spite of a low ceruloplasmin level, there is no evidence that defective iron mobilization exists. Anemia is not a feature of Wilson's disease except when liver failure is severe or during transient episodes of hemolysis. In the copper-deficient swine, hypoferremia was not observed until the ceruloplasmin levels fell below 1% of normal. The fact that hypoceruloplasminemia of this severity is uncommon in untreated Wilson's disease may explain the absence of defects in iron metabolism.

Kinetic evidence for an iron-ceruloplasmin compound that might serve as an intermediate in the transfer of iron to transferrin by way of a specific exchange reaction has been reported.[32] This supports the theory of Lee et al.[61] that ferrous iron occupies specific sites on the reticuloendothelial cell surface and that ceruloplasmin interacts with these sites to remove and transfer iron to transferrin.

Clinical reports have not as yet provided strong consistent evidence to fully support either hypothesis. The observations of Graham and Cordano[60] support the concept of two effects of copper deficiency on iron metabolism in man. The first, an impairment of iron absorption or mobilization occurs early as a result of copper deficiency. They suggested that the impairment of iron metabolism may be due to the loss of ferroxidase activity, not necessarily ceruloplasmin. The second, later, effect was an inadequate erythropoiesis, even in the presence of abundant iron stores. The anemia responded promptly to copper administration. The accompanying neutropenia responded better to free copper than to ceruloplasmin.

According to O'Dell,[3] the primary role of copper in hemopoesis is unclear. Copper is essential for iron release from the intestinal mucosa and from storage sites, and for development of reticulocytes. The facts that the occurrence of anemia and hypoferremia are rare, and that newborn normal infants with the lowest ceruloplasmin levels had the highest levels of iron-saturated transferrin while those with the highest ceruloplasmin had the lowest transferrin levels, were quoted as evidence against the concept that ceruloplasmin is a link between copper and iron metabolism.

It is, therefore, clear that in spite of many good observations and supporting data of the important role of ferroxidase activity of ceruloplasmin in iron metabolism, the functions of ceruloplasmin and the mechanism whereby ceruloplasmin exerts its effects under normal and pathological conditions are still not completely understood.

Cytochrome c oxidase is the terminal oxidase in the mitochondrial electron transport system catalyzing the transfer of electrons from cytochrome c to oxygen. This transfer is obviously a key reaction in energy metabolism, but it is difficult to associate the enzyme with a specific pathology.[3,68,97,98] Cytochrome c oxidase is a membrane-bound heme protein composed of cytochromes a and a3, protein, and lipid. It contains 1 g-atom of copper per mole of heme and a molecular weight of 270,000, of which 20% is estimated to be lipid.[5,68,97-100] It is found in particularly high concentrations in active tissues such as heart muscle, flight muscles of birds and insects, liver mitochondria, and brain gray matter.[3,5] It is estimated that more than 90% of the energy required for muscle contraction is provided by the respiratory chain involving cytochrome c oxidase.[98,100] The defective utilization of iron to form heme within normoblasts has also been linked to cytochrome oxidase deficiency. It is suggested that cytochrome oxidase maintains a flow of electrons essential to the availability of ferrous iron within the mitochondrion and heme synthesis.[5,61] Low brain cytochrome oxidase levels have been associated with neonatal ataxia, but a casual relationship has not as yet been established.[1,3]

Amine oxidases are enzymes catalyzing the oxidative deamination of amines by molecular oxygen to give an aldehyde and hydrogen peroxide.[3,5] The many different amine oxidases found in vertebrate tissues can be divided into two major types based on the cofactors involved.[3,5,101,102] Mitochondrial membrane-bound monoamine oxidases are flavoproteins that do not contain copper, but iron is present in some of them. The other general class of amine oxidases contain copper and pyrodoxal phosphate, or a closely related compound.[3,5] Among the amine oxidases of which copper is the primary factor are benzylamine oxidase isolated from pig plasma,[103,104] diamine oxidase from pig kidney,[105] spermine oxidase isolated from bovine plasma,[104,106,107] and lysyl oxidase isolated from connective tissues.[3,108-111] Lysyl oxidase has an important function in collagen and elastin metabolism.[3,112,113] Connective tissues also contain amine oxidases whose properties are highly similar to those found in plasma of the same species.[114-118] Copper deficiency is shown to have an adverse effect on connective tissues such as collagen and elastin, resulting in bone abnormalities and cardiovascular defects.[3,109-113] This is believed to be due to an impaired synthesis of cross-linking compounds of elastin or collagen, i.e., impaired peptidyl aldehyde (allysine), a step catalyzed by copper enzyme lysyl oxidase.[3,109-113]

The mammalian enzyme tyrosinase catalyzes the oxidation of tyrosine to 3,4-dihydroxyphenylanaline (dopa) and the oxidation of dopa to the quinone, which is ultimately used for melanin synthesis. Copper deficiency results in failure of pigmentation of hair and wool in animals, which is believed to be caused by a decrease in the tyrosinase level, which, in turn, reduces melanin formation.[1,3,5]

In 1938 Mann and Keilin[119] isolated two copper proteins from bovine erythrocytes and liver, which they named hemocuprein and hepatocuprein, respectively. In 1969, McCord and Fridovich[120] isolated an enzyme catalyzing the dismutation of superoxide ion from bovine blood and showed that this enzyme was identical to hemocuprein (erythrocuprein); the name superoxide dismutase was proposed. Later, Porter and Floch[121] isolated a similar protein from human brain tissue which was named cerebrocuprein. Since no apparent enzymatic activity was found for these proteins, and since they accounted for as high as 60% of copper found in the tissue, these proteins were thought to be copper storage sites.[86,88,121] Carrico and Deutsch[122] showed that erythrocuprein, hepatocuprein, and cerebrocuprein were, in fact, identical. They suggested that the name cytocuprein be used.

Superoxide dismutases (cytocupreins) are metalloenzymes that catalyze the dismutation of the superoxide anion free radical into hydrogen peroxide and oxygen and scavenge oxygen atoms in certain reactions.[3,123-125] They comprise a disparate class of metalloproteins containing either copper, iron, or manganese as essential cofactors for catalytic reactions.[8] The copper/zinc superoxide dismutase consists of two identical subunits, each containing one atom of copper and one atom of zinc. The copper participates in the catalytic activity of the enzyme, while the zinc plays only a structural role. The molecular weight of cytocuprein is about 33,000 and contains 0.34% of copper.[120,123-125] It is present in cytosol and in the intermembraneous space mitochondria. It has been suggested that superoxide dismutases provide protetion against oxygen toxicity, ionizing radiation, and also against oxygen toxicity, ionizing radiation, and also against the damaging sequelae of prolonged inflammation. They, also, may prevent injury to hepatocytic components from superoxide radicals generated by aerobic metabolic reactions.[124,125] The real biological function(s) of these proteins remains to be established.[8] No pathology in higher animals has been attributed to lack of dismutase activity.[3,125] Several reviews are available which cover various aspects of superoxide dismutases.[3,5,8,123-125]

There are several other cuproenzymes unassociated with specific pathologies besides cytochrome c oxidase and superoxide dismutase; among these are uricase and dopamine-beta-hydroxylase.[3,5,53,126] This lack of association to the pathology of copper deficiency may be due to several reasons beside the possibility that an intensive search for the association has simply not been made. Perhaps these enzymes bind copper very tightly, or have a slow rate of turnover, or they may be present in such excess that a slight decrease has no notable effect.[3]

The final supernatant or cytosol obtained during differential centrifugation contains the major portion of the total hepatic copper in the adult mammal. Although a small percentage of the copper in the hepatic cytosol is contained within copper-dependent enzymes, the predominant fraction of the metal is associated with a specific metal-binding protein.[53,68]

There are two hepatic low-molecular-weight copper-binding cytosolic proteins.[53,68] Metallothionein or (Cd,Zn)-thionein has a molecular weight of 6000 to 10,000 daltons and a cysteine content of approximately 30%, and no aromatic amino acids. It contains 0.1% copper by weight, but there is as much as 7% of zinc and cadmium.[68,127-131] L-6-D (Cu-LP), with a molecular weight of approximately 10,000, comprises about 2 mg/g of wet human liver and contains up to 3% of copper and little zinc, and much less cysteine content than metallothionein.[53,68,131] A third protein, copper-chelatin, with a molecular weight of 8000 daltons, has been isolated from rats injected with $CuCl_2$, but not from human liver.[132]

Although metallothionein is most likely one of the major copper-binding proteins in the liver, its function in the control of hepatic copper metabolism is still a matter of conjecture.[133] Since metallothionein is multifunctional with respect to metal binding, the protein probably functions in detoxification and storage of trace metals in liver as well as other tissues.[68,131] Kagi and Valee[127] found no evidence for a specific enzymatic function associated with metallothionein. These authors indicated that the physical characteristics of the protein are consistent with a role in various homeostatic mechanisms, including storage, immune phenomena, and detoxification, although the exact function remains obscure. According to Lerch,[134] metallothionein appears to play a role in copper absorption, transport, or detoxification. In 1962, Porter et al.[135] isolated a copper-rich protein from the mitochondria of immature bovine liver which was rich in cysteine (25%), had a 4% copper content, and a molecular weight between 5000 and 10,000. They named it neonatal hepatomitochondrocuprein. Later on, a similar protein was isolated from human fetal liver.[136] Hepatomitochondrocuprein is present in the hepatic lysosomes of neonates and patients with Wilson's disease.[137] Mitochondrial monoamine oxidase, present in liver mitochondria, forms hydrogen peroxide enzymatically. It forms ammonia and aldehydes from several monoamines, in particular epinephrine and serotonin, through oxidative deamination.[138,139] Tryptophan-2,3-dioxygenase is a heme protein present in liver cytosol. It has a molecular weight of 167,000 daltons and catalyzes the insertion of molecular oxygen into the pyrrole ring of L-tryptophan.[140]

Two pale-yellow copper-containing proteins, albocuprein I and II, have also been isolated from the human brain.[141] Albocuprein I has a molecular weight of 72,000 and contains 0.25% copper or 2.8 g-atoms/mol, while albocuprein II is smaller (molecular weight is 14,000) and contains 1.4% copper, or 3.1 g-atom/mol.

Our knowledge of the structure and function of copper-containing proteins and cuproenzymes is still very incomplete. There are many excellent articles, reviews, and books — some of which are listed here — which have appeared recently and should be consulted for more detailed information concerning copper biochemistry and metabolism.[142,143]

REFERENCES

1. **Underwood, E. J.,** *Trace Elements in Human and Animal Nutrition,* 4th ed., Academic Press, New York, 1977, 56-108.
2. **McMurray, C. H.,** Copper deficiency in ruminants, in *Biological Roles of Copper (Ciba Found. Symp. 79),* Excerpta Medica, Amsterdam, 1980, 183-207.
3. **O'Dell, B. L.,** Biochemistry and physiology of copper in vertebrates, in *Trace Elements in Human Health and Disease,* Vol. 1, Zinc and Copper, Prasad, A. S., Ed., Academic Press, New York, 1976, chap. 24.
4. **Sass-Kortsak, A.,** Copper metabolism, *Adv. Clin. Chem.,* 8, 1-67, 1965.
5. **Hsieh, S. H. and Hsu, J. M.,** Biochemistry and metabolism of copper, in *Zinc and Copper in Medicine,* Karcioglu, Z. A. and Sarper, R. M., Eds., Charles C Thomas, Springfield, Ill., 1980, chap. 4.
6. **Frieden, E.,** The complex nature of copper, in *Horizons in Biochemistry,* Kasha, M. and Pullman, B., Eds., Academic Press, New York, 1962, 461.
7. **Frieden, E., Osaki, S., and Kobayashi, H.,** Copper proteins and oxygen, *J. Gen. Physiol.,* 49(2), 213, 1965.
8. **Fee, J. A.,** The copper/zinc superoxide dismutase, in *Metal Ions in Biological Systems,* Vol. 13, Copper Proteins, Sigel, H., Ed., Marcel Dekker, New York, 1981, chap. 8.
9. **Holmberg, C. G. and Laurell, C. B.,** Investigations in serum copper. I. Nature of serum copper and its relation to the iron-binding protein in human serum, *Acta Chem. Scand.,* 1, 944, 1947.
10. **Holmberg, C. G. and Laurell, C. B.,** Investigations in serum copper. II. Isolation of the copper-containing protein and the description of some of its properties, *Acta Chem. Scand.,* 2, 550, 1948.
11. **Holmberg, C. G. and Laurell, C. B.,** Investigations in serum copper. III. Ceruloplasmin as an enzyme, *Acta Chem. Scand.,* 5, 476, 1951.
12. **Holmberg, C. G. and Laurell, C. B.,** Oxidase reactions in human plasma caused by ceruloplasmin, *Scand. J. Clin. Lab. Invest.,* 3, 103, 1951.
13. **Gubler, C. J., Lahey, M. E., Chase, M. S., Cartwright, G. E., and Wintrobe, M. M.,** Studies on copper metabolism. III. The metabolism of iron in copper deficient swine, *Blood,* 7, 1075, 1952.
14. **Gubler, C. J., Lahey, M. E., Ashenbrucker, H., Cartwright, G. E., and Wintrobe, M. M.,** Studies on copper metabolism. I. A method for the determination of copper in whole blood, red blood cells and plasma, *J. Biol. Chem.,* 196, 209, 1952.
15. **Lahey, M. E., Gubler, C. J., Chase, M. S., Cartwright, G. E., and Wintrobe, M. M.,** Studies on copper metabolism. II. Hematologic manifestations of copper deficiency in swine, *Blood,* 7, 1053, 1952.
16. **Chase, M. S., Gubler, C. J., Cartwright, G. E., and Wintrobe, M. M.,** Studies on copper metabolism. IV. The influence of copper on absorption of iron, *J. Biol. Chem.,* 199, 757, 1952a.
17. **Chase, M. S., Gubler, G. J., Cartwright, G. E., and Wintrobe, M. M.,** Studies on copper metabolism. V. Storage of iron in liver of copper deficient rats, *Proc. Soc. Exp. Biol. Med.,* 80, 749, 1952.
18. **Scheinberg, I. H., Cook, C. D., and Murphy, J. A.,** The concentration of copper and ceruloplasmin in maternal and infant plasma at delivery, *J. Clin. Invest.,* 33, 963, 1954.

19. **Bearn, A. G. and Kunkel, H. G.,** Localization of Cu in serum fractions following oral administration: an alteration in Wilson's disease, *Proc. Soc. Exp. Biol. Med.*, 85, 44, 1954.
20. **Markowitz, H., Gubler, C. J., Mahoney, J. P., Cartwright, G. E., and Wintrobe, M. M.,** Studies on copper metabolism. XIV. Copper, ceruloplasmin and oxidase activity in serum of normal human subjects, pregnant women and patients with infection, hepatolenticular degeneration and nephrotic syndrome, *J. Clin. Invest.*, 34, 1498, 1955.
21. **Scheinberg, I. H., Dubin, D., and Harris, R. S.,** The survival of normal ceruloplasmin in patients with hepatolenticular degeneration, *J. Clin. Invest.*, 34, 961, 1955.
22. **Bush, J. A., Jensen, W. N., Athens, J. W., Ashenbrucker, H., Cartwright, G. E., and Wintrobe, M. M.,** Studies on copper metabolism. XIX. The kinetics of iron metabolism and erythrocyte life span in copper-deficient swine, *J. Exp. Med.*, 103, 701, 1956.
23. **Bush, J. A., Mahoney, J. P., Gubler, C. J., Cartwright, G. E., and Wintrobe, M. M.,** Studies on copper metabolism, *J. Lab. Clin. Med.*, 47, 898, 1956.
24. **Scheinberg, I. H. and Morell, A. G.,** Exchange of ceruloplasmin copper with ionic Cu with reference to Wilson's disease, *J. Clin. Invest.*, 36, 1193, 1957.
25. **Scheinberg, I. H., Harris, R. S., Morrell, A. G., and Dubin, D.,** Some aspects of the relation of ceruloplasmin to Wilson's disease, *Neurology*, 8 (Suppl. 1), 44, 1958.
26. **Laurell, C. B.,** Metal-binding plasma proteins and cation transport, in *The Plasma Proteins*, Putnam, F. W., Ed., Academic Press, New York, 1960, 1 and 349.
27. **Sternlieb, I., Morell, A. G., Tucker, W. D., Greene, M. W., and Scheinberg, I. H.,** The incorporation of copper into ceruloplasmin in vivo: studied with copper^{64} and copper^{67}, *J. Clin. Invest.*, 40, 1834, 1961.
28. **Cartwright, G. E., Markowitz, H., Shields, G. S., and Wintrobe, M. M.,** Studies on copper metabolism. XXIX. A critical analysis of serum copper in normal subjects, patients with Wilson's disease and relatives of patients with Wilson's disease, *Am. J. Med.*, 28, 555, 1962.
29. **Scheinberg, I. H.,** Ceruloplasmin, a review, in *The Biochemistry of Copper*, Peisach, J., Aisen, P., and Blumberg, W., Eds., Academic Press, New York, 1966, 513.
30. **Owen, C. A., Jr. and Hazelrig, J. B.,** Metabolism of Cu^{64} labeled copper by the isolated rat liver, *Am. J. Physiol.*, 210, 1059, 1966.
31. **Osaki, S., Johnson, D. A., and Frieden, E.,** The possible significance of the ferrous oxidase activity of ceruloplasmin in normal human serum, *J. Biol. Chem.*, 241, 2746, 1966.
32. **Osaki, S. and Wallace, O.,** Kinetic studies of ferrous iron oxidation with crystalline human ferroxidase. II. Rate constants at various steps and formation of a possible enzyme-substrate complex, *J. Biol. Chem.*, 242, 5053, 1967.
33. **Evans, G. W. and Wiederanders, R. E.,** Pituitary-adrenal regulation of ceruloplasmin, *Nature (London)*, 215, 766, 1967.
34. **Evans, G. W. and Wideranders, R. E.,** Effect of hormones on ceruloplasmin and copper concentrations in the plasma of the rat, *Am. J. Physiol.*, 214, 1152, 1968.
35. **Trip, J. A.,** Ceruloplasmin, Clinical and Experimental Investigations in Origin, Presence and Behavior of Fractions, thesis, University Hospital, Gronigen, 1968.
36. **Lee, G. R., Cartwright, G. E., and Wintrobe, M. M.,** Heme biosynthesis in copper-deficient swine, *Proc. Soc. Exp. Biol. Med.*, 127, 977, 1968.
37. **Lee, G. R., Nacht, S., Lukens, J. N., and Cartwright, G. E.,** Iron metabolism in copper-deficient swine, *J. Clin. Invest.*, 47, 2058, 1968.
38. **Lee, G. R., Nacht, S., Christensen, D., Hansen, S. P., and Cartwright, G. E.,** The contribution of citrate to the ferroxidase activity of serum, *Proc. Soc. Exp. Biol. Med.*, 131, 918, 1969.
39. **Simons, K. and Bearn, A. G.,** Isolation and partial characterization of the polypeptide chains in human ceruloplasmin, *Biochim. Biophys. Acta*, 175, 260, 1969.
40. **Evans, G. W., Myron, D. R., and Wiederanders, R. E.,** Effect of protein synthesis inhibitors on plasma ceruloplasmin in the rat, *Am. J. Physiol.*, 216, 340, 1969.
41. **Evans, G. W., Majors, P. F., and Cornatzer, W. E.,** Induction of ceruloplasmin synthesis by copper, *Biochem. Biophys. Res. Commun.*, 41, 1120, 1970.
42. **Evans, G. W., Cornatzer, N. F., and Cornatzer, W. E.,** Mechanism for hormone induced alterations in serum ceruloplasmin, *Am. J. Physiol.*, 218, 613, 1970.
43. **Topham, R. W. and Frieden, E.,** Identification and purification of a nonceruloplasmin ferroxidase of human serum, *J. Biol. Chem.*, 245, 6698, 1970.
44. **Frieden, E.,** Ceruloplasmin, a link between copper and iron metabolism, *Adv. Chem. Ser.*, 100, 292, 1971.
45. **Owen, C. A., Jr.,** Metabolism of copper^{67} by the copper-deficient rat, *Am. J. Physiol.*, 221, 1722, 1971.
46. **Ryden, L.,** Evidence for proteolytic fragments in commercial samples of human ceruloplasmin, *FEBS Lett.*, 18, 321, 1971.
47. **Ryden, L.,** Comparsion of polypeptide-chain structure of four mammalian ceruloplasmins by gel filtration in guanidine hydrochloride solutions, *Eur. J. Biochem.*, 28, 46, 1972.

48. **Danks, D. M., Champbell, P. E., Stevens, B. J., Mayne, V., and Cartwright, E.,** Menkes' kinky hair syndrome: an inherited defect in copper absorption with widespread effects, *Pediatrics,* 50, 188, 1972.
49. **Vanngard, T.,** Copper proteins, in *Biological Applications of Electron Spin Resonance,* Swartz, H. M., Bolton, J. R., and Berg, B. C., Eds., John Wiley & Sons, New York, 1972, 411-477.
50. **Marceau, N. and Aspin, N.,** Distribution of ceruloplasmin-bound ^{67}Cu in the rat, *Am. J. Physiol.,* 222, 106, 1972.
51. **Marceau, N. and Aspin, N.,** The intracellular distribution of the radio-copper derived from ceruloplasmin and from albumin, *Biochim. Biophys. Acta,* 328, 338, 1973.
52. **Freeman, S. and Daniet, E.,** Dissociation and reconstruction of human ceruloplasmin, *Biochemstry,* 12, 4806, 1973.
53. **Evans, G. W.,** Copper homestasis in mammalian system, *Phys. Rev.,* 53, 535, 1973.
54. **Scheinberg, I. H. and Morell, A. G.,** Ceruloplasmin, in *Inorganic Chemistry,* Eichorn, G. I., Ed., Elsevier, New York, 1973.
55. **Frieden, E.,** The ferrous to ferric cycles in iron metabolism, *Nutr. Rev.,* 31, 41, 1973.
56. **Frieden, E.,** The evolution of metals as essential elements (with special reference to iron and copper), *Adv. Exp. Med. Biol.,* 48, 1, 1974.
57. **Frieden, E. and Osaki, S.,** Ferroxidases and ferrireductases: their role in iron metabolism, *Adv. Exp. Med. Biol.,* 48, 235, 1974.
58. **Hsieh, H. S. and Frieden, E.,** Evidence for ceruloplasmin as copper transport protein, *Biophys. Res. Commun.,* 67, 1326, 1975.
59. **Fee, J. A.,** Copper proteins-systems containing the "blue" copper centre, *Struct. Bond,* 23, 1, 1975.
60. **Graham, G. G. and Cordano, A.,** Copper deficiency in human subjects, in *Trace Elements in Human Health and Diseases,* Vol. 1, Prasad, S. A., Ed., Academic Press, New York, 1976, chap. 22.
61. **Lee, G. R., Williams, D. M., and Cartwright, G. E.,** Role of copper in iron metabolism and heme biosynthesis in *Trace Elements in Human Health and Disease,* Vol. 1, Prasad, A. S., Ed., Academic Press, New York, 1976, 373-390.
62. **Ryden, L. and Björk, I.,** Reinvestigation of some physiochemical and chemical properties of human ceruloplasmin (ferroxidase), *Biochemistry,* 15, 3411, 1976.
63. **Frieden, E.,** Ceruloplasmin, a multifunctional protein, in *Trace Element Metabolism in Man and Animals,* Vol. 3, Arbeitkreis für Tierernährungsforschung, Kirchgessner, M., Ed., Weihenstephan, 1978, 36.
64. **Frieden, E.,** Ceruloplasmin: the serum copper transport protein with oxidase activity, in *Copper in the Environment, Part II,* Nriagu, J. O., Ed., John Wiley & Sons, New York, 1979, 241-284.
65. **Frieden, E. and Hsieh, H. S.,** Ceruloplasmin: the copper transport protein with essential oxidase activity, *Adv. Enzymol.,* 44, 187, 1976.
66. **Frieden, E.,** Ceruloplasmin: a multifunctional metalloprotein of vertebrate plasma, in *Biological Roles of Copper (Ciba Found. Symp. 79),* Excerpta Medica, Amsterdam, 1980, 93.
67. **Frieden, E.,** Ceruloplasmin: a multifunctional metalloprotein of vertebrate plasma, in *Metal Ions in Biological Systems,* Vol. 13, Copper Proteins, Sigel, H., Ed., Marcel Dekker, New York, 1981, chap. 4.
68. **Sternlieb, I.,** Copper and liver, *Gastroenterology,* 78, 1615, 1980.
69. **Owen, C. A.,** Copper and hepatic function, in *Biological Roles of Copper (Ciba Found. Symp. 79),* Excerpta Medica, Amsterdam, 1980, 267-282.
70. **Broman, L.,** Separation and characterization of two ceruloplasmin from human serum, *Nature (London),* 182, 1655, 1958.
71. **Johannsen, R., Haupt, H., Bohn, H., Heide, K., Seiler, F. R., and Schwick, H. G.,** Inhibition of the mixed leukocyte culture (MLC) by proteins: mechanism and specificity of the reaction, *Z. Immun. Forsch.,* 152, 280, 1976.
72. **Larsen, B. and Heron, I.,** Modification of lymphocyte response to phytomitogens by polycations and polyanions, *Experientia,* 34, 1224, 1978.
73. **Linder, M. C., Moor, J. R., and Wright, K.,** Ceruloplasmin assays in diagnosis and treatment of human lung, breast and gastrointestinal cancer, *J. Natl. Cancer Inst.,* 57, 263, 1981.
74. **Fisher, G. F. and Shifrine, M.,** Hypothesis for the mechanism of elevated serum copper in cancer patients, *Oncology,* 35, 22, 1978.
75. **Weissmann, G., Smolen, J. E., and Hoffstein, S.,** Polymorphonuclear leukocytes as secretory organs of inflammation, *J. Invest. Dermatol.,* 71, 95, 1978.
76. **Weissmann, G.,** *Prostaglandin in Acute Inflammation,* The Upjohn Company, Kalamazoo, Mich., 1980.
77. **Itoh, O., Torikai, T., Satoh, M., Okumura, O., and Osawa, T.,** Antitumor and toxohormone-neutralizing activities of human ceruloplasmin, *Gann,* 12, 370, 1981.
78. **Al-Timimi, D. J. and Dormandy, T. L.,** Inhibition of lipid autoxidation by human ceruloplasmin, *Biochem. J.,* 168, 283, 1977.
79. **Hart, E. B., Steenbock, H., Waddell, J., and Elvehjem, C. A.,** Iron in nutrition: copper as a supplement to iron for hemoglobin building in the rat, *J. Biol. Chem.,* 77, 797, 1928.

80. **Cartwright, G. E., Lauritsen, M. A., Jones, P. J., Merrill, I. M., and Wintrobe, M. M.,** The anemia of infection. Hypoferremia, hypercupremia, and alterations in porphyrin metabolism in patients, *J. Clin. Invest.,* 25, 65, 1946.
81. **Cartwright, G. E., Huguley, C. M., Ashenbrucker, H., Fay, J., and Wintrobe, M. M.,** Studies on free erythrocyte protoporphyrin, plasma iron and plasma copper in normal and anemic subjects, *Blood,* 3, 501, 1948.
82. **Cartwright, G. E.,** Copper metabolism in human subjects, in *Copper Metabolism: A Symposium on Animal, Plant and Soil Relationships,* McElroy, W. D. and Glass, B., Eds., Johns Hopkins University Press, Baltimore, 1950, 274-312.
83. **Lahey, M. E., Gubler, C. J., Cartwright, G. E., and Wintrobe, M. M.,** Studies on copper metabolism. VI. Blood copper in normal human subjects, *J. Clin. Invest.,* 32, 322, 1953a.
84. **Lahey, M. E., Gubler, C. J., Cartwright, G. E., and Wintrobe, M. M.,** Studies on copper metabolism. VII. Blood copper in pregnancy and various pathologic states, *J. Clin. Invest.,* 32, 329, 1953b.
85. **Lahey, M. E., Gubler, C. J., Brown, D. M., Smith, E. L., Jager, B. V., Cartwright, G. E., and Wintrobe, M. M.,** Studies on copper metabolism. VIII. The correlation between the serum copper level and various serum protein fractions, *J. Lab. Clin. Med.,* 41, 829, 1953.
86. **Markowitz, H., Cartwright, G. E., and Wintrobe, M. M.,** Studies on copper metabolism. XXVII. The isolation and properties of an erythrocyte cuproprotein (erythrocuprein), *J. Biol. Chem.,* 234, 40, 1953.
87. **Cartwright, G. E., Gubler, C. J., and Wintrobe, M. M.,** Studies on copper metabolism. XI. Copper and iron metabolism in the nephrotic syndrome, *J. Clin. Invest.,* 33, 685, 1954.
88. **Shields, G. S., Markowitz, H., Klassen, W. H., Cartwright, G. E., and Wintrobe, M. M.,** Studies on copper metabolism. XXXI. Erythrocyte copper, *J. Clin. Invest.,* 40, 2007, 1961.
89. **Cartwright, G. E. and Wintrobe, M. M.,** Copper metabolism in normal subjects, *Am. J. Clin. Nutr.,* 14, 224, 1964a.
90. **Cartwright, G. E. and Wintrobe, M. M.,** The question of copper deficiency in man, *Am. J. Clin. Nutr.,* 15, 94, 1964b.
91. **Williams, D. M., Christensen, D. D., Lee, G. R., and Cartwright, G. E.,** Serum acide-resistance ferrioxidase activity, *Biochem. Biophys. Acta,* 350, 129, 1974a.
92. **Cordano, A., Baertl, J. M., and Graham, G. G.,** Copper deficiency in infancy, *Pediatrics,* 34, 324, 1964.
93. **Cordano, A. and Graham, G. G.,** Copper deficiency complicating severe chronic intestinal malabsorption, *Pediatrics,* 38, 596, 1966.
94. **Graham, G. G. and Cordano, A.,** Copper depletion and deficiency in the malnourished infant, *Johns Hopkins Med. J.,* 124, 139, 1969.
95. **Bates, G. W. and Schlabach, M. R.,** The reaction of ferric salts with transferrin, *J. Biol. Chem.,* 248, 3228, 1973.
96. **Elvehjem, C. A.,** The biological significance of copper and its relation to iron metabolism, *Physiol. Rev.,* 15, 471, 1935.
97. **Cohen, E. and Elvejhem, C. A.,** The relation of iron and copper to the cytochrome and oxidase content of animal tissues, *J. Biol. Chem.,* 107, 97, 1934.
98. **Brunori, M., Antonini, E., and Wilson, M. T.,** Cytochrome C oxidase: an overview of recent work, in *Metal Ions in Biological Systems,* Vol. 13, Copper Proteins, Sigel, H., Ed., Marcel Dekker, New York, 1981, chap. 6.
99. **Kelin, D. and Hartee, E. F.,** Cytochrome and cytochrome oxidase, *Proc. R. Soc. (London) Ser. B,* 127, 167, 1939.
100. **Beinert, H.,** Cytochrome C oxidase, a review, in *Biochemistry of Copper,* Peisach, J., Eisen, P., and Blumberg, N. E., Eds., Academic Press, New York, 1966, 213-230.
101. **Sourkes, T. L.,** Copper, biogenic amines, and amine oxidases, in *Biological Roles of Copper (Ciba Found. Symp. 79),* Excerpta Medica, Amsterdam, 1980, 143-156.
102. **Cass, A. E. G. and Hill, H. A. O.,** Copper proteins and copper enzymes, in *Biological Roles of Copper (Ciba Found. Symp. 79),* Excerpta Medica, Amsterdam, 1980, 71-91.
103. **Corper, W. R., Stoddard, D. D., and Martin, D. F.,** Pig liver monoamine oxidase. I. Isolation and characterization, *Biochem. Biophys. Acta,* 334, 287, 1974.
104. **Blaschko, H.,** The amine oxidase of mammalian blood plasma, *Adv. Comp. Physiol. Biochem.,* 1, 67, 1962.
105. **Mondovi, B., Rotilio, G., Costa, M. T., Finazzi-Agro, A., Chiancone, E., Hansen, R. E., and Beinert, H.,** Diamine oxidase from pig kidney. Improved purification and properties, *J. Biol. Chem.,* 242, 1160, 1967.
106. **Yamada, H. and Yasunobu, K. T.,** Monoamine oxidase. II. Copper, one of the prosthetic groups of plasma monoamine oxidase, *J. Biol. Chem.,* 237, 3077, 1962.
107. **Hirsch, J. G.,** Spermine oxidase: an amine oxidase with specificity for spermine and spermidine, *J. Exp. Med.,* 97, 345, 1953.

108. **Harris, E. D., Gonnerman, W. A., Savage, J. E., and O'Dell, B. L.,** Connective tissue amine oxidase. II. Purification and partial characterization of lysyl oxidase from chick aorta, *Biochim. Biophys. Acta,* 341, 332, 1974.
109. **Siegel, R. C., Pinnell, S. R., and Martin, G. R.,** Cross-linking of collagen and elastin: properties of lysyl oxidase, *Biochemistry,* 9, 4486, 1970.
110. **Narayanan, A. S., Siegel, R. C., and Martin, G. R.,** Stability and purification of lysyl oxidase, *Biochim. Biophys. Acta,* 162, 231, 1974.
111. **O'Dell, B. L., Hardwick, B. C., Reynolds, G., and Savage, J. E.,** Connective tissue defect resulting from copper deficiency, *Proc. Soc. Exp. Biol. Med.,* 108, 402, 1961.
112. **Shields, G. S., Coulson, W. F., Kimball, D. A., Carnes, W. H., Cartwright, G. E., and Wintrobe, M. M.,** Studies on copper metabolism. XXXII. Cardiovascular lesions in copper deficient swine, *Am. J. Pathol.,* 41, 603, 1962.
113. **O'Dell, B. L., Bird, D. W., Ruggles, D. L., and Savage, J. E.,** Composition of aortic tissue from copper deficient chicks, *J. Nutr.,* 88, 9, 1966.
114. **Rucker, R. B. and Goettlich-Rieman, W.,** Properties of rabbit aorta amine oxidase, *Proc. Soc. Exp. Biol. Med.,* 139, 286, 1972.
115. **Rucker, R. B. and O'Dell, B. L.,** Connective tissue amine oxidase. I. Purification of bovine aorta amine oxidase and its comparison with plasma amine oxidase, *Biochim. Biophys. Acta,* 235, 32, 1971.
116. **Rucker, R. B., Parker, H. E., and Rogler, J. C.,** Effect of copper deficiency on chick bone collagen and selected bone enzymes, *J. Nutr.,* 98, 57, 1969.
117. **Bird, D. W., Savage, J. E., and O'Dell, B. L.,** Effect of copper deficiency and inhibitors on the samine oxidase activity of chick tissues, *Proc. Soc. Exp. Biol. Med.,* 123, 250, 1966.
118. **Chou, W. S., Savage, J. E., and O'Dell, B. L.,** Role of copper in biosynthesis of intramolecular cross-links in chick tendon collagen, *J. Biol. Chem.,* 244, 5785, 1969.
119. **Mann, T. and Keilin, D.,** Hemocuprein and hepatocuprein. Copper-protein compounds of blood and liver of mammals, *Proc. R. Soc. (London) Ser. B,* 126, 303, 1938.
120. **McCord, J. M. and Fridovich, I.,** Superoxide dismutase, and enzymic function for erythrocuprein (hemocuprein), *J. Biol. Chem.,* 244, 6049, 1969.
121. **Porter, H. and Floch, J.,** Brain copper-protein fractions in the normal and in Wilson's disease, *Arch. Neurol. Psychiatry.,* 77, 8, 1957.
122. **Carrico, R. J. and Deutsch, H. F.,** The presence of zinc in human cytocuprein and some properties of the apoprotein, *J. Biol. Chem.,* 245, 723, 1970.
123. **Fridovich, I.,** Superoxide dismutases, *Adv. Enzymol.,* 41, 35, 1974.
124. **Fridovich, I.,** Superoxide dismutases, *Ann. Rev. Biochem.,* 44, 147, 1975.
125. **Hassan, H. M.,** Superoxide dismutase, in *Biological Roles of Copper (Ciba Found. Symp. 79),* Excerpta Medica, Amsterdam, 1980, 125-145.
126. **Lerch, K.,** Copper monooxigenases: tyrosinase and dopamine beta-monooxigenase, in *Metal Ions in Biological Systems,* Vol. 13, Copper Proteins, Sigel, H., Ed., Marcel Dekker, New York, 1981, chap. 5.
127. **Kagi, J. H. R. and Vallee, B. L.,** Metallothionein: a cadmium and zinc-containing protein from equine renal cortex. II. Physico-chemical properties, *J. Biol. Chem.,* 236, 2435, 1961.
128. **Bühler, R. H. O. and Kagi, J. H. R.,** Human hepatic metallothionein, *Fed. Eur. Biochem. Soc. Lett.,* 39, 229, 1974.
129. **Evans, G. W., Dubois, R. S., and Hambridge, K. M.,** Wilson's disease; identification of an abnormal copper-binding protein, *Science,* 181, 1175, 1973.
130. **Bremner, I. and Young, B. W.,** Isolation of (copper, zinc) — thioneins from pig liver, *Biochem. J.,* 155, 631, 1976.
131. **Morell, A. G., Shapiro, J. R., and Scheinberg, I. H.,** Copper binding protein from human liver, in *Wilson's Disease: Some Current Concepts,* Walshe, J. M. and Cumings, J. N., Eds., Charles C Thomas, Springfield, Ill., 1961, 36-41.
132. **Winge, D. R., Premakumar, R., Wiley, R. D. E., and Rajagopalan, K. V.,** Copper-chelatin: purification and properties of a copper-binding protein from a rat liver, *Arch. Biochem. Biophys.,* 170, 253, 1975.
133. **Bremner, I.,** Absorption, transport and distribution of copper, in *Biological Roles of Copper (Ciba Found. Symp. 79),* Excerpta Medica, Amsterdam, 1980, 32.
134. **Lerch, K.,** The chemistry and biology of copper metallothioneins, in *Metal Ions in Biological Systems,* Vol. 13, Copper Proteins, Sigel, H., Ed., Marcel Dekker, New York, 1981, chap. 9.
135. **Porter, H., Johnson, J., and Porter, E. M.,** Neonatal hepatic mitochondrocuprein. I. Isolation of a protein fraction containing more than 4% copper from mitochondria of immature bovine liver, *Biochim. Biophys. Acta,* 65, 66, 1962.
136. **Porter, H., Sweeney, M., and Porter, E. M.,** Human hepatocuprein: Isolation of a copper-protein from the subcellular soluble fraction of adult human liver, *Arch. Biochem. Biophys.,* 104, 97, 1964.

137. **Porter, H.,** The particulate half-cysteine-rich copper protein of newborn liver. Relationship to metallothionein and subcellular location on non-mitochondrial particules possibly representing heavy lysosomes, *Biochem. Biophys. Res. Commun.,* 56, 661, 1974.
138. **Ryden, L. and Deutsch, H. F.,** Preparation and properties of the major copper-binding component in human fetal liver, *J. Biol. Chem.,* 253, 519, 1978.
139. **Nostrand, I. F. and Glantz, M. D.,** Purification and properties of human liver monoamine oxidase, *Arch. Biochem. Biophys.,* 158, 1, 1973.
140. **Schultz, G. and Feigelson, P.,** Purification and properties of rat liver tryptophan oxygenase, *J. Biol. Chem.,* 247, 5327, 1972.
141. **Flushimi, H., Hamison, C. R., and Ravin, H. A.,** Two new copper proteins from human brain isolation and properties, *J. Biochem. (Tokyo),* 69, 1041, 1971.
142. **Frieden, E.,** The copper connection, in *Seminars in Hematology,* Vol. 20, Metal Metabolism in Hematologic Disorders, Miescher, P. A. and Jaffe, E. R., Eds., Grune & Stratton, New York, 1983, 114-117.
143. **Deur, C. J., Stone, M. J., and Frenkel, E. P.,** Trace metals in hematopoiesis, *Am. J. Hematol.,* 11, 309, 1981.

Chapter 5

METABOLISM OF COPPER

I. INTRODUCTION

Copper metabolism in man has been the subject of many excellent papers,[1-10] reviews,[11-22] symposia,[23-26] and books.[27-29] Its homeostasis in the mammalian system is regulated through a variety of interrelated mechanisms. Copper is absorbed from the upper gastrointestinal[4-6,30-32] system, probably bound to amino acids and small peptides.[13,16,22] The forms in which copper passes to the serosal side of the intestinal mucosa have not been established.[23] Copper absorption, in general, is influenced by the amount and chemical form of the copper ingested, dietary level of several other metal ions and organic substances, and by the age of the animal species.

Copper antagonists, chemically similar transition elements, inhibit absorption as a result of competition for binding sites on intestinal metallothionein.[29,33-35]

Evans[13] hypothesized that metallothionein functions in copper absorptions consist of providing binding sites within the intestinal mucosa to ensure adequate supply of the element and preventing excessive absorption of copper and other metals. Following intestinal absorption, copper is distributed throughout the body (Figure 1). In the liver, copper undergoes the metabolic interrelationship involved in the maintenance of copper homeostasis (Figure 2), whereas in the extra hepatic tissues the metabolism of copper is confined mainly to the normal synthesis and degradation of copper-dependent enzymes.[13] Copper absorption, tissue, cellular and subcellular distribution, storage, detoxification, structure and function of copper-containing proteins, enzymes, and regulative mechanisms of copper homeostasis are still not well understood.[29] These and other aspects of copper metabolism will be discussed in subsequent sections.

II. NUTRITIONAL ASPECTS (COPPER DEFICIENCY)

In many mammalian species, consumption of a low-copper diet results in decreased hepatic copper concentrations.[13-29] A wide variety of disease states have been associated with a dietary deficiency of copper or with response to copper therapy.[11-14,23-29] These include anemia, growth retardation, bone disorders — many of which are common to a scorbutic process, depigmentation of hair and wool, abnormal wool growth, alopecia, neonatal ataxia, impaired reproduction, cardiovascular defects, heart failure, and gastroenterological disturbance. The extent to which one or more of these functions is actually revealed depends upon the species, its age and sex, environment, and the severity and duration of the copper deficiency.[13,14,26,27,29]

Nutritional copper deficiency was recognized in the 1950s in infants suffering from malabsorption syndrome. A syndrome characterized by anemia, hypocupremia, and hypoferremia responding to combined iron and copper therapy was reported.[36] The hypocupremia is believed to result from insufficient copper absorption and increased loss of copper-protein into the bowel. Copper deficiency has been also reported in severly malnourished infants rehabilitated on a milk-based low-copper diet.[24,37,38]

Graham and Cordano[39] described a disorder characterized by hypocupremia, hypoferremia, neutropenia, and hypoproteinemia in 62 out of 173 malnourished Peruvian infants with histories of poor nutrition and recurrent diarrhea, who were fed on a low-copper, modified milk diet. These findings as well as depigmentation of hair and skin, bone changes, defective elastin formation, and central nervous system abnormalities reminiscent of Menke's syn-

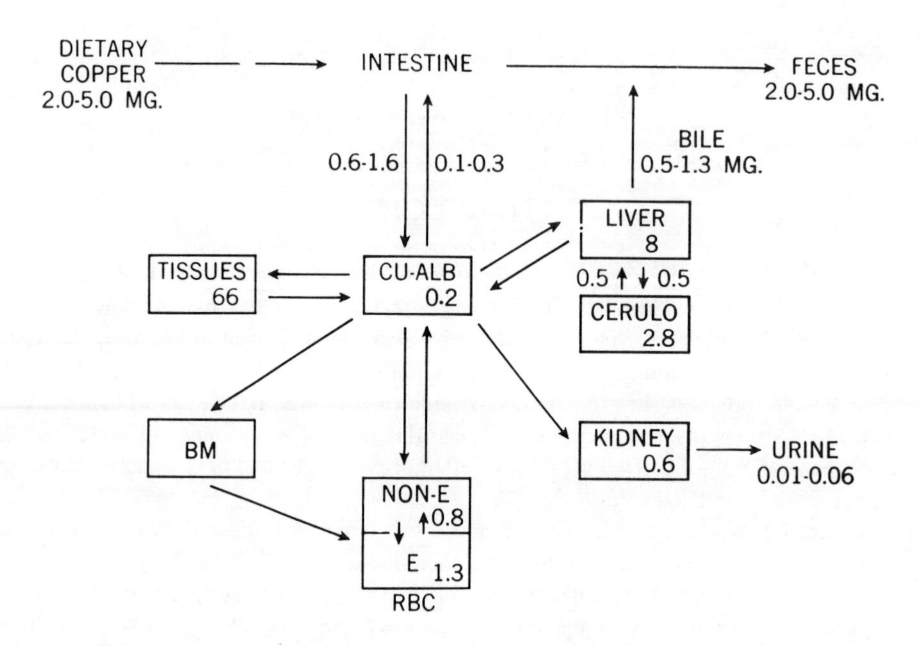

FIGURE 1. Schematic representation of some metabolic pathways of copper in man. The numbers in the boxes refer to milligrams of copper in the pool. The numbers next to the arrow refer to milligrams of copper traversing the pathway each day: Cu-ALB = direct-reacting fraction; cerulo = ceruloplasmin: NON-E = nonerythrocuprein; BM = bone marrow; and RBC = red blood cell. (From Cartwright, G. E. and Wintrobe, M. M., *Am. J. Clin. Nutr.*, 14, 224, 1964. © Am. J. Clin. Nutr., American Society for Clinical Nutrition, Bethesda, Md. With permission.)

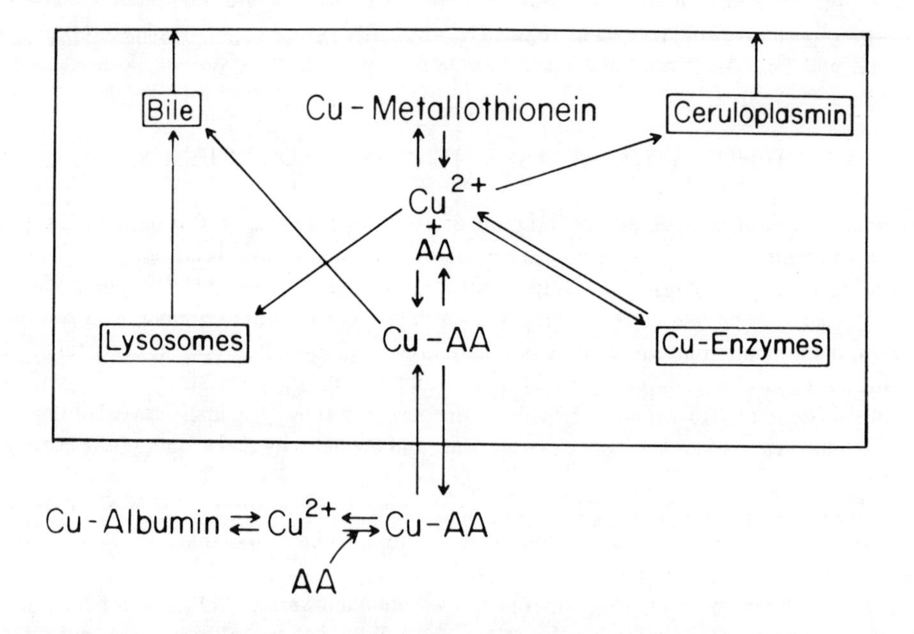

FIGURE 2. Schematic diagram illustrating copper homeostasis in normal hepatocyte: AA = amino acids. (From Evans, G. W., *Physiol. Rev.*, 53, 535, 1973. With permission.)

drome which were reported in association with the above-mentioned syndrome responded to oral copper supplementation.[39,40-45] Anemia has also been related to copper deficiency at 3 to 4 months of age in premature infants and babies of low birth weight in Western countries.[27-46] Malnourished infants who were rehabilitated on a high-caloric, low-copper milk diet[37-41] and patients of all age groups on total parenteral nutrition without copper supplementation are likely to suffer from copper deficiency.[40,42,43,47,48] In addition to dietary lack of copper, decreased absorption and increased excretion caused by several factors such as iron-fortified milk formulas, malabsorption syndromes, chronic infantile diarrhea, and extensive intestinal resection are likely to result in copper deficiency.[24,27,37-44,46-49]

The earliest detectable manifestation of copper depletion was found to be hypocupremia and hypoceruloplasminemia, the latter reaching undetectable levels (by *p*-phenylenediamine oxidase) when serum copper was still measurable.[24,50] In addition to the above, early findings of copper deficiency include persistent neutropenia with vacuolation of circulating neutrophils, not seen in Menkes' syndrome, was described in most reports.[37,51,52] The ability of the deficient patients to respond to infection with a significant neutrophil count was unimpaired.[37,51,52] A maturation arrest of the erythrocytic and granulocytic series in the bone marrow was described in some of the most severe and prolonged copper-deficient cases.[37,38,41,42] Bone marrow examination revealed megaloblastosis and vacuolation.[38,41,42,52] Hypochromic, normocytic anemia with depressed reticulocyte response has been observed.[24,37] The studies of the bone marrow stem cells of a copper-deficient patient suggested that a relative erythropoetin deficiency is important in red cell hypoplasia and that a defective granulocytic proliferation secondary to hypocupremia is responsible for neutropenia.[51] A deficiency of the ferroxidase activity of ceruloplasmin may, in part, explain the anemia. Lack of iron availability does not seem a sufficient explanation for anemia and neutropenia, because of normal bone marrow iron stores. Based on decreased serum iron and a normal bone marrow iron storage, observed in copper deficient cases, a defective iron mobilization was suspected.[48] The skeletal abnormalities reported in children suffering from copper deficiency include cortical thinning, metaphyseal flaring and cupping, epiphyseal widening and periosteal reaction, and bone demineralization in addition to pathological fractures.[37,40,41]

Graham and Cordano[39] noticed that the peak incidence in infants suffering from copper deficiency was just under 1 year of age. Beyond that age, consumption of foods rich in copper resulted in a decreased incidence of clinically detectable copper deficiency.

A full term infant, or even a premature infant or more than 1200 g at birth, does not become copper-deficient without first experiencing prolonged and significant body losses of the element. Therefore, with this single exception, no instance of copper deficiency has been reported that has not resulted from repeated and prolonged diarrhea, poor dietary copper intake, or prolonged total parenteral nutrition without copper supplementation.[24,27,45-49]

In humans, prolonged therapy with large doses (5 g/day) of zinc have been associated with drastic reduction in serum copper levels and an associated anemia.[53,54] High levels of dietary zinc resulted in increased fecal losses of copper in adolescent girls.[55] The fortification of a low-copper diet with iron also appears to precipitate copper deficiency in infants.[46] In view of the known antagonism between copper and zinc,[33,34,53-55] the public health consequences of marginal copper intake are not fully understood, but it has been suggested that the zinc-copper ratio in the diet is determinant of cholesterol levels and may have a role in the genesis of arteriosclerosis.[56,57]

Copper antagonists, zinc, cadmium, silver, mercury, and excess of iron and its salts affect the utilization of copper by reducing its solubility within the intestinal lumen, by competing with copper for binding sites during its absorption or transport, or by modifying its distribution between receptors in body tissues.[29,33-36] Metabolic interactions between copper and other trace elements influence the susceptibility of animals and humans to deficiency or toxicity of copper.[36,58]

The laboratory indices of copper status in humans include serum copper and ceruloplasmin determinations, erythrocyte copper, hair copper, and 24-hr urinary copper.[49] Serum copper levels and ceruloplasmin determinations are simple and useful tests for assessment of copper nutriture. Red cell superoxide dismutase, serum amine oxidase, leukocyte cytochrome c oxidase, and fingernail copper are not reliable, in general, in the assessment of copper nutriture.[12,49] Red cell copper concentrations represent a tissue index of this metal. Normal red cell copper content has been found in many pathological conditions such as Menkes' syndrome,[59] acute and chronic liver disease,[60] pulmonary tuberculosis,[61] sickle-cell disease,[62] and infantile gastroenteritis.[63] Hypocuperemic patients with nephrotic syndrome have depressed erythrocyte copper.[5] Female patients with rheumatoid arthritis[64] also have reduced red cell copper levels, but the nutritional significance of this observation is uncertain.[49] Elevated serum copper levels in infections,[7,65-68] pregnancy,[69-74] and estrogen therapy[75-80] are probably due to increased ceruloplasmin synthesis and can mask depleted copper status. Sanstead[81] believes that even in conditions such as Kwashiorkor, which is usually accompanied by copper deficiency, increased serum copper level attributable to a superimposed infection may be seen. Hypercupremia or hypocupremia, per se, do not signify copper excess or deficiency. Copper in hair and nail does not reflect the accumulation or depletion of copper in the body, thus the analysis of these easily accessible tissues is of little value for daignostic purposes.[12,49,82,83] The use of hair copper determination in the assessment of nutritional status has been called into serious question by the demonstration by Hambidge[83] of a progressive increase of copper concentration in the hair shaft with increasing distance from the scalp, caused by contamination. It has, therefore, been suggested that only copper in the most proximal hair segments be analyzed in assessment of copper nutritures.[49] Due to the low urinary excretion of copper (less than 60 μg/24 hr) and the inadequacy of many testing procedures, its determination in the assessment of copper nutriture is of limited value.[84] In some patients undergoing total parenteral nutrition, lowered urinary copper excretion was noted.[49] However, daily urinary copper excretion in infants with Menkes' kinky hair syndrome was found to be increased.[52,59] Copper concentration in fingernails has been measured in Wilson's disease[85] and in cystic fibrosis[86] and could be applied to the study of presumed copper-deficiency states.

Measurement of the serum ceruloplasmin concentrations before and after moderate copper repletion is suggested as a method of detecting mild copper deficiency.[52] Valuable information concerning the assessment of copper nutriture can be obtained by the use of metabolic balance techniques, in which the total dietary intake and total fecal and urinary excretions of copper are measured and the net absorption and retention are calculated. The net absorption is the difference between dietary intake and fecal excretion and the retention is the difference between dietary intake and the sums of the fecal and urinary excretions.[58]

From the above, it is obvious that diagnosing copper deficiency is rather difficult because of a lack of sensitive laboratory tests and interference with other factors which can mask copper nutriture. Clinical findings and awareness of predisposing factors such as malabsorption, or copper losses, in addition to anemia, neutropenia and hypocupremia or hypoceruloplasminemia, provide a high index of suspicion of depleted copper stores. Careful interpretation of clinical and laboratory data is required before a final therapeutic decision can be made.

As the result of the foregoing consideration of human needs for dietary copper and of possible ill effects on humans, it seems probable that the normal homeostatic mechanism (in which the liver plays a key role) is capable of absorbing and conserving sufficient copper for physiological needs, and to prevent an accumulation that could be toxic, thus maintaining a zero copper balance. Therefore, and under normal conditions, neither a deficiency nor excess of copper occurs despite wide ranges of dietary copper intake. Specific genetic defects characterized by copper deficiency or accumulation of copper, resulting in liver and other organ damages and dysfunctions, will be discussed later.

III. ABSORPTION

Little is known about the mechanism of copper absorption in higher animals and humans.[11,29,52] Absorption of copper, as mentioned previously, occurs in the upper gastrointestinal system,[4-6,30,31,87,88] although precise definition of the site and mechanism of absorption is not well understood.[29] The transfer of copper across the intestinal mucosa is believed to be associated with superoxide dismutase or metallothionein. Whether the role of mucosal metallothionein is to promote or to prevent excessive uptake and copper absorption is not known. The precise role of this protein in copper metabolism (intestinal and hepatic homeostasis) is still a matter of conjecture.[26] After oral administration of copper to humans, the isotope appears in the blood within 1 to 2 hr, which suggests that absorption occurs in the stomach and upper small intestine.[9,89] In normal individuals approximately one third (32%) of an oral radiocopper is absorbed from the upper gastrointestinal tract and 61.8 to 94.6% of the ^{64}copper is recovered in the stool.[9,89] Recent studies using intestinal perfusion techniques confirmed the proximal absorption site for dietary copper.[90] They suggested an even greater efficiency of net copper absorption than the radiocopper experiments had shown, as identical values for dietary copper intake, biliary copper output, and fecal copper excretion were detected. It has been estimated that of the 2 to 5 mg copper ingested daily by adult man, 0.6 to 1.6 mg is absorbed, 0.5 to 1.3 mg is excreted in the bile, 0.1 to 0.3 mg passes directly into the bowel, and 0.01 to 0.06 mg appears in the urine[3,4,27] (Figure 1).

Several hypotheses concerning the mechanism of copper absorption have been quite lucidly summarized by Evans.[13] Experimental animal data indicate that copper absorption is, in part, protein mediated.[13] A fraction of copper may be transported from the mucosal side to the serosal side by an energy-dependent mechanism such as a copper-amino acid complex.[30] Transferral of copper from the intestinal mucosa in the form of organic complexes as well as the ionic form (passive simple diffusion) has also been suggested.[91] Two distinct copper-binding proteins, superoxide dismutase and metallothionein, isolated from the cytosol of duodenal mucosa of experimental animals have been implicated with copper absorption.[13] Evans and LeBlanc[92] isolated and characterized a copper-binding protein from rat intestine which seems to be different from the conventional metallothionein and they suggested that this intestinal copper-binding protein plays a role in copper absorption as a ''mucosal block'', regulating the passage of copper from the mucosa of the blood.

According to Evan's[13] hypothesis, copper dissociates from ingested food and is released in either ionic form or a copper-amino acid complex. In the intestinal lumen or on the mucosa, a fraction of the ionic copper combines with available amino acids and in complexed form is actively transported to the serosal side. An uncomplexed copper ion combines with metallothionein. After dissociation from intestinal metallothionein, the element either diffuses directly into the plasma or becomes complexed for subsequent transport to the serosal side.

There is evidence to suggest that the form in which copper is present in food greatly influences its availability for absorption.[11,12,29] Mills[91,93,94] has shown that copper in herbage is in the form of water-soluble, negatively charged organic complexes and that copper administered in this form is more effective in increasing copper stores in the copper-deficient rat than inorganic copper sulfate. It is suggested that copper may in fact be transported through the intestinal mucosa in the form of a negatively charged organic complex as well as in the ionic form.[13] While absorption of the copper contained in foods of plant origin is seemingly efficient, it has been reported that rats fed on raw meat developed copper deficiency. This suggests that the copper present in raw meat is not available for absorption.[95] Guggenheim,[96] on the other hand, presented evidence to the effect that the ''meat anemia'' in mice is due to the high zinc content of meat in the presence of low (but by itself, adequate) amounts of copper, and that a concomitant lack of calcium further aggravates the situation. Sulfides in the diet form insoluble salts of copper sulfide, reducing copper absorption.[13,97,98] Undenatured proteins, as in raw meat or isolated soy protein, form insoluble complexes with copper which are unavailable for absorption;[95,99] denaturation of meat protein by cooking improves the availability of copper absorption.[12,13] Transition metals as mentioned earlier, such as zinc, cadmium, silver, and mercury interfere with copper absorption by competing for binding sites in the copper-binding protein in the intestine.[13,33-36] Ascorbic acid also affects copper availability adversely.[100-102] In addition to the above, antagonists of copper absorption include excesses of iron and its salts, phytate, alkaline pH (calcium carbonate), and excesses of molybdenum.[13,29,36,97,103] Oral doses of potassium sulfide and certain ion exchange resins are used to reduce the absorption of copper in certain pathological conditions.[12,104]

IV. COPPER TRANSPORT

The major portion of copper in whole blood is divided among five separate fractions: erythrocytic superoxide dismutase, a unidentified copper complex in the erythrocyte-designated "labile pool", plasma albumin, ceruloplasmin, and amino acids. Each of these has unique functions in copper metabolism. Administration of radiocopper to normal persons, either orally or intravenously, results in a transient rise in serum activity followed by a rapid fall during the following 4 hr, and followed again by a slower secondary rise observable for about 48 to 72 hr.[6,9,104] Initial radioactivity of serum is associated with the albumin fraction and the slow secondary rise with ceruloplasmin.[3,6,9] Copper in plasma reversibly bound to serum albumin, constituting 5 to 10% of the total plasma copper, is distributed widely to the tissues and can pass readily into the erythrocytes[9] (Figure 1). This so-called "direct reacting" fraction of plasma copper, though small in relation to ceruloplasmin, occupies a pivotal position in copper metabolism. From this pool copper is distributed to the liver, bone marrow, and other tissues. Albumin-bound copper also receives copper from the tissues[3,4,105,106] (Figure 1). Albumin-bound copper circulating in human plasma is promptly cleared by the liver, where a highly efficient membrane recognition system for albumin-bound copper exists.[15] This is evident since, within hours, 60 to 90% of the absorbed radiocopper is deposited in the liver,[106] and 10 to 20% of liver activity is released into the blood as ceruloplasmin incorporated into proteins such as superoxide dismutase and copper-binding proteins (both Cu-thionein and Cu-chelatin types).[107-110] About 80% of the copper in the normal adult human liver is present in these proteins.[15] The remainder is incorporated into specific copper proteins such as cytochrome c oxidase and ceruloplasmin,[111,112] mitochondria, microsomes, nuclei, and the soluble fraction of the parenchymal cells,[113-118] or is taken up by lysosomes before being excreted in the bile.[13,119,120] The copper is either stored in these sites or is released for incorporation into erythrocuprein and ceruloplasmin and the various other copper-containing enzymes of the cells.[27,120] Ceruloplasmin is synthesized in the liver and secreted into the serum, and it has been shown that ceruloplasmin can serve as a vehicle carrying copper from the liver to other tissues.[121-124]

Studies with ^{64}Cu and ^{67}Cu show that the copper is incorporated irreversibly into ceruloplasmin at synthesis and remains incorporated throughout the life of the protein.[119] Injected radiocopper is handled similarly by the liver and the 2-hr uptake is approximately 34 to 87% of the injected dose.[125] The whole-body retention of radiocopper has a biological half-life reported to be 17.0 days,[126] whereas other investigations found it to be 27.1 days in the normal subject.[127] Hazelrig et al.[128] postulated at least three distinct hepatic copper processes; (1) preparation of copper for excretion in bile, (2) temporary storage of copper, and (3) incorporation of copper into ceruloplasmin and its release into the plasma (Figure 2). Since ceruloplasmin is synthesized in the liver, the key organ of this metaloenzyme, copper is not incorporated into the globulin until after the metal permeates the hepatic cells. Furthermore, the copper in ceruloplasmin does not exchange with nonceruloplasmin in vivo.[25,129] Thus, ceruloplasmin does not function in transporting ingested copper through the portal blood to the liver.[13] Owen[123] observed that after the intravenous injection of radioactive copper, the isotope did not accumulate in extrahepatic organs until after the emergence of ceruloplasmin-Cu^{64}, suggesting that ceruloplasmin is a copper donor for the tissues.

These results have recently been confirmed using ceruloplasmin-Cu.[124,130] Since the copper in ceruloplasmin is not readily dissociated, the exchange of copper between ceruloplasmin and the extrahepatic tissues probably involves a degradative mechanism either on the cellular membrane or within the cell. Considering the dissociability of albumin-bound copper and the facilitory action of amino acid-bound copper, the copper in ceruloplasmin is probably not the sole source for the metal in extrahepatic tissues.[13]

There are two main routes by which copper is mobilized from the liver. First, during hepatic synthesis of the blue copper-glycoprotein, ceruloplasmin, copper is incorporated into this protein which is then released into the blood.[6,108,111] In normal subjects the available pool of copper utilized for this process seems to be small, turning over rapidly, since the specific activity of radiocopper in newly synthesized ceruloplasmin is very high only during the first hours after the deposition of the isotope in the liver.[15] The turnover of ceruloplasmin copper accounts for about 0.5 to 1.0 mg daily.[131,132] Catabolism of ceruloplasmin also involves the liver.[133,134] Second, the maintenance of copper balance appears to depend principally on the copper that leaves the liver via the bile. This amounts to about 1.5 mg/day[90,135] and is bound to a carrier that seems to prevent or inhibit its reabsorption from the intestinal tract.[136] The identity of this specific carrier for biliary copper and physiological variations of biliary copper excretion are not well understood.[13,15] Copper transport through the gastrointestinal tract, in the blood, and its excretion in physiological and pathological conditions has been well summarized in Sarkar's excellent review.[137]

V. COPPER EXCRETION

The route of copper excretion is primarily through the biliary tract (80%) with a small amount excreted via the intestine (16%) and the urinary tract (4%).[13,27,106,120,137-140] Using a duodenal perfusion technique, Van Berge Henegowen et al.[90] recently found that in healthy man the amount of copper excreted in the bile was 25 ± 13 mg/kg/day or about 1.7 mg for a 70-kg man (Figure 1). The biliary copper has been shown to occur in low-molecular-weight components such as amino acids and small peptides, as well as macromolecules and bile pigments.[13,15,139,141] Different workers have isolated bile components of low-molecular-weight proteins or polypeptides of about 5000 daltons, macromolecular complexes as large as 800,000 daltons,[137] and bile acids[138] or bile pigment[139] — all with bound copper. Golan et al.[141] indicated that the low-molecular-weight components are more prominent in hepatic bile and the high-molecular-weight fraction predominates in gall bladder bile. Farrer and Mistilis[142] demonstrated that one fraction of copper is associated with amino acids and small peptides, whereas a second fraction is associated nonspecifically with high-molecular-weight molecules. The low-molecular-weight fraction or amino acid-bound copper probably represents one of the forms in which copper is transported across the bile canaliculus, since amino acids are known to facilitate the membrane transport of copper.[13] There are at least two probable sources of macromolecularly bound biliary copper: one of which permeates the bile canaliculus and subsequently combines nonspecifically with proteins in the bile, and one which is deposited in the bile as a result of protein catabolism and pinocytosis by the hepatic lysosomes.[13,141-144] Gregoriadis et al.[145] demonstrated that desialylated ceruloplasmin is catabolized principally by the hepatic lysosomes. Biliary copper is poorly absorbed, probably because it is bound to a specific protein macromolecular ligand that impairs absorption and, therefore, the enterohepatic circulation of biliary copper is insignificant, but dependent on the degree of protein binding.[13,143,144]

Experimental data seem to suggest that the metal enters the bile in the form of low-molecular-weight complexes during the initial phase of biliary copper excretion. The proportion of copper complexes bound to macromolecules increases as time goes on due to increased association of ionic copper with biliary protein, and increased intrabiliary secretion of copper-binding proteins and enzymes.[13,143,144] Since protein-bound biliary copper is apparently unavailable for reabsorption, copper homeostasis depends, in part, on adequate protein synthesis.[13] Gregoriadis and Sourkes[146] have demonstrated that inhibitors of protein synthesis severely retard the removal of hepatic copper. The amount of copper in feces depends to a great extent on the dietary intake. Copper output via feces is not a reliable measure of excretion of copper from the body since some of the dietary copper may pass through the gastrointestinal tract without being absorbed and thus may constitute a significant portion of fecal copper.[27,146] It has been shown, however, that copper given by the intravenous route is excreted very efficiently, and almost exclusively via feces.[147]

Only a small amount of copper (0.01 to 0.06 mg) is excreted in the urine[3,4] (Figure 1). Since most of the copper is circulating blood is bound to ceruloplasmin or confined within the erythrocytes, very little copper (approximately 7 to 10%) permeates the glomerular capillaries. This ultrafiltrable or "free" copper is in balance with the protein-bound metal. The small fraction of ingested copper that does appear in urine emanates from the amino acids, nicotinic acid, small peptides-copper complexes, and possibly some other small-molecular-weight ligands.[148] After filtration by the glomeruli, less than 1% of the filtered copper appears in the urine, the remainder being reabsorbed by the tubules. Therefore, urinary copper excretion contributes little to the maintenance of copper homeostasis under normal physiological conditions. The daily urinary excretion of copper in normal men is less than 50 μg, but it varies widely and changes from day to day in any individual.[3,4] This cyclic variation in urinary copper excretion more likely reflects cyclic variation in "free" (or nonceruloplasmin-bound) copper than changes in ceruloplasmin since the ceruloplasmin-bound copper complex is too large to filter through the normal glomerulus.[7,10] Other routes of copper losses include sweat, in negligible amounts,[149] and loss by menstruation (0.5 mg/cycle).[27,150] Copper losses in milk at the height of human lactation average 0.4 mg/day.[29]

VI. COPPER TOXICITY

Although copper is essential to maintaining life, an excess can be fatally toxic. Because copper is so ubiquitous and is so efficiently metabolized, deficiencies and excesses rarely develop in humans. There are several situations in which copper toxicity requires special consideration.

Wilson's disease, or hepatolenticular degeneration, is a hereditary disease of abnormal copper accumulation in various organs, particularly the liver, resulting in hepatic and cerebral dysfunction. It will be discussed in detail later.

Acute and chronic forms of copper toxicity can occur in man following the ingestion of gram quantities of copper salts accidently, or with suicidal intent. Acute copper poisoning occurs in man when gram quantities of copper sulfate or "bluestone" are ingested.[151,152] Although unusual, acute and chronic forms of copper toxicity can occur in man due to a variety of reasons.[153] Acute poisoning due to copper salts, though not common in the Western world, is often seen in some of the countries of Asia and Europe, particularly in India. Cases with copper levels in excess of 265 μg/100 mℓ of plasma are generally fatal.[151] The fatal period ranges from 4 to 12 hr. Patients usually complain of a sensation of metallic taste, nausea, vomiting, pain in the mouth and esophagus, epigastric burning, colicky abdominal pains, diarrhea, and sometimes cramps of the limbs.[151,154] In chronic poisoning, the main indications are on a parallel with salts of lead. There is evidence of progressive emaciation, gastrointestinal irritation, vomiting, loss of appetite, diarrhea, or constipation.[154,155]

In severe poisoning, intravascular hemolysis, jaundice, hepatic necrosis and failure, gastrointestinal ulcerations and bleeding, oliguria, azotemia, hemoglobinuria, hematuria, proteinuria, hypotension, tachycardia, convulsions, kidney failure, coma, and death are not infrequent.[151-158] Hemolysis has been reported following the application of solutions of copper salts to large areas of burned skin.[159] In excess, copper inhibits the catalytic activity of many enzymes, including microsomal membrane ATPase and a variety of glycolytic enzymes.

The most striking autopsy findings are greenish or bluish colorations of gastrointestinal tract mucosa with congestion and ulcerations and hepatic cell necrosis, and occasionally jaundice.[124-156]

Various therapeutic modalities have been tried for copper poisoning. In acute copper poisoning, besides the gastric lavage, potassium ferrocyanide as antidote, the chelating agents, penicillamine and dimercaprol, exchange transfusion, and hemodialysis or peritoneal dialysis have all been utilized.[154,155,159-161] Adding albumin, which binds copper to the peritoneal dialysate, results in a greater than tenfold increase in copper removal. Exchange transfusion with supplemented albumin results in a substantial reduction in the serum copper concentration.[162]

Copper toxicity can occur in hemodialysis in which water is left overnight, if the pH of the water or dialysis fluid is low (below 6.5).[163-168] Twice-weekly dialysis for a year or two can elevate the hepatic copper concentration to an abnormally high level.[163,165] Concentrations of copper in drinking water range from 0.09 to 0.30 ppm.[29] Some copper-lined hot-water systems deliver up to 1 mg/ℓ of copper in areas with soft water supplies.[52] The storage of acidic food or drink such as vinegar, carbonated beverages, and citrus juices in copper containers for an extended period of time can cause an increased content of this metal.[153,169-173] Overnight contact of carbonated water with copper check valves of drink-dispensing machines has proved capable of increasing the copper content of the first drink of the day sufficiently to cause metallic taste, ptyalism, nausea, vomiting, epigastric burning, and diarrhea.[153,169-173] Whiskey sours, mixed in copper-lined cocktail shakers, have induced similar effects.[174,175] In these instances the vomiting and diarrhea induced by the copper generally protect the patient from serious systemic toxic effects.

Another form of copper toxicity is the occupational hazard occurring in vineyard sprayers using copper sulfate with hydrated lime (Bordeaux mixture), or using copper sulfate as an algicide, or molluscacide.[176-178] Copper toxicity may present itself as a local granulomatous reaction in isolated organs, such as lung and liver (vineyard sprayer's lung or liver). The copper in the Bordeaux mixture has been incriminated in the pathogenesis of interstitial-pulmonary lesions in the lungs of some vineyard workers, actually turning the lungs blue.[153,179] Pulmonary copper deposition with fibrosis and granulomatous reactions has also been described.[179] Diffuse pulmonary fibrosis with multiple histiocytic granulomas containing abundant inclusions of copper were found during autopsy.[178,179] Increased incidence of alveolar cell carcinoma of the lung have been reported in workers with "vineyard sprayer's lung".[176] The copper-containing granulomata were also found in the nasal mucosa, liver, kidney, spleen, and lymph nodes.[178,180]

The addition of copper salts to animal feeds may cause the abnormal accumulation of copper in man and animals.[153] Swine fed rations containing 250 ppm copper for the purpose of accelerating increases in carcass weight have significant elevations in their hepatic copper concentrations from the normal mean of 24 to 220 μg/gm of dry tissue.[153,181] One-quarter pound of liver from pigs fed rations with 250 ppm copper contains about 10 mg of copper, which is two to three times the average daily supply of the metal in the Western diet. There are no data on the adverse effects of eating this amount of copper for long periods.[153] Manure from pigs and poultry raised on copper-supplemented soil used for fertilizing human food crops, and water run-offs from dressed fields also constitute a potential problem in increasing the copper content of vegetation, with consequent ill effects on the environment.[153]

The use of copper-containing intrauterine contraceptive devices is another potential cause of local and possibly systemic toxicity. The intrauterine presence of copper is associated with polymorphonuclear leukocytic infiltration of the endometrium, changes in the endometrial enzyme levels, liquifaction of mucus, and possibly with the production of free radicals.[182-188] Analysis of copper IUDs after their presence in utero for varying durations, show that about 25 to 50 mg of copper can be dissolved in a year.[151] Some of the dissolved copper is lost through menstrual secretion, but experiments in rats show that at least some copper is absorbed within hours, and is found in many extrauterine organs and tissues, principally liver and serum.[182] There is reason for concern that chronic copper toxicity may develop over the years or decades that a woman is likely to use an intrauterine contraceptive device.[153] Unfortunately, no applicable laboratory tests are available to detect sublinical toxicity.

Hefnavi et al.[183] studied copper levels in endometrial tissues and plasma in women using intrauterine devices (Cu-T 200 or Lippes Loop®), or oral contraceptives. In the control group not using contraception, copper levels of the endometrium were significantly lower during the secretory phase compared to those in the proliferative phase. Plasma copper levels were similar in both phases (78 ± 20 μg/100 mℓ). Likewise, the copper levels of the endometrial tissues of oral contraceptive users were reduced in the secretory phase, although the baseline levels were significantly higher in this group than in the control. Women using intrauterine devices (Cu-T 200 or Lippes Loop®) had higher endometrial tissues in the secretory than in the proliferative phase. The copper level of endometrial tissue obtained from patients with long-term use of a Lippes Loop® was, surprisingly, as high as those levels found in patients using the copper device. The plasma copper levels in both groups using IUDs were elevated during both phases as compared to the control group. The observations are consistent with the viewpoint that local inflammation is associated with IUD use.[184,185] Increased endometrial tissue copper content and plasma copper levels are the result of a local as well as generalized body reaction to the intrauterine foreign body. This is in agreement with the report of elevated serum immunoglobulins, IgG, and IgM levels in IUD users.[186,187] Hagenfeldt[188] reported a similar rise in endometrial copper levels during both phases of the menstrual cycle in the presence of the Cu-T device; however, the change in the proliferative phase was not significant after the first few months. This rise was explained by the uptake of released copper by the endometrium. However, this author reported no change in plasma copper after the insertion of Cu-T device.

Rubinfeld et al.[189] reported a progressive rise in serum copper levels in women taking oral contraceptives, particularly Neogynone® (Shering) for an extended period of time. This increase of 137% over control values is worrisome when compared with results in chronically copper-loaded rats, in which serum copper levels exceeded normal values by 130%.[190] According to this author, the potential hazard of chronic copper intoxication should be considered, particularly in patients taking Neogynone®, and should be further investigated.

Schenker et al.[73] reported increased serum copper levels of over 275 μg/100 mℓ in 31 (6.5%) of the 502 females using oral contraceptives which may point to liver dysfunction. This was supported by the fact that this group included six subjects suffering from jaundice associated with abnormal liver function tests indicative of intrahepatic choleostasis.

During the last few years a growing concern with problems of human environmental health has given renewed vigor to studies with trace elements. The possibility of chronic deleterious effects upon human health from contamination of the air, water supply, and food with elements arising from modern agricultural, industrial, and medical practices, and from the increasing motorization and urbanization of large sections of the community has stimulated investigation on many trace elements, including copper[191,192] (Figure 3).

Fortunately, homeostatic mechanisms were developed in man and at least some other mammals to regulate metal absorption and excretion in such a way that specific metabolic needs may be met and toxicity avoided in spite of fluctuations in the amount of copper in the environment. In addition, almost all the organism's copper is tightly bound as a prosthetic element of a number of proteins and very little is present in the ionic, toxic form.[153,193,194]

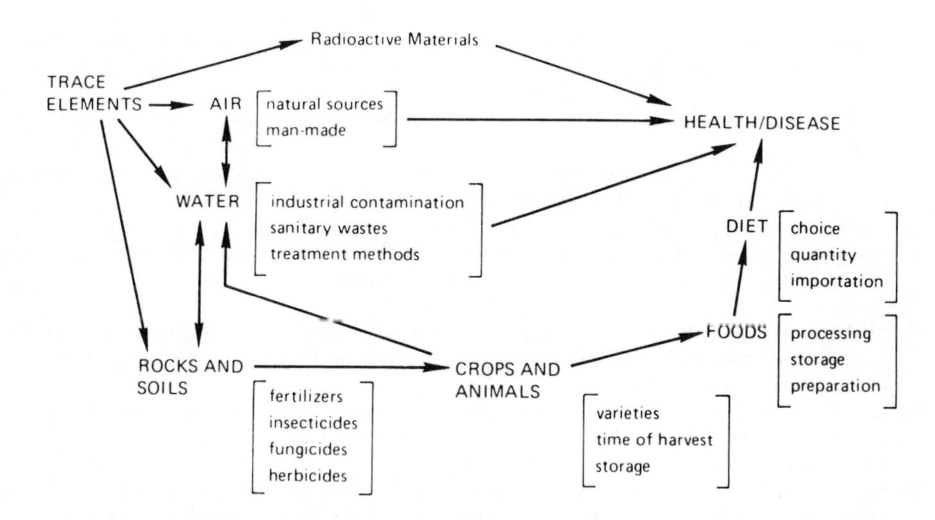

FIGURE 3. Schema showing mechanisms by which trace elements find their way to man, influencing the quality of his health or producing disease. (From Subcommittee on the Geochemical Environment in Relation to Health and Disease, in *Geochemistry and Environment,* Vol. 1, National Academy of Sciences, Washington, D.C., 1974. With permission.)

REFERENCES

1. **Cartwright, G. E.,** Copper metabolism in human subjects, in *A Symposium on Copper Metabolism,* McElroy, W. D. and Glass, B., Eds., Johns Hopkins University Press, Baltimore, 1950, 274-314.
2. **Schroeder, H. A.,Nason, A. P., Tipton, I. H., and Balassa, J. J.,** Essential trace metals in man: copper, *J. Chronic Dis.,* 19, 1007, 1966.
3. **Wintrobe, M. M., Cartwright, G. E., and Gubler, C. J.,** Studies on function and metabolism of copper, *J. Nutr.,* 50, 395, 1953.
4. **Cartwright, G. E. and Wintrobe, M. M.,** Copper metabolism in normal subjects, *Am. J. Clin. Nutr.,* 14, 224, 1964a.
5. **Cartwright, G. E. and Wintrobe, M. M.,** The question of copper deficiency in man, *Am. J. Clin. Nutr.,* 15, 94, 1964b.
6. **Bearn, A. G. and Kunkel, H. G.,** Metabolic studies in Wilson's disease using ^{64}Cu, *J. Lab. Clin. Med.,* 45, 623, 1955.
7. **Markowitz, H., Gubler, C. J., Mahoney, J. P., Cartwright, G. E., and Wintrobe, M. M.,** Studies on copper metabolism. XIV. Copper ceruloplasmin and oxidase activity in sera of normal human subjects, pregnant women and patients with infection, hepatolenticular degeneration and nephrotic syndrome, *J. Clin. Invest.,* 34, 1498, 1955.
8. **Lahey, M. E., Gubler, C. J., Cartwright, G. E., and Wintrobe, M. M.,** Studies on copper metabolism. VI. Blood copper in normal human subjects, *J. Clin. Invest.,* 32, 322, 1953.
9. **Bush, J. A., Mahoney, J. P., Markowitz, H., Gubler, C. J., Cartwright, G. E., and Wintrobe, M. M.,** Studies on copper metabolism. XVI. Radioactive copper studies in normal subjects and in patients with hepatolenticular degeneration, *J. Clin. Invest.,* 34, 1766, 1955.
10. **Gubler, C. J., Lahey, M. E., Cartwright, G. E., and Wintrobe, M. M.,** Studies on copper metabolism. IX. The transportation of copper in blood, *J. Biol. Chem.,* 224, 405, 1957.
11. **Adelstein, S. J. and Vallee, B. L.,** Copper, in *Mineral Metabolism,* Vol. 2B, Comar, C. L. and Bronner, F., Eds., Academic Press, New York, 1962, 371-401.
12. **Sass-Kortsak, A.,** Copper metabolism, *Adv. Clin. Chem.,* 8, 1, 1965.
13. **Evans, G. W.,** Copper homeostasis in the mammalian system, *Physiol. Rev.,* 53, 535, 1973.
14. **O'Dell, B. L.,** Biochemistry and physiology of copper in vertebrates, in *Trace Elements in Human Health and Disease,* Vol. 1, Prasad, A. S., Ed., Academic Press, New York, 1976, 391-413.
15. **Sternlieb, I.,** Copper and the liver, *Gastroenterology,* 78, 1615, 1980.

16. **Evans, G. W.,** New aspects of biochemistry and metabolism of copper, in *Zinc and Copper in Clinical Medicine,* Hambidge, K. M. and Nichols, B. L., Eds., S. P. Medical and Scientific Books, New York, 1976, 113-118.
17. **Linder, M. C. and Munro, H. N.,** Iron and copper metabolism during development, *Enzyme,* 15, 111, 1973.
18. **Widowson, E. M.,** Trace elements in human development, in *Mineral Metabolism in Pediatrics,* Barltrop, D. and Burland, W. L., Eds., Blackwell Scientific, Oxford, 1969, 85-98.
19. **Dowdy, R. P.,** Copper metabolism, *Am. J. Clin. Nutr.,* 22, 887, 1969.
20. **Hurley, L. S., Keen, C. L., and Lönnerdal, B. O.,** Copper in fetal and neonatal development, in *Biological Roles of Copper,* Excerpta Medica, Amsterdam, 1980, 227-245.
21. **Jacob, R. A.,** Zinc and copper, *Clin. Lab. Med.,* 1, 743, 1981.
22. **Evans, G. W. and Johnson, P. E.,** Copper and zinc-binding ligands in the intestinal mucosa, in *Trace Element Metabolism in Man and Animals,* Vol. 3, Arbeitskreis für Tierenährungsforschung, Kirchgessner, M., Ed., Weihenstephan, 1978, 98.
23. **Lee, G. R., Williams, D. M., and Cartwright, G. E.,** Role of copper in iron metabolism and heme biosynthesis, in *Trace Elements in Human Health and Disease,* Vol. 1, Zinc and Copper, Prasad, A. S., Ed., Academic Press, New York, 1976, chap. 23.
24. **Graham, G. G. and Cordano, A.,** Copper deficiency in human subjects, in *Trace Elements in Human Health and Disease,* Vol. 1, Zinc and Copper, Prasad, A. S., Ed., Academic Press, New York, 1976, chap. 22.
25. **Bremner, I.,** Absorption, transport and distribution of copper, in *Biological Roles of Copper (Ciba Found. Symp. 79),* Excerpta Medica, Amsterdam, 1980, 23-48.
26. **O'Dell, B. L.,** Biochemistry of Copper. Symposium on Trace Elements, *Med. Clin. North Am.,* 60, 687-849, 1976.
27. **Prasad, A. S.,** *Trace Elements and Iron in Human Metabolism,* Plenum Press, New York, 1978, chap. 2.
28. **Hsieh, S. H. and Hsu, J. M.,** Biochemistry and metabolism of copper, in *Zinc and Copper in Medicine,* Karcioglu, Z. A. and Sarper, R. M., Eds., Charles C Thomas, Springfield, Ill., 1980, chap. 4.
29. **Underwood, E. J.,** *Trace Elements in Human and Animal Nutrition,* 4th ed., Academic Press, New York, 1977, 56-108.
30. **Crampton, R. F., Matthews, E. M., and Poisner, R.,** Observations on the mechanism of absorption of copper by small intestine, *J. Physiol. London,* 178, 111, 1965.
31. **Van Campen, D. R. and Mitchell, E. A.,** Absorption of ^{64}Cu, ^{65}Zn, ^{99}Mo, and ^{59}Fe from ligated segments of the rat gastrointestinal tract, *J. Nutr.,* 86, 120, 1965.
32. **Solomons, N. W.,** Zinc and copper in gastrointestinal system, in *Zinc and Copper in Medicine,* Karcioglu, Z. A. and Sarper, R. H., Eds., Charles C Thomas, Springfield, Ill., 1980, chap. 9.
33. **VanCampen, D. R.,** Effects of zinc, cadmium, silver, and mercury on the absorption and distribution of copper-64 in rats, *J. Nutr.,* 88, 125, 1966.
34. **Hill, C. H., Starcher, B., and Matrone, G.,** Mercury and silver inter-relationships with copper, *J. Nutr.,* 83, 107, 1964.
35. **Mills, C. F.,** Metabolic interactions with other trace elements, in *Biological Roles of Copper (Ciba Found. Symp. 79),* Excerpta Medica, Amsterdam, 1980, 49-69.
36. **Sturgeon, P. and Brubaker, C.,** Copper deficiency in infants: a syndrome characterized by hypocupremia, iron deficiency anemia, and hypoproteinemia, *J. Dis. Child.,* 92, 254, 1956.
37. **Cordano, A., Baertl, J. M., and Graham, G. G.,** Copper deficiency in infancy, *Pediatrics,* 34, 324, 1964.
38. **Cordano, A. and Graham, G. G.,** Copper deficiency complicating severe chronic intestinal malabsorption, *Pediatrics,* 38, 596, 1966.
39. **Graham, G. G. and Cordano, A.,** Copper depletion and deficiency in the malnourished infant, *Johns Hopkins Med. J.,* 124, 139, 1969.
40. **Ashkenazi, A., Levin, S., Djaldetti, M., Fishel, E., Benvenisti, D.,** The syndrome of neonatal copper deficiency, *Pediatrics,* 52, 525, 1973.
41. **Al-Rashid, R. A. and Spangler, J.,** Neonatal copper deficiency, *N. Engl. J. Med.,* 285, 841, 1971.
42. **Karpel, J. T. and Peden, V. H.,** Copper deficiency in long-term parenteral nutrition, *J. Pediatr.,* 80, 32, 1972.
43. **Griscom, N. T., Craig, J. N., and Newhouser, E. B. D.,** Systemic bone disease developing in small premature infants, *Pediatrics,* 48, 883, 1971.
44. **Heller, R. M., Kirchner, S. G., O'Neill, J. A., Hough, A. J., Howard, L., Kramer, S. S., and Green, H. L.,** Skeletal changes of copper deficiency in infants receiving prolonged total parenteral alimentation, *J. Pediatr.,* 92, 947, 1978.
45. **Danks, D. M.,** Copper transport and utilization in Menkes' syndrome and in mottled mice, *Inorg. Perspect. Biol. Med.,* 1, 73, 1977.

46. **Seely, J. R., Humprey, G. B., and Matter, B. J.,** Copper deficiency in premature infant fed on iron-fortified formula, *N. Engl. J. Med.*, 286, 109, 1972.
47. **Dunlap, W. M., James, G. W., III, and Hume, D. M.,** Anemia and neutropenia caused by copper deficiency, *Ann. Intern. Med.*, 80, 470, 1974.
48. **Vilter, R. W., Bozian, R. C., Hess, E. V., Zellner, D. C., and Petering, H. G.,** Manifestations of copper deficiency in a patient with systemic sclerosis on intravenous hyperalimentation, *N. Engl. J. Med.*, 291, 188, 1974.
49. **Solomons, N. W.,** Zinc and copper in human nutrition, in *Zinc and Copper in Medicine*, Karcioglu, Z. A. and Sarper, R. F., Eds., Charles C Thomas, Springfield, Ill., 1980, chap. 8.
50. **Holtzman, N. A., Charache, P., Cordano, A., and Graham, G. G.,** Distribution of serum copper in copper deficiency, *Johns Hopkins Med. J.*, 126, 34, 1970.
51. **Zidar, B. L., Shadduck, R. K., Ziegler, Z., and Winkelstein, A.,** Observations on anemia and neutropenia of human copper deficiency, *Am. J. Hematol.*, 3, 177, 1977.
52. **Danks, D. M.,** Copper deficiency in humans, in *Biological Roles of Copper (Ciba Found. Symp. 79)*, Excerpta Medica, Amsterdam, 1980, 209-225.
53. **Prasad, A. S., Brewer, G. J., Schoomaker, E. B., and Rabbani, P.,** Hypocupremia induced by large doses of zinc therapy in adults, *J. Am. Med. Assoc.*, 235, 2396, 1978.
54. **Pfeifer, C. C. and Jenny, E. H.,** Excess oral zinc in man lowers copper levels (abstr.), *Fed. Proc.*, 37, 324, 1978.
55. **Greger, J. L., Zaikis, S. C., Bennet, O. A., Abernathy, R. P., and Huffman, J.,** Mineral and nitrogen balance of adolescent females fed two levels of zinc, *Fed. Proc.*, 37, 254, 1978.
56. **Klevay, L. M.,** Hypercholesteremia in rats produced by an increase in the ratio of zinc to copper ingested, *Am. J. Clin. Nutr.*, 26, 1060, 1973.
57. **Klevay, L. M.,** Coronary heart disease: the zinc/copper hypothesis, *Am. J. Clin. Nutr.*, 28, 764, 1975.
58. **Delves, H. T.,** Dietary sources of copper, in *Biological Roles of Copper (Ciba Found. Symp. 79)*, Excerpta Medica, Amsterdam, 1980, 5-22.
59. **Williams, D. M., Atkins, C. L., Ferns, D. B., and Bray, P. F.,** Menkes' kinky hair syndrome: studies of copper metabolism and long term copper therapy, *Pediatr. Res.*, 11, 823, 1977.
60. **Versiek, J., Hoste, J., and Barbier, F.,** Determination of manganese, copper and zinc in serum and packed red cells during acute hepatitis, chronic hepatitis and post necrotic cirrhosis, *Clin. Chem.*, 20, 1141, 1974.
61. **Bogden, J. D., Lintz, D. I., Joselow, M. M., and Salak, C. J.,** Effect of pulmonary tuberculosis on blood concentrations of copper and zinc, *Am. J. Clin. Pathol.*, 67, 251, 1977.
62. **Prasad, A. S., Ortega, J., Brewer, G. J., Oberleas, D., and Schoomaker, E. B.,** Trace elements in sickle cell disease, *J. Am. Med. Assoc.*, 235, 2396, 1976.
63. **Khalil, M., Elkhateeb, S., Aref, K., Jahin, S., and Ellozy, M.,** Plasma and red cell copper and elements in infantile diarrhea, *Gaz. Egypt. Pediatr. Assoc.*, 22, 105, 1974.
64. **Scudder, P., Stocks, J., and Dormandy, T. L.,** The relationship between erythrocyte superoxide dismutase activity and erythrocyte copper levels in normal subjects and in patients with rheumatoid arthritis, *Clin. Chim. Acta*, 69, 397, 1976.
65. **Munch-Petersen, S.,** On serum copper in angina simplex and in infectious mononucleosis, *Acta Med. Scand.*, 131, 588, 1948.
66. **Beisel, W. R.,** Trace elements in infectious processes, *Med. Clin. North Am.*, 60, 831, 1976.
67. **Brenner, W.,** Beutrage zur Kentnisse des Eisen und Kupferstoffwechels im Kindersalter: Serum Eisen und Serum Kupfer bei Akuten und Chronischen Infektionen, *Z. Kinderheilk.*, 66, 14, 1949.
68. **Brendstrup, P.,** Serum copper, serum iron and total iron binding capacity of serum in acute and chronic infections, *Acta Med. Scand.*, 145, 315, 1953.
69. **Krebs, H. A.,** Über das Kupfer im menschlichen Blutserum, *Klin. Wochenschr.*, 7, 584, 1928.
70. **Nielsen, A. L.,** On serum copper. IV. Pregnancy and parturition, *Acta Med. Scand.*, 118, 92, 1944.
71. **Fay, J., Cartwright, G. E., and Wintrobe, M. M.,** Studies on free erythrocyte protoporphyrin, serum iron, serum iron-binding capacity and plasma copper during normal pregnancy, *J. Clin. Invest.*, 28, 487, 1949.
72. **Effkemann, G. and Rottger, H.,** Über den Kupferhaushalt während der Schwangerschaft, *Klin. Wschr.*, 28, 13, 1950.
73. **Schenker, J. G., Jungreis, E., and Polishuk, W. Z.,** Serum copper levels in normal and pathologic pregnancies, *Am. J. Obstet. Gynecol.*, 105, 933, 1969.
74. **Burrows, S. and Pekala, B.,** Serum copper and ceruloplasmin in pregnancy, *Am. J. Obstet. Gynecol.*, 109, 907, 1971.
75. **Eisalo, A., Jarvinen, P. A., and Luukkainen, T.,** Liver function tests during intake of contraceptive tablets in premenopausal women, *Br. Med. J.*, 1, 1416, 1965.
76. **Carruthers, M. E., Hobbs, B. C., and Warren, L. R.,** Raised serum copper and ceruloplasmin levels in subjects taking oral contraceptives, *J. Clin. Pathol.*, 19, 498, 1966.

77. **Halsted, J. A., Hackley, B. M., and Smith, J. C., Jr.,** Plasma zinc and copper in pregnancy and after oral contraceptives, *Lancet,* 2, 278, 1968.
78. **Russ, E. M. and Raymunt, J.,** Influence of estrogens on total serum copper and ceruloplasmin, *Proc. Soc. Exp. Biol. Med.,* 92, 465, 1956.
79. **Briggs, M., Austin, J., and Staniford, M.,** Oral contraceptives and copper metabolism, *Nature (London),* 225, 81, 1970.
80. **Eisalo, A., Jarvinen, P. A., and Luukkainen, T.,** Hepatic impairment during the intake of contraceptive pills. Clinical trial with postmenopausal women, *Br. Med. J.,* 2, 426, 1964.
81. **Sandstead, H. H.,** Some trace elements which are essential for human nutrition: Zinc, copper, manganese, and chromium, *Prog. Food Nutr. Sci.,* 1, 371, 1975.
82. **Martin, G. M.,** Copper content of hair and nails of normal individuals and in patients with hepatolenticular degeneration, *Nature (London),* 202, 903, 1964.
83. **Hambidge, K. M.,** Increase in hair copper concentrations with increasing distance from the scalp, *Am. J. Clin. Nutr.,* 26, 1212, 1973.
84. **Li, T. K. and Valle, B. L.,** The biochemical and nutritional role of trace elements, in *Modern Nutrition in Health and Disease,* Godhardt, R. S. and Shils, M. E., Eds., Lea & Febiger, Philadelphia, 1973, 372.
85. **Rice, E. W. and Goldstein, H. D.,** Copper content of hair and nails in Wilson's disease (hepatolenticular degeneration), *Metabolism,* 10, 1085, 1961.
86. **Van Stekebenberger, G. J., Van de Larr, A. J. B., and Van der Laag, J.,** Copper analysis of nail clippings: an attempt to differentiate between normal children and patients suffering from cystic fibrosis, *Clin. Chim. Acta,* 59, 233, 1975.
87. **Sacks, A., Levine, V. E., Hill, F. C., and Hughes, R. C.,** Copper and iron in human blood, *Arch. Intern. Med.,* 71, 489, 1943.
88. **Bowland, J. P., Braude, R., Chamberlain, A. G., Glascock, R. F., and Mitchell, K. G.,** The absorption, distribution and excretion of labelled copper in young pigs given different quantities, as sulfate or sulfide, orally or intravenously, *Br. J. Nutr.,* 15, 59, 1961.
89. **Vierling, J. M., Shrager, R., Rumble, W. F., Aamodt, R., Berman, M. D., and Jones, E. A.,** Incorporation of radiocopper into ceruloplasmin in normal subjects and in patients with primary biliary cirrhosis, *Gastroenterology,* 74, 652, 1978.
90. **Van Berge Henegowen, G. P., Tangedahl, T. N., Hofmann, A. F., Northfield, T. C., LaRusso, N. F., and McCall, J. T.,** Biliary secretion of copper in healthy man: quantitation by an intestinal perfusion technique, *Gastroenterology,* 72, 1228, 1977.
91. **Mills, C. F.,** The dietary availability of copper in form of naturally occurring organic complexes, *Biochem. J.,* 63, 190, 1956.
92. **Evans, G. W. and LeBlanc, F. N.,** Copper binding protein in rat intestine: amino acids composition and function, *Nutr. Rep. Int.,* 14, 281, 1976.
93. **Mills, C. F.,** Copper complexes in grassland herbage, *Biochem. J.,* 57, 603, 1954.
94. **Mills, C. F.,** Availability of copper in freeze-dried herbage and herbage extracts to copper-deficient rats, *Br. J. Nutr.,* 9, 398, 1955.
95. **Moore, T., Constable, B. J., Day, K. C., Impey, S. G., and Symonds, K. F.,** Copper deficiency in rats fed upon raw meat, *Br. J. Nutr.,* 18, 135, 1964.
96. **Guggenheim, K.,** The role of zinc, copper and calcium in the etiology of ''meat anemia'', *Blood,* 23, 786, 1964.
97. **Dick, A. T.,** Studies on the assimilation and storage of copper in crossbred sheep, *Aust. J. Agric. Res.,* 5, 511, 1954.
98. **McMurray, C. H.,** Copper deficiency in ruminants, in *Biological Roles of Copper (Ciba Found. Symp. 79),* Excerpta Medica, Amsterdam, 1980, 182.
99. **Davis, P. N., Norris, L. C., and Kratzer, F. H.,** Interference of soybean proteins with the utilization of trace minerals, *J. Nutr.,* 77, 217, 1962.
100. **Evans, G. W., Majors, P. F., and Cornatzer, W. E.,** Ascorbic acid interaction with metallothionein, *Biochem. Biophys. Res. Commun.,* 41, 298, 1970.
101. **Carlton, W. W. and Henderson, W.,** Studies in chickens fed a copper deficient diet supplemented with ascorbic acid, reserpine and diethylstilbesterol, *J. Nutr.,* 85, 67, 1965.
102. **Hill, C. H. and Starcher, B.,** Effect of reducing agents on copper deficiency in the chick, *J. Nutr.,* 85, 271, 1965.
103. **Seelig, M. S.,** Proposed role of copper-molybdenum interaction in iron-deficiency and iron storage disease, *Am. J. Clin. Nutr.,* 26, 657, 1973.
104. **Cartwright, G. E., Hodge, R. E., Gubler, C. J., Mahoney, J. P., Daum, K., Wintrobe, M. M., and Bean, W. B.,** Studies on copper metabolism. XIII. Hepatolenticular degeneration, *J. Clin. Invest.,* 33, 1487, 1954.
105. **Wintrobe, M. M.,** *Clinical Hematology,* 7th ed., Lea & Febiger, Philadelphia, 1974, 150-152.

106. **Sternlieb, J. and Scheinberg, I. H.,** Radiocopper in diagnosing liver disease, *Semin. Nucl. Med.*, 2, 176, 1972.
107. **Bloomer, L. D. and Lee, G. R.,** Normal hepatic copper metabolism, in *Metals and the Liver,* Powell, I. W., Ed., Marcel Dekker, New York, 1978, 179-239.
108. **Terao, T. and Owen, C. A., Jr.,** Nature of copper compounds in liver supernate and bile of rats: studies with ^{67}Cu, *Am. J. Physiol.*, 224, 682, 1973.
109. **Evans, G. W., Wolenetz, M. L., and Grace, C. I.,** Copper-binding proteins in the neonatal and adult rat liver soluble fraction, *Nutr. Rep. Int.*, 12, 261, 1975.
110. **Terao, T. and Owen, C. A., Jr.,** Copper in supernatant fractions of various tissues. Studies with ^{67}Cu, *Mayo Clin. Proc.*, 49, 376, 1974.
111. **Sternlieb, I., Morell, A. G., Scheinberg, I. H.,** The uniqueness of ceruloplasmin in the study of plasma protein synthesis, *Trans. Assoc. Am. Phys.*, 75, 228, 1962.
112. **Neifakh, S. A., Monakhov, N. K., Shaponshnikov, A. M., and Zubhitski, N. Yu.,** Localization of ceruloplasmin biosynthesis in human and monkey liver cells and its copper regulation, *Experientia,* 25, 337, 1969.
113. **Thiers, R. E. and Vallee, B. L.,** Distribution of metals in subcellular fractions of rat liver, *J. Biol. Chem.*, 226, 911, 1957.
114. **Herman, G. E. and Kun, D.,** Intracellular distribution of copper in rat liver and its response to hypophysectomy and growth hormone, *Exp. Cell. Res.*, 22, 257, 1961.
115. **Gregoriadis, G. and Sourkes, T.,** Intracellular distribution of copper in the liver of the rat, *Can. J. Biochem.*, 45, 1841, 1967.
116. **Milne, D. B. and Weswig, P. H.,** Effects of supplementary copper on blood and liver copper-containing fractions in rats, *J. Nutr.*, 95, 429, 1968.
117. **Porter, H.,** Neonatal hepatic mitochondriacuprein: The nature, submitochondrial localization, and possible function of the copper accumulating physiologically in the liver of newborn animals, in *Trace Element Metabolism in Animals,* Mills, C. F., Ed., E. & S. Livingstone, Edinburgh, 1970, 237-247.
118. **Porter, H., Wiener, W., and Barker, M.,** The intracellular distribution of copper in immature liver, *Biochim. Biophys. Acta,* 52, 419, 1961.
119. **Sternlieb, I., Van Den Hamer, C. J. A., Morell, A. G., Alpert, S., Gregoriadis, G., and Scheinberg, I. H.,** Lysosomal defect of hepatic copper excretion in Wilson's disease (hepatolenticular degeneration), *Gastroenterology,* 64, 99, 1973.
120. **Owen, C. A.,** Copper and hepatic function, in *Biological Roles of Copper (Ciba Found. Symp. 79),* Excerpta Medica, Amsterdam, 1980, 267-282.
121. **Frieden, E. and Hsieh, S. H.,** Ceruloplasmin: The copper transport protein with essential oxidase activity, *Adv. Enzymol.*, 44, 187, 1976.
122. **Frieden, E.,** Ceruloplasmin: a multifunctional metalloprotein of vertebrate plasma, in *Biological Roles of Copper Ciba Found. Symp. 79),* Excerpta Medica, Amsterdam, 1980, 93-124.
123. **Owen, C. A., Jr.,** Metabolism of radiocopper (^{64}Cu) in the rat, *Am. J. Physiol.*, 209, 900, 1965.
124. **Owen, C. A., Jr.,** Metabolism of copper-67 by the copper-deficient rat, *Am. J. Physiol.*, 221, 1722, 1971.
125. **Osborn, S. B. and Walshe, J. M.,** Studies with radioactive copper (^{64}Cu and ^{67}Cu) in relation to the natural history of Wilson's disease, *Lancet,* 1, 346, 1967.
126. **Dekaban, A. S., O'Neilly, S., Aamondt, R., and Rumble, F. W.,** Study of copper metabolism in kinky hair disease (Menkes' disease) and in hepatolenticular degeneration (Wilson's disease) utilizing ^{67}Cu and radioactivity counting in the total body and various tissues, *Trans. Am. Neurol. Assoc.*, 99, 106, 1974.
127. **Hamamoto, K., Tauxe, W. M., Novak, L. P., and Goldstein, N. P.,** Use of whole-body counter to study body retention of radiocopper in Wilson's disease, *J. Lab. Clin. Med.*, 72, 754, 1968.
128. **Hazelrig, J. B., Owen, C. A., Jr., Ackerman, E.,** A mathematical model for copper metabolism and its relation to Wilson's disease, *Am. J. Physiol.*, 211, 1075, 1966.
129. **Sheinberg, I. H. and Morell, A. E.,** Exchange of ceruloplasmin copper with ionic Cu^{64} with reference to Wilson's disease, *J. Clin. Invest.*, 33, 963, 1954.
130. **Marceau, N., Aspin, N., and Sass-Kortsak, A.,** Absorption of copper 64 from gastrointestinal tract of the rat, *Am. J. Physiol.*, 218, 377, 1970.
131. **Sternlieb, I., Morell, A. G., Tucker, W. D., Greene, M. W., and Scheinberg, I. H.,** The incorporation of copper into ceruloplasmin in vivo: studies with copper^{64} and copper^{67}, *J. Clin. Invest.*, 40, 1834, 1961.
132. **Sternlieb, I.,** Gastrointestinal copper absorption in man, *Gastroenterology,* 52, 1038, 1967.
133. **Waldman, T. A., Morell, A. G., Wochner, R. D., Strober, W., and Sternlieb, I.,** Measurement of gastrointestinal protein loss using ceruloplasmin labeled with copper^{67}, *J. Clin. Invest.*, 46, 10, 1967.
134. **Gregoriadis, G., Morell, A. G., Sternlieb, I., and Scheinberg, I.,** Catabolism of desialylated ceruloplasmin in the liver, *J. Biol. Chem.*, 245, 5833, 1970.
135. **Frommer, D. J.,** Defective biliary excretion of copper in Wilson's disease, *Gut,* 15, 125, 1974.
136. **Gollan, J. L.,** Studies on the nature of complexes formed by copper with human alimentary secretions and their influence on copper absorption in the rat, *Clin. Sci. Mol. Med.*, 49, 237, 1975.

137. **Sarkar, B.,** Transport of copper, in *Metal Ions in Biological Systems,* Vol. 12, Properties of Copper, Sigel, H., Ed., Marcel Dekker, New Yrok, 1981, chap. 6.
138. **Frommer, D. J.,** Biliary copper excretion in man and the rat, *Digestion,* 15, 390, 1977.
139. **Lewis, K. O.,** The nature of the copper complexes in bile and their relationship to the absorption and secretion of copper in normal subjects and in Wilson's disease, *Gut,* 14, 221, 1973.
140. **McCullars, G. M., O'Reilly, S., and Brennan, M.,** Pigment binding of copper in human bile, *Clin. Chim. Acta,* 74, 33, 1977.
141. **Gollan, J. L., Davis, P. S., and Deller, D. J.,** Binding of copper by human alimentary secretions, *Am. J. Clin. Nutr.,* 24, 1025, 1971.
142. **Farrer, P. A. and Mistilis, S. P.,** Copper metabolism in the rat: studies of the biliary excretion and intestinal absorption of ^{64}Cu-labeled copper, *Birth Defects Orig. Artic. Ser.,* 4(2), 14, 1968.
143. **Farrer, P. and Mistilis, S. P.,** Absorption of exogenous and endogenous biliary copper in the rat, *Nature (London),* 213, 291, 1967.
144. **Mistilis, S. P. and Farrer, P. A.,** The absorption of biliary and nonbiliary radiocopper in the rat, *Scand. J. Gastroenterol.,* 3, 586, 1968.
145. **Gregoriadis, G., Morell, A. G., Sternlieb, I., and Scheinberg, I. H.,** Catabolism of desialylated ceruloplasmin in the liver, *J. Biol. Chem.,* 245, 5833, 1970.
146. **Gregoriadis, G. and Sourkes, T. L.,** Role of protein in removal of copper from the liver, *Nature (London),* 218, 290, 1968.
147. **Gitlin, D., Hughes, W. J., and Janeway, C. A.,** Absorption and excretion of copper in mice, *Nature (London),* 198, 150, 1960.
148. **Mertz, W. and Roginski, E. E.,** *Newer Trace Elements in Nutrition,* Marcel Dekker, New York, 1971, 123.
149. **Mitchell, H. H. and Hamilton, T. S.,** The dermal excretion under controlled environmental conditions of nitrogen and minerals in human subjects, with particular reference to calcium and iron, *J. Biol. Chem.,* 178, 345, 1949.
150. **Leverton, R. M. and Brinkley, E. S.,** The copper metabolism in young women, *J. Nutr.,* 27, 43, 1944.
151. **Chuttani, H. K., Gupta, P. S., Gulati, S., and Gupta, D. N.,** Acute copper sulfate poisoning, *Am. J. Med.,* 39, 849, 1965.
152. **Wahal, P. K., Mehrota, M. P., Kishore, B., Patney, N. L., Mital, V. P., Harza, D. K., Raizada, M. N., and Tiwari, S. R.,** Study of whole blood, red cell and plasma copper levels in acute copper sulfate poisoning and their relationship with complications and prognosis, *J. Assoc. Physicians India,* 24, 153, 1976.
153. **Scheinberg, I. H. and Sternlieb, I.,** Copper toxicity and Wilson's disease, in *Trace Elements in Human Health and Disease,* Vol. 1, Prasad, A. S., Ed., Academic Press, New York, 1976, 415-438.
154. **Glaister, J. and Rentoul, E.,** *Medical Jurisprudence and Toxicology,* E. & S. Livingstone, London, 1966, 530.
155. **Moeschlin, S.,** *Poisoning: Diagnosis and Treatment,* (transl. from the 4th Germ. ed.), Grune & Stratton, New York, 1965, 124.
156. **Davenport, S. J.,** Health Hazards of Metals. I. Copper, U.S. Department of Interior, U.S. Government Printing Office, Washington, D.C., 1953, 114.
157. **Chugh, K. S., Sharma, B. K., Singhal, P. C., Dass, K. C., and Datta, B. N.,** Acute renal failure following copper sulphate intoxication, *Postgrad. Med. J.,* 53, 18, 1977.
158. **Wahal, P. K., Mittal, V. P., and Bansal, O. P.,** Renal complications in acute copper sulphate poisoning, *Indian Pract.,* 18, 807, 1965.
159. **Holtzman, N. A., Elliott, D. A., and Heller, R. H.,** Copper intoxication. Report of a case with observations on ceruloplasmin, *N. Engl. J. Med.,* 275, 347, 1966.
160. **Agarwal, B. N. and Agrawal, P.,** Zinc and copper in nephrology, in *Zinc and Copper in Medicine,* Karcioglu, Z. A. and Sarper, R. M., Eds., Charles C Thomas, Springfield, Ill., 1980, chap. 13.
161. **Zelkowitz, M., Verghese, J. P., and Antel, J.,** Copper and zinc in the nervous system, in *Zinc and Copper in Medicine,* Karcioglu, A. Z. and Sarper, R. F., Eds., Charles C Thomas, Springfield, Ill., 1980, chap. 14.
162. **Cole, D. E. C. and Lyrenman, D. S.,** Treatment with albumin-enriched peritoneal dialysate in acute copper poisoning, *J. Pediatr.,* 92, 955, 1978.
163. **Blomfield, J., McPherson, J., and George, C. R. P.,** Active uptake of copper and zinc during hemodialysis, *Br. Med. J.,* 2, 141, 1969.
164. **Manzler, A. D. and Schreiner, A. W.,** Copper-induced acute hemolytic anemia, *Ann. Intern. Med.,* 73, 409, 1970.
165. **Blomfield, J., Dixon, S. R., and McCredie, D. A.,** Potential hepatotoxicity of copper in recurrent hemodialysis, *Arch. Intern. Med.,* 128, 555, 1971.
166. **Lyle, W. H.,** Chronic dialysis and copper poisoning, *N. Engl. J. Med.,* 276, 1209, 1967.

167. **Lyle, W. H., Payton, J. E., and Hui, M.,** Haemodialysis and copper fever, *Lancet,* 1, 1324, 1976.
168. **Matter, B. J., Pederson, J., Psimenos, G., and Lindeman, R. D.,** Lethal copper intoxication in hemodialysis, *Trans. Am. Soc. Artif. Intern. Organs,* 15, 309, 1969.
169. **Hopper, S. H. and Adams, H. S.,** Copper poisoning from vending machines, *Public Health Rep.,* 73, 910, 1958.
170. **Semple, A. B., Parry, W. H., and Phillips, D. E.,** Acute copper poisoning; outbreak traced to contaminated water from corroded geyser, *Lancet,* 2, 700, 1960.
171. **LeVan, J. H. and Perry, E. L.,** Copper poisoning on shipboard, *Public Health Rep.,* 76, 334, 1961.
172. **Bohre, G. F., Huisman, J., and Lifferink, H. F. L.,** Acute copper poisoning aboard a ship, *Ned. Tijdschr. Geneesk.,* 109, 978, 1965.
173. **Paine, C. H.,** Food poisoning due to copper, *Lancet,* 2, 520, 1968.
174. **McMullen, W.,** Copper contamination of soft drinks from bottle pourers, *Health Bull. (Edinburgh),* 29, 94, 1971.
175. **Wyllie, J.,** Copper poisoning at cocktail party, *Am. J. Public Health,* 47, 615, 1957.
176. **Pimentel, J. C. and Marques, F.,** "Vineyard Sprayers' Lung." A new occupational disease, *Thorax,* 24, 678, 1969.
177. **Pimentel, J. C. and Menezes, A. P.,** Lung cancer and copper inhalation in vineyard sprayers, Proc. 5th Cong. Eur. Soc. Pathol., Vienna, 1975, 140.
178. **Pimentel, J. C., Menezes, A. P., and Avila, R.,** Doencas Pulmonares Inhalatorias Provacados por Fungicidas. Relatorio annul. Lisbon, 1974.
179. **Pimentel, J. C. and Menezes, A. P.,** Liver granulomas containing copper in vineyard sprayer's lung. A new etiology of hepatic granulomatosis, *Am. Rev. Respir. Dis.,* 111, 189, 1975.
180. **Pimentel, J. C. and Menezes, A. P.,** Liver disease in vineyard sprayers, *Gastroenterology,* 72, 275, 1977.
181. **Braude, R., Mitchell, K. G., and Pittman, R. J.,** A note on cuprous chloride as a feed additive for growing pigs, *Anim. Prod.,* 17, 321, 1973.
182. **Okereke, T., Sternlieb, I., Morell, A. G., and Scheinberg, I. H.,** Systemic absorption of intrauterine copper, *Science,* 177, 358, 1972.
183. **Hefnawi, F., Kandil, O., and Askalani, H.,** Copper levels in women using intrauterine devices or oral contraceptives, *Fertil. Steril.,* 25, 556, 1974.
184. **Mishell, D. R., Jr. and Moyer, D. L.,** Association of pelvic inflammatory disease with the intrauterine device, *Clin. Obstet. Gynecol.,* 12, 179, 1969.
185. **Cuadros, A. and Hirsch, I. G.,** Copper on intrauterine devices stimulates leukocyte exudation, *Science,* 175, 175, 1972.
186. **Gump, D. W., Mead, P. B., Horton, E. L., Lambron, K. R., and Forsyth, B. R.,** Intrauterine contraceptive device and increased serum immunoglobulin levels, *Obstet. Gynecol.,* 41, 259, 1973.
187. **Holub, W. R., Reyner, F. C., and Forman, G. H.,** Increased levels of serum immunoglobulins G and M in women using intrauterine contraceptive devices, *Am. J. Obstet. Gynecol.,* 110, 326, 1971.
188. **Hagenfeldt, K.,** Intrauterine contraception with the copper device. I. Effect on trace elements in the endometrium, cervical mucus and plasma, *Contraception,* 6, 37, 1972.
189. **Rubinfeld, Y., Maor, Y., Simon, D., and Modai, D.,** A progressive rise in serum copper levels in women taking oral contraceptives: a potential hazard? *Fertil. Steril.,* 32, 599, 1976.
190. **Owen, C. A., Jr.,** Similarity of copper toxicity in rats of copper deposition of Wilson's disease, *Mayo Clin. Proc.,* 49, 368, 1974.
191. **Subcommittee on the Geochemical Environment in Relation to Health and Disease,** The relationship of selected trace elements to health and disease, in *Geochemistry and Environment,* Vol. 1, National Academy of Sciences, Washington, D.C., 1974.
192. **Subcommittee on the Geochemical Environment in Relation to Health and Disease,** Distribution of trace elements related to the occurrence of certain cancers, cardiovascular diseases and urolithaisis, in *Geochemistry and Environment,* Vol. 3, National Academy of Sciences, Washington, D.C., 1978.
193. **Deur, C. J., Stone, M. J., and Frenkel, E. P.,** Trace metals in hematopoiesis, *Am. J. Hematol.,* 11, 309, 1981.
194. **Williams, D. M.,** Copper deficiency in humans, in *Seminars in Hematology,* Vol. 20, Metal Metabolism in Hematologic Disorders, Miescher, P. A. and Jaffe, E. R., Eds., Grune & Stratton, New York, 118-128.

Chapter 6

FACTORS INFLUENCING COPPER HOMEOSTASIS

I. INTRODUCTION

Alterations in copper metabolism have been observed in a variety of normal and disease states.[1-34] Excluding copper poisoning and the two genetically inherited diseases discussed below,[22-24] very few pathological conditions result from impaired copper homeostasis per se.[3,7,10,16] In most diseases where an alteration in copper metabolism is involved, the disease has affected one or more homeostatic mechanisms and copper imbalance appears only as a secondary manifestation of the disease.[16] For instance, the hypocupremia and depressed hepatic copper levels observed in Kwashiorkor syndrome probably result from a low dietary copper intake, the unavailability of suitable ligands for transporting the metal across the intestinal mucosa, and lack of other essential nutrients required for maintaining the normal function of the mucosal cells and increased loss into the bowels.

Diet,[1,3,34,35] age,[15,17] race,[17,18] sex,[1,30] hormonal imbalance,[2-6] smoking,[8] obesity,[20] and pathological conditions of various etiology[7-29,31,36] have been reported to influence serum copper and ceruloplasmin concentrations.

Logical explanations and probable causative mechanisms whereby the homeostasis of copper is altered in various physiological and pathological conditions can be given in a number of instances. Sex hormone imbalance seen in pregnancy,[1-3,10,14] estrogen and oral contraceptive administration,[5,6] and androgen therapy[35] are associated with a markedly increased serum copper level and are believed to be induced by increased synthesis and release of ceruloplasmin.[3,6,36] Also, administration of copper itself in small doses has been shown to increase the rate of ceruloplasmin synthesis, while repeated injections of large doses of copper results in depressed copper and ceruloplasmin levels.[37,38]

Since bile is in the main excretory route of copper, it is not surprising to find elevated serum copper levels in obstructive hepatobiliary diseases, although no good correlation between hyperbilirubinemia and hyperceruloplasminemia was found suggesting that the enzyme is excreted by a different mechanism and that other factors are involved in the production of elevated ceruloplasmin levels in these conditions.[39] Cirrhosis of the liver is associated with impaired inactivation of estrogens via glucoronide conjugation. Elevated circulating levels of these hormones in addition to intrahepatic obstruction are assumed to account for the increased copper concentrations.

Hyperfunction or hypofunction of pituitary-adrenal axis hormones has been associated with changes in the copper concentration of blood, bile, and urine.[16,40] The same holds true for other target organs like thyroid and parathyroid glands.[16,40] The role of hypothalmus influence on copper concentration is not known.[16] The mechanism by which pituitary-adrenal axis hormones imbalance affects copper homeostasis is not quite clear, but this probably is the result of some redistribution of body copper and shifting of the metal to tissue compartments or promoting copper excretion.[16]

Little is known as to how these hormonal changes might influence, mediate, or contribute to the observed changes in copper metabolism during an infection or inflammation.[21] In many infections and inflammatory conditions moderately elevated serum copper and ceruloplasmin concentrations have been observed.[10,11,14] The mediation of this rise is thought to be caused by products of polymorphonuclear leukocytes.[20,21,41-48] It has been observed that a substance with hormone-like properties is released from phagocytizing white blood cells into the serum during infectious illnesses. This substance has been independently recognized and given a variety of names reflecting its function: interleukin-1, lymphocyte activating factor, leukocyte endogenous mediator (LEM), and endogenous pyrogen. The mechanism by which interleukin-1 causes copper elevation is believed to be through the stimulation of liver synthesis and the release of acute phase reactants including ceruloplasmin. The efflux of copper and ceruloplasmin from the liver in association with numerous other acute phase reactants, as a host defense system response, is probably the result of several factors, some of which have not as yet been completely defined. Striking correlations between percentage of blast cells in bone marrow and serum copper levels in patients with untreated acute childhood leukemia have been reported.[28] Also, it has been observed that serum copper levels reflect disease activity in malignant lymphoma (Hodgkin's or non-Hodgkin's type)[27-29,31] and some other malignancies.[32,33,49] These observations are restricted in most cases to a description of alterations in the blood levels of copper and ceruloplasmin in these patients. Efforts to find an explanation for these changes and attempts to sort out the reasons for their development in the course of malignant diseases are made infrequently.

Drugs like chelating agents,[50,51] corticosteroids,[16,52] and prolonged zinc therapy[25,53] will result in reduced serum copper levels. On the other hand administration of sex hormones,[5,6] psoralens,[54] triiodothyronine,[55] and diphenylhydantoin[56] have been reported to increase serum copper levels.

Incidence of hypocupremia (decreased serum copper) is much less often encountered and is better understood. Dietary lack of copper,[34,35] impaired absorption and/or transport,[23,24] and increased excretion[34,57] are the most common causes of copper depletion.

The various conditions in which altered serum copper homeostasis has been observed associated with increased or decreased serum, liver, or urinary copper concentrations can be seen in Tables 1, 2, 3, and 4. The underlying factors causing alterations of copper metabolism in some conditions are not defined clearly. It is not possible, for this reason, to place these factors in logical order or to explain fully their effects or mechanism of action. For these reasons a pragmatic classification has been adopted, and we will discuss under the two subsequent headings the alterations of copper homeostasis in physiological and pathological conditions.

II. ALTERATION OF COPPER HOMEOSTASIS IN PHYSIOLOGICAL CONDITIONS

Variations of serum copper and ceruloplasmin levels and most known causative factors altering copper metabolism are related to age and hormonal status of the subjects studied, as can be seen from Table 5. Pregnant women (last trimester) and parturient mothers contain approximately six times the serum copper and ceruloplasmin concentration of the newborn and two times the copper of normal adults. One- to two-year-old children have about five times the serum copper and ceruloplasmin concentration than that of newborns and approximately two times that of the adult values. The black population has higher serum copper and ceruloplasmin levels than their white or Hispanic counterparts. The serum copper levels and ceruloplasmin concentrations in women taking estrogens or oral contraceptives are $2^1/_2$ times higher than that of females who are not using drugs to control fertility. Slightly higher serum copper levels and ceruloplasmin concentrations are found in females as compared to males. Finally, smoking[8] and eating[20] habits are also found to affect the serum copper and ceruloplasmin levels, being higher in smokers and obese persons vs. the non-smoking or lean population. These and other factors altering copper homeostasis will be discussed in more detail in subsequent sections.

A. Diet

Little is known of the chemical form in which copper exists in food.[1] Because of the ubiquitous presence of copper in food constituents and even in drinking water (0.09 to 0.30

Table 1
CONDITIONS WHICH MAY BE ASSOCIATED WITH ELEVATED SERUM COPPER (HYPERCUPREMIA)

Pregnancy

Hormonal administration (estrogens — oral contraceptives, progesterones, androgens, thyroid, and parathyroid hormone), diphenylhydantoin, triiodothyronine, and psoralens

Obstructive (intra- or extrahepatic) hepatobiliary disorders: primary biliary cirrhosis, Indian childhood cirrhosis, extrahepatic obstruction for any reason, and most acute and chronic liver diseases

Various malignant diseases (acute childhood leukemia, Hodgkin's disease, non-Hodgkin's lymphoma, and ? chronic leukemia), hepatic neoplasms, lung, breast, gastric carcinoma, and osteogenic sarcoma

Acute and chronic infections or inflammations (viral, bacterial, fungal, or parasitic origin), collagen diseases, rheumatoid arthritis, rheumatic fever, glomerulonephritis, and psoriasis

Endocrine disorders (hypopituitarism, Addison's disease, hypophysectomy, or adrenalectomy for any reason, and ? hyperthyroidism and hyperparathyroidism).

Acute myocardial infarction, congestive heart failure, and atherosclerotic peripheral vascular disease with claudication

Postoperative states following major surgery

Anemias (aplastic, sideropenic, pernicious, and sickle cell anemia)

Young age group (children 1—14 years)

Copper toxicity

IgG myeloma (rare)

Smoking

ppm), diets composed of natural foods contain adequate amounts of copper. Diets abnormally low or high in copper generally cause corresponding changes in the concentration of this metal in the liver and frequently in the blood in mammalian system.[1,3,15]

Copper is found in various foodstuffs, the amount in agricultural products depending upon the copper content of the soil. The richest dietary sources of copper are animal livers and kidneys, Crustacea, shellfish, dried fruit, nuts, and chocolate.[61,62] The copper content of processed baby food has been determined; the concentration in strained beef and cereals are highest and those in fruits and vegetables are next. Cow's milk and dairy products, cereals, white sugar, and honey are the poorest sources of this element. Human milk contains 450 μg/ℓ, cow's milk 150 μg/ℓ, and homogenized milk 15 to 140 μg/ℓ. It is clear that some artificial milk diets will provide less than the recommended allowance for copper. The refining of cereals and sugar for human consumption results in significant losses of copper, although it is not as severe as it is with Fe, Zn, and Mn. In addition to the absolute quantity of copper found in the diet, the tissue stores, and particularly those of the liver, are influenced by conditioning factors including the chemical form of copper presented to the organism, as well as the quantity of phytate, fiber, iron, zinc, and other anions (sulfide) in the dietary complement.[61] Dietary protein protects against accumulation of toxic levels of copper.[16,63] The effects of dietary protein on copper absorption probably result from the formation of copper complexes of macromolecules, which does not render the copper suitable for intestinal absorption.

Since milk is particularly known to be deficient in copper with less than 20 μg/100 kcal,

Table 2
CONDITIONS WHICH MAY BE ASSOCIATED WITH DECREASED SERUM COPPER (HYPOCUPREMIA)

Conditioned copper deficiency
- Decreased copper dietary intake
  - Premature neonates of low birth weight fed on milk-based low-copper infant formula
  - Prolonged total parenteral nutrition without copper supplementation
- Decreased copper absorption
  - Iron-fortified milk formula
  - Zinc and ascorbic acid long-term therapy
  - Malabsorption syndromes
    - Extensive jejunoileal bypass?
    - Chronic diarrhea in infancy

Excessive loss of copper (states associated with hypoproteinemia)
- Kwashiorkor
- Celiac disease
- Sprue—tropical and nontropical
- Idiopathic hypoproteinemia
- Active ulcerative colitis
- Nephrotic syndrom
- Burns and scalds
- Cystic fibrosis with severe hypoproteinemia

Hereditary diseases
- Menkes' kinky hair syndrome
- Wilson's disease

Endocrine disorders
- Cushing's syndrome
- Adrenal cortical carcinoma
- Adrenal corticosteroid therapy

Infancy (newborn)

Table 3
CONDITIONS ASSOCIATED WITH INCREASED LIVER COPPER CONTENT

Hepatolenticular degeneration (Wilson's disease)

Heterozygotes for Wilson's disease

Obstructive hepatobiliary diseases: extrahepatic biliary obstruction for any cause (primary biliary cirrhosis, various cholestatic and noncholestatic syndromes, Indian childhood cirrhosis, extrahepatic biliary atresia, and cirrhosis of $alpha_1$ antitrypsin deficiency)

Hepatic granulomatosis ("vineyard sprayer's liver")

Copper toxicity

Newborn infants

copper deficiency in severely malnourished infants rehabilitated on milk-based, low-copper diets has been reported.[34,35] Preterm infants of low birth weight, pregnant and nursing women, patients of all age groups on prolonged (3 to 17 weeks) total parenteral nutrition or zinc therapy, and children with various types of malabsorption syndromes or protein-loosing enteropathies fed with synthetic diets require more copper than the recommended daily

Table 4
CONDITIONS ASSOCIATED WITH INCREASED URINARY COPPER (HYPERCUPRIURIA)

Wilson's disease

Obstructive (intra- or extrahepatic) hepatobiliary disorders

Nephrotic syndrome

Endocrine disorders (Cushing's syndrome, adrenal cortical carcinoma, and primary hyperparathyroidism)

Administration of adrenal corticosteroids, parahormone, growth hormone, chelating agents (D-penicillamine, triethylenetetramine)

Burns and scalds

Water diuresis

Menkes' kinky hair syndrome

Table 5
VARIATIONS OF SERUM COPPER AND CERULOPLASMIN LEVELS ACCORDING TO AGE, SEX, RACE, AND HORMONAL STATUS

	No. of persons	Copper (μg/100 mℓ)	No. of subjects	Ceruloplasmin (mg/100 mℓ)	Ref.
Adult males	40	105(68—134)	100	32.3 ± 4.9	58, 59
Adult females	23	116(84—143)			
Adult black men	21	118.1 ± 32.7	21	30.0 ± 6.6	18
Adult white men	28	96.5 ± 11.5	28	25.6 ± 5.3	18
Adult black women	27	118.4 ± 16.5	27	27.9 ± 4.80	18
Adult white women	35	118.5 ± 25.5	35	30.1 ± 8.4	18
Healthy nonsmokers	40	99 (52—191)	35	30.3(18—42.5)	8
Smokers	25	116(71—160)	25	41.1(22.1—58.6)	
Women taking oral contraceptives	14	211(101—211)	14	52.5(30—61.7)	8
Women taking oral contraceptives	3	275	3	72	6
Pregnant women (6—9 months)	4	279(207—332)	4	48.9(32.9—57.4)	60
Parturient mothers	12	216(118—302)	12	55.5(39.4—89.0)	15
Newborns (cord blood)	12	36(12—26)	12	6.5(1.8—13.1)	15
Children (years)					
2	133	140(95—186)	134	42.6(31.4—53.9)	15
6	133	129(83—174)	134	38.1(26.9—49.2)	15
10	133	117(72—162)	134	33.5(22.4—44.6)	15
11 (white)	77	124.9 ± 24.11	—	—	17
11 (black)	19	143.7 ± 32.71	—	—	17
12 (white)	52	120.9 ± 19.68	—	—	17
12 (black)	13	136.8 ± 19.38	—	—	17

allowance suggested by the National Academy of Science (U.S.A.) or the World Health Organization (WHO).[34,61] Adult Western-type diets are reported to provide 2 to 5 mg/day.[1,61,62] Few adult diets provide the recommended daily allowance of 2 mg/day or 30 μg/kg, and many contain less than half of this value.[64,65] Those who rely heavily upon prepared foods from supermarket shelves are at risk to suffer from subclinical copper deficiency.[62] Food processors are designed to eliminate copper because its oxidant properties limit shelf-life. The suggested WHO copper intakes are 80 μg/kg/day for infants and young children and 40 μg/kg/day for older children.[61] Extrapolating from data obtained in patients receiving total parenteral nutrition, Dunlap et al.[66] calculated the copper requirement for adults to be 400 μg/day.

If a patient is to be maintained on parenteral hyperalimentation or zinc therapy on a long-term basis, serum copper levels should be monitored.[53,66] Copper should be added either parenterally or orally, depending on the clinical circumstances. It would seem that 1.25 mg elemental copper (as 5 mg copper sulfate) is the proper initial dosage, with 0.4 mg (1.6 mg copper sulfate) as maintenance.[66]

Plasma copper does not increase following meals, nor decrease during short-term fasting.[1] Weight loss and negative nitrogen balance have not been significant features in the accounts of copper deficiency in man.[67] On the other hand, obesity has been reported to be associated with increased serum copper levels.

B. Age, Sex, and Race

The serum copper and ceruloplasmin levels vary quite markedly with age, as shown in Table 5. It has been known for some time that the serum copper content of newborn infants is very low - about one third of the normal adult range and, more significantly, one sixth of the values in parturient mothers. Towards the end of pregnancy, copper is extensively transferred to the fetus, much being retained in the liver by specific fetal protein-mitochondrocuprein, which is present in early life and is lost soon after birth.[68,69] Increased copper store in the liver of newborn (300 to 400 μg/g dry weight) is utilized during the suckling period, since milk is relatively low in copper. The low serum copper levels in the newborn are paralleled by similarly low serum levels of ceruloplasmin. The placenta does not seem to permit the transfer of ceruloplasmin from mother to fetus, or only to a very small extent, and perhaps most or all of the fetal ceruloplasmin is synthesized by the fetus.[3,68,69] The increased copper stores in the liver and the other tissues of the newborn are apparently transferred from the mother via the albumin-bound fraction of serum copper, which seems to be in equilibrium on both sides of the placenta.[3,15] The equilibration is probably further aided by the third, small molecular-sized amino acid-bound fraction of the serum copper which has been shown to exist.[63] On the other hand, it has been reported that nonceruloplasmin copper of the umbilical cord represents a much larger proportion (23%) of total plasma copper, which is only 8% or less in the plasma of adults.[68]

Interesting age-dependent changes in the concentration of both serum copper and ceruloplasmin levels in children were observed by Cox and Sass-Kortsak.[15] In a group of 134 healthy children from 2 to 15 years of age, with close to equal numbers of males and females, there was no significant difference related to sex, but a highly significant ($p > 0.001$) negative regressing coefficient with increasing age was found. From this it seems that by about 1 year of age the initially rather low levels of copper and ceruloplasmin rise above the normal adult range and then gradually decrease in a linear fashion to the normal adult level which is reached about the age of puberty. This age-related pattern of serum copper levels has been confirmed in our large series of 681 subjects (2 to 12 years old) and will be discussed later.[17,28] The reasons for and the physiological meaning of the marked changes in the tissue copper content, serum copper, and ceruloplasmin levels with age are not known. A slight linear increase in serum copper levels between the 3rd and 7th decade was reported,[49,70] whereas Sinha[14] found no change with age in adult life. Lahey et al.[58] found very little change in plasma copper from day to day or week to week. On the other hand, considerable variation from time to time, though not outside normal limits for individual determinations, has been reported.[15,70] A circadian pattern of serum copper and ceruloplasmin variation as well as urinary copper excretion has been observed[71,72] indicating an inverse relationship between plasma copper levels and adrenal steroids activity.[16,40] Slightly higher levels of both serum and ceruloplasmin levels in normal adult females compared to males were reported, although no statistically significant difference was found.[30,58] There has been disagreement regarding variations in plasma copper during the menstrual cycle. Lahey et al.[58] found no correlation whereas von Studnitz and Berezin[73] reported ±15% variations of serum copper during the menstrual cycle. Sass-Kortsak[15] observed a change in serum copper levels of as much as 33% in a single individual during the menstrual cycle; the variation was considered not to be outside the normal limits established for the female. Values for red blood cell copper were also found to be relatively constant during the menstrual cycle.[58] From the above, it is obvious that variations of serum copper throughout the menstrual cycle were sought, but are still not well documented. As can be seen from Table 5, the most significant factor influencing the levels are hormonal imbalance and children's age group.

Williams[18] studied 47 black and 63 white subjects and reported hemoglobin levels in blacks are consistently lower than in whites. Copper levels were lower in white men than in other groups and zinc levels were similar between racial groups. The red blood cells of blacks were also smaller than those in whites. These differences do not appear to be related to socioeconomic dietary differences or differences of hereditary disorders such as sickle cell disease. Race-related differences in serum copper levels was also noted in our studies and will be discussed later.[17]

C. Stress

In most mammalian species plasma copper and ceruloplasmin levels are elevated as a result of a variety of stress conditions. Several investigations demonstrated that both serum copper levels and ceruloplasmin concentrations are significantly elevated in experimental animals after violent exercise.[1,16] Haralambie and Kreul[74] observed an increase in serum copper and ceruloplasmin in humans after 2 hr of physical exercise. On the other hand, Prasad[67] stated that physical exertion in humans produces no change in serum copper levels. In various experiments an inverse relationship was found between plasma copper and corticosteroid levels, indicating that hormones of the adrenal cortex are not responsible for the rise in plasma copper during physical stress.[16]

D. Pregnancy

One of the earliest indications that hormones influence copper metabolism was the observation of Krebs[75] in 1928, that the serum copper level of pregnant females is nearly twice that of control subjects. von Studnitz and Berezin[73] suggested that this elevation of the serum copper level was due to the action of increased endogenous estrogen occurring during pregnancy. It is well documented that in normal pregnant women there is a gradual increase, beginning at the 8th week of pregnancy, in both serum copper and ceruloplasmin levels, culminating in levels approximately twice normal at term.[2-4,10,14] The copper content of the liver is also increased in pregnancy, but urine copper is normal suggesting that the increased serum copper levels are due to increased ceruloplasmin concentration which is unable to cross glomerulus. The high levels of serum copper and ceruloplasmin decrease rapidly after delivery and reach the normal adult range within the first 6 to 8 weeks postpartum. Nielsen,[76] in a study of 31 pregnant women, found serum copper to increase from the 3rd month to an average of 2.7 μg/mℓ, compared with a normal nonpregnant level of 1.2 μg/mℓ. These observations have been confirmed by many investigators.[2-4,10,14,58,73] According to Lahey et al.[58] increased serum copper levels in pregnant women range from 150 to 285 μg/100 mℓ. Red blood cell copper remains normal during pregnancy indicating that elevation of whole blood copper is entirely due to an increase in serum copper, mainly through induction of the copper-carrying protein, ceruloplasmin.

Markowitz et al.[10] reported a high correlation between serum copper and ceruloplasmin levels in pregnant women. The mechanism by which copper increases in blood following increases in blood estrogen level lies in the induction of the major copper-bound protein, ceruloplasmin. Sinha and Gabrieli[14] reported serum copper levels in 138 pregnant subjects ranging from 129 to 410 μg/100 mℓ, with a mean value of 227 ± 50 μg/100 mℓ. In their series, 92% of the pregnant women had values more than 2 SD above normal, and 8% had normal serum copper levels.

E. Sex Hormones

Russ and Raymunt[77] in 1956 demonstrated that estrogen administration produces an increase in serum copper and copper-binding protein. It has been reported on many occasions that the serum copper levels and ceruloplasmin concentrations are elevated in patients following the administration of estrogen or an estrogen-progestin oral contraceptive pill:[5,6,78-80] 1 mg of ethinyl estradiol per day has raised serum copper concentrations to over 300 μg/100 mℓ in a number of cases.[3] It is believed that estrogens induce *de novo* synthesis of ceruloplasmin, although the precise mechanism or process by which estrogen increases the ceruloplasmin concentrations is not known.[15,16] Abnormal liver function, pruritus and icterus in contraceptive users suffering from reversible noninflammatory intrahepatic cholestasis have been reported.[5] Halsted et al.[81] reported a mean of 3.0 ± 0.7 μg/mℓ in women on contraceptive medication, compared with 1.18 ± 0.2 μg/mℓ in the control group.

Carruthers et al.[6] found a high degree of correlation between serum copper and ceruloplasmin levels in 25 female subjects taking oral contraceptives. The coefficient of correlation of the serum copper and ceruloplasmin concentrations was 0.94. Normal urinary copper findings in these subjects were consistent with the fact that ceruloplasmin is too large a protein to cross the renal glomeruli freely. In no case was the serum copper or ceruloplasmin level within the normal range, the lowest copper level in this series was 180 μg/100 mℓ. Schenker et al.,[5] in a large series of 502 women taking oral contraceptives, found the mean serum copper level of 207 μg/100 mℓ. The rise in serum copper level appears after the first two cycles of treatment and normalizes in 4 to 6 weeks after discontinuation of therapy. The rise in serum copper levels to over 275 μg/100 mℓ in 31 women (6.5%) indicated liver dysfunction (cholestasis). The authors suggested that serum copper determinations may aid in selecting suitable subjects for oral contraceptive medications, and in early detection of liver injury before the appearance of clinical signs. There were no differences noted in the serum copper levels of women taking contraceptives of different steroid compositions. On the other hand, Rubinfeld et al.[82] reported that females using Neogynon® (Schering) had significantly higher serum copper levels as compared to other estrogen-progestin preparations, suggesting possible potential hazard of excessive copper accumulation in these patients. Administration of androgen also has been reported to increase serum copper levels associated with increased ceruloplasmin concentrations.[83] A number of elderly men and women have been given 25 mg of testosterone propionate twice a week, and after a few weeks the serum copper concentrations increased from 114 to 153 μg/100 mℓ in the women, whereas that of men increased from 106 to 137 μg/100 mℓ. Androgen therapy can also cause liver dysfunction as a result of defective hepatobiliary excretion.

Numerous adverse effects of estrogen therapy have been noted. The most important are increased incidence of thrombembolic phenomena, elevated blood pressure, fluid retention, menstrual irregularities, migraine headache, depression, and hypercalcemia in patients with destructive bone lesions.[84] The prolonged estrogen therapy in postmenopausal women has also been associated with increased risk ratio of endometrial carcinoma.[85,86] Benign hepatic adenomas have been reported to be associated with the use of oral contraceptives.[87-89] The possibility of induction of other malignancies (breast, vagina, and cervix) has been raised but it has not been confirmed or refuted.[90-94] The use of estrogens or androgens in patients with tumors known to be hormone-dependent (breast or prostatic carcinoma) is contraindicated.

F. Habits

Davidoff et al.[8] reported elevation in serum copper, red blood cell copper, and ceruloplasmin concentrations in smokers. Significant increase of serum copper ($p < 0.0053$) and ceruloplasmin ($p < 0.0001$) were found in the smoking vs. the nonsmoking population. No correlation was observed between duration of smoking (pack-years smoked) or frequency of usage (packs per day) and elevations of serum copper and ceruloplasmin levels or erythrocytic copper. This interesting and useful information should be kept in mind when interpreting serum or ceruloplasmin alterations in lymphoma patients. It is conceivable that lymphoma patients who discontinue smoking in the middle of their treatment may have serum or ceruloplasmin levels showing a decrease suggestive of favorable response to therapy. On the other hand, lymphoma patients in remission who begin to smoke may show an increase in serum or ceruloplasmin levels indicative of relapse.

Linder and Moor[49] independently also found increased serum copper and ceruloplasmin concentrations in heavy smokers. Two or three smokers with elevated ceruloplasmin levels subsequently developed lung cancer.

Similarities between serum copper levels and carcino-embryonic antigen (CEA) are obvious. The CEA level is also known to be elevated in heavy smokers. The cause of blood copper elevation in the smoking population is not known.

III. ALTERATIONS OF COPPER HOMEOSTASIS IN PATHOLOGICAL NONMALIGNANT CONDITIONS

Most of the knowledge concerning alterations of copper metabolism in human disease derives from studies in the blood and urine concentration of this element in afflicted patients. The only exceptions are Wilson's disease and Menkes' kinky hair syndrome, in which tissue copper also has been extensively studied. Such studies have their obvious limitations in clarifying the mechanism of pathogenesis. Little is known of the metabolic fate of elevated copper and/or ceruloplasmin levels or the distribution of this metaloenzyme in different tissues during the course of various diseases. A more profound understanding of the biochemical pathology may be anticipated when tissues, cells, and subcellular particles as well as body fluids can be studied in their native states. Since serum copper and ceruloplasmin alterations are found in a variety of pathological conditions of different etiopathogenesis, probably more than one causative mechanism responsible for these changes is involved. For these reasons we will discuss serum copper and ceruloplasmin changes in the three succeeding sections separately under headings:

1. Hereditary Diseases
2. Inflammatory and Endocrine Disorders
3. Malignant Diseases

A. Hereditary Diseases

Modern medicine has identified two genetic syndromes whose effects are a direct result of abnormal copper metabolism. In one, Wilson's disease, patients suffer from the deleterious effects of excess tissue copper;[51,95-99] in the other, Menkes' kinky hair syndrome, the features of the disease result from copper deficiency.[23,24,100,101]

1. Wilson's Disease

In Wilson's disease (hepatolenticular or hepatocerebral degeneration) the genetic metabolic defect leads to the abnormal accumulation of copper first in the liver resulting in cirrhosis, and later on in the central nervous system causing movement disorders and intellectual deterioration.[51,96,99] Excessive copper accumulation in corneal tissue leads to Kayser-Fleisher ring and sunflower cataract formation and in the kidneys to glomerular and tubular defects.[51,96,97] The disease follows a homozygous recessive pattern of transmission. This abnormal gene is distributed throughout the world, with a prevalence in heterozygotes of about 1 in 200 and in abnormal homozygotes patients with Wilson's disease of about 5 in 1 million.[95] The age of onset of clinical manifestations of disease varies widely, most commonly occurring in older children (above 6 years of age), adolescents, and young adults, and occasionally in individuals aged 40 to 60 years.[51,95,99] Some of the symptoms are the results of the primary toxic effects of copper on various tissues, and some such as the endocrine disturbances probably reflect changes in the hormonal metabolism by the diseased liver.[95] Etiopathogenesis of Wilson's disease remains enigmatic. Many hypotheses have been suggested to explain the subcellular defects. One of the primary defects in Wilson's disease seems to be associated with an increased synthesis of copper-binding protein and probably defective ceruloplasmin synthesis and biliary copper excretion.[51,95-99] Evans[16] proposed that copper-binding hepatic proteins bind the copper atoms which would normally be used for ceruloplasmin synthesis and biliary excretion. As the disease progresses, the binding sites of the protein become saturated, resulting in a decrease of hepatic copper uptake and an increase of serum nonceruloplasmin copper. Evans et al.[102] extracted the metal-binding protein, metallothionein, from liver tissue from two patients with primary biliary cirrhosis and two patients with hepatolenticular degeneration. The number of copper-binding sites was identical for all specimens, but a fourfold increase in binding affinity for copper was characteristic of the protein isolated from Wilson's disease patients. These observations of enhanced protein binding await further confirmation.[103]

Sternlieb et al.[104] suggested that a lysosomal defect might account for the reduction of biliary copper excretion and the consequent hepatic accumulation of the metal in Wilson's disease. The molecular abnormality producing hepatic copper accumulation as yet has not been precisely defined.

Hypocupremia (decreased serum copper) and hypoceruloplasminemia are seen in 90 to 95% of patients with Wilson's disease. A total of 10% of heterozygotes have decreased circulating copper and ceruloplasmin levels, but have intermediary values (117 ± 51 μg/g dry liver) for hepatic copper.[105] A histochemical determination of hepatic copper concentrations is essential in suspected cases of hepatolenticular degeneration accompanied by hypocupremia and hypoceruloplasminemia. A normal hepatic copper concentration (32 ± 7 μg/g dry liver) excludes the diagnosis of untreated Wilson's disease, in which copper concentrations in excess of 250 μg/g dried liver are commonly found.[95]

Since several forms of chronic liver disease may result in abnormal accumulations of copper, an increased liver-copper content alone cannot confirm the diagnosis of Wilson's disease and must be correlated with clinical, biochemical, and histological findings.[95] The distinction between primary biliary cirrhosis or prolonged cholestatic syndromes with Kayser-Fleischer rings and Wilson's disease is primarily made by a serum ceruloplasmin determination; the concentration is almost invariably greater than 30 mg/dℓ in the former.[95] In all forms of liver disease (other than hepatolenticular degeneration) the concentration of ceruloplasmin and copper is almost always either normal or increased, with the exception of fulminent massive hepatic necrosis where it is rarely decreased.[106] Occasionally, in patients with normal ceruloplasmin levels, it is necessary to resort to radiocopper loading tests in order to resolve the diagnostic dilemma.[107] A linear correlation between hepatic copper content and urinary copper output has been observed in Wilson's disease patients receiving chelation therapy. The urinary copper content is a useful index of hepatic copper content and a good monitor of the efficacy of therapy with chelating agents. The increased urinary copper level (100 μg/24 hr) is pathognomic, but not a diagnostic test of Wilson's disease for several reasons:[108] inaccurate assays are frequent, the absence of hypercupriuria (increased urinary copper) during the early stages of Wilson's disease, and presence of hypercupriuria observed in other pathological conditions such as advanced primary biliary cirrhosis and nephrotic syndrome. Penicillamine therapy-induced cupriuria cannot discriminate between Wilson's and non-Wilsonian liver disease.

D-Penicillamine, at a daily dosage of 0.75 to 2g, is the drug of choice for the treatment of any stage of Wilson's disease.[51,109-112] The earlier this therapy is instituted, the better the results.[95,112] This is particularly true for asymptomatic patients who are being treated prophylactically. The treatment of Wilson's disease should be coupled with a low copper diet, small doses of pyridoxine, and possible zinc supplementation. Long-term therapy for several years is required before hepatic copper content returns to the normal range. Side effects of D-penicillamine therapy include rash, bone marrow depression, nephrotic and Goodpasture's syndrome, disseminated lupus erythematodes, optic neuritis, peripheral neuropathy and myasthenia gravis. During penicillamine therapy the patient should be closely observed with frequent urinalyses and blood counts.

2. Menkes' Kinky Hair Syndrome

Menkes et al.[23] reported in 1962 a genetically determined defect in copper absorption and utilization associated with hypothermia, uncontrollable seizures and progressive cerebral and cerebellar degeneration, growth and mental retardation, and abnormally textured hair resulting in death within the first 2 years of life. In addition to the above listed findings, scorbut-like changes in skeletal mineralization, altered elastic layers in the most severely affected major vessels, and depigmentation of hair and skin were also noted.[24,100,101,113-116] The disorder has been termed trichopoliodystrophy, or simple kinky hair syndrome, because of the unique hair abnormalities.

The similarity between the clinical manifestations of kinky hair syndrome and copper deficiency in sheep was noticed by Danks et al.[100,101] who documented low serum copper and ceruloplasmin concentration in affected infants and suggested altered copper metabolism leading to deficiency as the etiopathogenetic defect.[100,101] Since normal levels of copper and ceruloplasmin are found in cord blood at birth, it is presumed that symptoms occur early in life after depletion of maternal copper storage sites. Serum copper and ceruloplasmin levels and hepatic copper content are universally low beyond the first 2 weeks of life and, in addition to clinical findings, microscopic examination of hair confirm the diagnosis of kinky hair syndrome. The red blood cell copper is normal and 24-hr urinary copper is increased. The anemia and neutropenia common to patients with copper deficiency have not been observed in kinky hair syndrome.[114,115] Studies using radiocopper have confirmed that intestinal malabsorption of copper is a primary defect in kinky hair syndrome.[100,101,113] The defect seems not to be in the failure of copper absorption from the intestinal lumen but rather impairment of the transport of copper out of the intestinal cell,[117] as the intestinal concentration of copper is markedly increased in specimens of duodenal mucosa from intestinal biopsy and from autopsy.[100,118] The subcellular location of copper is in the brush border membrane.[119] Whole-body retention and biological half-life of radiocopper is abnormally increased.[113] Most of the features of this disease appear to be a direct or indirect consequence of copper deficiency and reduced cuproenzyme activity.[114] The cerebral dysfunction can be attributed to cytochrome c oxidase deficiency,[100] to dopamine beta-hydroxylase deficiency,[117] and to a poorly defined myelination defect.[120] The cause of the hypothermia is believed to be due to impaired energy metabolism resulting from cytochrome c oxydase deficiency.[116] The abnormality of the ''kinky hair'' is due to the unusual frequency of free sulfhydryl groups in hair protein that remain unoxidized because of defective copper-dependent amine oxidases, while tyrosinase deficiency is responsible for the depigmentation of skin.[116] Defective lysyl oxidase is presumably the cause of the ultrastructural arterial wall defect (arterialtortuosity, fragmentation and disintegration of the internal elastic lamina with irregular thickening of the intima).[100,121] Garnica et al.[118] have experimentally confirmed a copper metaloenzyme deficiency in patients with this syndrome; their leukocyte cytochrome c oxidase concentration was reduced 40%. Serum amine oxidase was also decreased to 6% of the normal. Attempts to treat Menkes' kinky hair syndrome patients with oral or parenteral copper administration were made, but the results were generally disappointing.[114]

Haas et al.[122] in 1981 reported a new x-linked disease of the nervous system associated with abnormal copper metabolism. This syndrome was similar to Menkes' disease in some respects: x-linked recessive inheritance, marked psychomotor retardation with seizures, low serum copper and ceruloplasmin levels, and a block in intestinal copper absorption. There were also some striking differences from Menkes' disease. Patients had normal birth weight at term, no hypothermia, and survived beyond the usual Menkes' age group with static neurologic disease including hypotonia and choreoathetosis. In addition, physical findings in these children were unremarkable apart from undescended testes and growth retardation. The hair, facies, and skin were normal and there was no radiologic evidence of bony changes. Whether or not this is a new disease of abnormal copper metabolism, as the authors claim it to be, or a variant of Menkes' kinky hair syndrome awaits further investigation and verification.

B. Inflammatory and Endocrine Disorders

Increased serum copper levels have been reported in a number of various pathological conditions. Viral, bacterial, fungal, and parasitic infections are frequently accompanied by elevated serum copper levels. Patients with acute and chronic infections, disseminated lupus erythematodes and other types of collagen diseases, rheumatoid arthritis, rheumatic fever, and glomerulonephritis have been reported to have both increased levels of copper and ceruloplasmin concentrations.[3,5,7-15,19-22,27-29,39,54,58,70,123,124]

A high degree of correlation between total serum copper levels and ceruloplasmin concentrations have been observed in most conditions associated with altered copper homeostasis.[6,8,10,31,36] This is not surprising, because over 90% of total serum or plasma copper is tightly bound to ceruloplasmin. Therefore, analysis of serum copper reflects the concentration of ceruloplasmin.[125] Other conditions associated with hypercuperemia (increased serum copper level) and hyperceruloplasminemia (increased ceruloplasmin level) are myocardial infarction[11,12] and various types of anemia,[58,67] and following major surgery.[126,127] Elevated serum copper levels and ceruloplasmin were also reported in various forms of acute[123,124,128] and chronic liver disease,[129,130] and particularly in hepatobiliary disease associated with defective biliary excretion (intra- or extrahepatic obstruction).[10,13,39] Plasma copper and ceruloplasmin are markedly affected by inflammatory agents. Attempts to increase serum copper by "activation" of the reticuloendothelial system have not been uniformly successful.[16] Turpentine, thorium dioxide, milk, and typhoid vaccine have each been employed.[16] Most of these inflammatory conditions are associated with a change in serum protein and erythrocyte sedimentation rate.[54,131,132]

Weissmann et al.[133] described the release of three classes of inflammatory mediators from human granulocytes exposed to a phagocytic stimulus. The generation of the products of molecular oxygen (O_2^-, OH, and H_2O_2), the secretion of lysosomal enzymes into the cells surrounding fluids, and the generation of biologically active products from membrane-derived arachidonic acid — which includes the thromboxanes and the stable prostaglandins, potent mediators of inflammation. There is also evidence that prostaglandins, released from phagocytizing granulocytes can mediate the synthesis of acute phase proteins, directly or indirectly through simulation of macrophages to release interleukin-1 or other mediator of acute phase.[20,21,41-48,131] The exact roles of prostaglandins and thromboxanes in inflammation are not clear. Data in the literature are complex and often conflicting.[133,134] The action of interleukin-1 (product of phagocytizing granulocytes) on hepatic cells consists of stimulating an accelerated synthesis of nuclear and ribosomal RNA and causes the rough endoplasmic reticulum of the hepatocytes to increase synthesis of various acute phase reactant proteins as host response to inflammation.[21,41] Increased protein synthesis accounts for the subsequent release of ceruloplasmin and copper from the liver, which in turn gives rise to their increased serum levels during inflammation. This substance (interleukin-1), when obtained from leukocytes of infected animal, can produce a response in noninfected animals.[21,47] Sokal and Shimaoka[48] have detected in the urine of patients with febrile Hodgkin's disease the occurrence of a pyrogen of the endogenous type usually associated with the breakdown of polymorphonuclear granulocytes.

According to common view, ceruloplasmin is considered as one of the numerous "acute phase reactants" a heterogeneous group of proteins, including serum transferrin, haptoglobin, alpha$_1$-acid glycoprotein, alpha$_1$-antitrypsin, C_3 (complement component), and C-reactive protein, fibrinogen, seromucoid — just a sampling of a much larger group.[21,54,131] It is argued that increased blood levels of these, and of ceruloplasmin, are found whenever there is inflammation or other type of tissue destruction in the body. However, the argument in favor of this contention is weakened by the lack of positive correlation between serum ceruloplasmin and C-reactive protein levels in some of these conditions. Although hypercupremia and C-reactive proteins are often seen together, a sufficient number of exceptions have been recorded to indicate that these two serum components are not interdependent.[54] Furthermore, according to Foster et al.[127] untreated polycythemia rubra vera is physiological stress and if ceruloplasmin was acting as an acute phase reactant only, one would expect an increase, which is not found. The increase in serum copper levels that occurs in acute and chronic infection begins soon after the onset of the infection and returns to normal during convalescence. Red blood cell copper is unchanged in a majority of cases studied indicating that hypercupremia is mainly the result of an increase in the plasma ceruloplasmin-bound copper.[135-138] Copper within the red blood cells fluctuates very little. Increased red blood cell copper is reported in iron deficiency and pernicious anemia.[58]

Free copper can be present in greater amounts than the copper contained in the ceruloplasmin, particularly in Wilson's disease.[98] The reverse is sometimes true in primary biliary cirrhosis in which PPD-oxidase activity (and therefore, presumably, the amount of ceruloplasmin) is usually higher than can be accounted for by the serum copper.[98]

In order to determine whether patients with chronic infections and hypercupremia are capable of removing copper from the plasma at a normal rate, a patient with chronic tuberculosis was given 100 mg of "cupralene" (19.3 mg of copper) i.v. and the subsequent changes in the plasma and red cell copper levels were measured. The same amount of organic copper compound was given to four normal controls. In the normal individual the plasma copper level increased within 10 min to between 545 and 740 μg/100 mℓ plasma, thereafter declining rapidly to reach approximately the initial values in 4 hr. A similar disappearance rate was observed in the patient with chronic pulmonary tuberculosis.[58] As can be seen from Table 1, most consistent hypercupremia and hyperceruloplasminemia have been observed in pregnancy,[2,4,10] women taking oral contraceptive medication or any form of sex hormone,[5,6,76-78] obstructive (intrahepatic or extrahepatic) biliary disease,[13,39] untreated acute leukemia in children,[28] Hodgkin's[29] and non-Hodgkin's lymphoma[30] in active phase, and some other malignancies[31,32] in decreasing order, followed by severe acute and chronic infection.[10,58] Much less consistent elevated serum copper levels are found in the other conditions listed in the table.

The secretions of the endocrine glands are known to affect a wide variety of metabolic functions in mammalian systems. Since the early observation of Krebs in 1928,[75] hormones of the pituitary gland, adrenal glands, thyroid gland, and sex glands have been reported to produce alterations in copper metabolism.[16,40] Little is known of how these hormonal changes might influence or contribute to the observed changes in copper homeostasis during an infection or inflammation. Some of the antiinflammatory effects of the corticosteroids can be attributed to inhibition of polymorphonuclear leukocyte locomotion, phagocytosis, degranulation, and membrane stabilization.[131] In adult mammals the pituitary adrenal axis apparently exerts a secondary effect on copper metabolism through the influence of the adrenal steroids on hepatic bile secretion.[16] Elevated endogenous secretions of adrenal corticosteroids, as occur in Cushing's syndrome or in patients with adrenal cortical carcinoma, has been associated with decreased plasma copper, increased urinary copper excretion, and decreased tissue retention of this metal.[40] Treatment of Cushing's syndrome, by surgery or irradiation, resulted in the reduction of excessive endogenous adrenal corticosteroid secretion and return of serum copper to normal levels, diminution in urinary excretion of copper, and its increase in tissue.[40] Changes similar to the above observed in patients with hypersecretion of adrenal corticosteroids were observed following oral administration of adrenal corticosteroids to normal volunteers. In these studies, following 5 days of prednisone 50 mg daily, reduced serum copper concentration and increased urinary copper excretions were found.[139] Adrenalectomy or adrenal cortical insufficiency in humans from several causes, including idiopathic Addison's disease and hypopituitarism, has been associated with increased serum copper concentration, decreased urinary copper, and increased retention of copper in the liver.[40,139-141] These observations indicate an inverse relationship between adrenal corticosteroid levels in blood and copper concentration and a direct relationship between blood levels of adrenal corticosteroids and biliary and urinary excretion of copper.[16,40] These changes may be related to direct effects of some adrenal corticosteroids on mobilization of hepatic copper and bile secretion, and the production of increased ultrafiltrable serum copper resulting in net body loss of this metal due to its increased excretion. Epinephrine injections result in significant elevation in serum copper and ceruloplasmin in experimental animals. The exact mechanism by which this hormone produces elevated serum copper levels is not known.[16] A decreased concentration of circulating growth hormone, as occurs in patients with untreated isolated growth hormone deficiencies, has been associated with elevated levels of serum copper and decreased excretion of urinary copper.[40] Treatment of these patients with exogenous growth hormone resulted in the lowering of their elevated serum copper levels and an increase of their previously reduced urinary copper excretion.[40]

Adrenocorticotropic hormone (ACTH) therapy reduces both ceruloplasmin and serum copper as a result of the hormone's stimulation of the adrenal glands.[16] The role of the thyroid hormones in copper metabolism remains an enigma. Increased serum copper concentrations have been reported in hyperthyroid patients and decreased serum copper levels accompany hypothyroidism.[142] Antithyroid drugs produce fall in serum copper levels as well as surgical correction of disease.[3,16] Patients with untreated hyperparathyroidism also show increased serum copper and urinary copper. Following surgical treatment of the hyperparathyroidism there was a decrease in the urinary excretion of copper among those patients in whom normalization of parathyroid hormone occurred.[139] The effects of sex hormones on serum copper and ceruloplasmin concentrations have been already discussed.

In view of the above, it appears that endocrine disorders are associated with increased or decreased hepatic, serum, and urinary copper, depending upon hyperfunction or insufficiency of responsible glands.

Although, the incidence of and the number of diseases associated with hypercupremia are far greater than those associated with decreased serum copper levels (hypocupremia), the mechanism which cause diminished serum copper concentrations in many instances may be traced to specific causes, as discussed previously. Most consistent decreased serum copper level and decreased ceruloplasmin level have been reported in patients with Wilson's disease[96-99] and in Menkes' kinky hair syndrome,[23-24] inborn errors of copper metabolism. Nephrotic syndrome is also associated with decreased serum copper levels.[10,57] Hypocupremia is specific for nephrotic syndrome and this has not been seen in other patients with hypoalbuminemia (Table 2).

Increased urinary copper (hypercupruria) generally occurs during the advanced stages of a pathological condition affecting the primary homeostatic mechanisms such as biliary copper excretion, copper storage, ceruloplasmin synthesis, or increased excretion. A good correlation between increased tissue copper, plasma copper and urinary copper have been noted. Wilson's disease, obstructive (intra- or extrahepatic) hepatobiliary disease and primary biliary cirrhosis have been demonstrated to have increased hepatic and urinary copper content.[50,51,95] In patients with hepatolenticular degeneration undergoing therapy with penicillamine, 24-hr urinary copper contents can exceed 1000 μg.[67,95] Cartwright et al.[57] found a high degree of correlation between the proteinuria and cupriuria in patients with nephrotic syndrome. Approximately 31 μg of copper was excreted per gram of protein. The copper excreted was undialyzable and, therefore, probably was protein bound. Immunochemical analysis demonstrated that 60 to 80% of the excreted copper can be accounted for as ceruloplasmin in patients with nephrotic syndrome[10] (Table 4).

Considerable differences were found in the copper concentrations in cerebrospinal fluid in patients with multiple sclerosis, convulsive disorders, myasthenia gravis, central nervous system involvement with carcinoma, and schizophrenia.[143] Consistently high values of copper in cerebrospinal fluid were found only in the patients with Wilson's disease, Huntington's chorea, portal cirrhosis, and central nervous system involvement with carcinoma. The significance of these findings is not known and measurements of serum copper or ceruloplasmin have little, if any, value in the diagnosis or management of patients with the above-mentioned disorders. In approximately 50% of oral contraceptive users, increased cerebrospinal fluid copper was found. A marked fall of the blood and cerebrospinal fluid copper content following electroshock therapy has been reported.[144] Bogden and Troiano[9] reported significantly higher plasma copper concentrations and significantly lower zinc levels in those patients who developed delirium tremens or a prolonged hallucinatory state during a period of alcohol withdrawal than the patients who recovered uneventfully. Liver function tests were abnormal in only 4 of 30 cases studied. Increased copper concentrations (84.8 ± 20.7 μg/100 g) are found in the synovial fluid of patients with rheumatoid arthritis.[145] Elevation of copper levels in aqueous humor have been found in hepatolenticular disease, intraocular copper foreign bodies, and after prolonged therapy of trachoma with copper sulfate and in a rare case of IgG myeloma.[146]

Liver concentrations in excess of 250 μg/g of dried liver are frequently found in patients with Wilson's disease, whereas copper content in heterozygotes was approximately 117 ± 51 μg/g of dry liver compared to normal copper hepatic content of 32 ± 7 μg/g of dry liver.[105] Hepatic copper concentration even higher than those seen in some patients with Wilson's disease may be encountered in patients with primary biliary cirrhosis, extrahepatic biliary atresia, various cholestatic and noncholestatic syndromes, chronic active liver disease, vineyard sprayers' liver (who inhale sprays containing copper salts), and in children with Indian childhood cirrhosis[95] (Table 3).

C. Copper and/or Ceruloplasmin in Malignant Diseases

Many biological markers in cancer patients have been studied in an attempt to improve clinical monitoring of disease activity and efficacy of therapy. Numerous studies, mainly from Europe, indicated serum copper concentrations were significantly elevated in patients with malignant diseases. Linder and Moor[49] in reviewing the literature from 1956 to 1969 found 27 reports, of which only 4 showed that cancer patients did not have copper values higher than controls. The majority of investigators believe that increase in serum copper is a general phenomenon in patients with neoplastic diseases.

In our pilot studies published in 1968, we reported the clinical significance and usefulness of serum copper changes in 70 patients with malignant lymphoma and acute leukemia with special reference to Hodgkin's disease.[147] Since that time, an increasing number of reports have been noted and alterations of serum copper and/or ceruloplasmin concentrations have been the subject of considerable investigation.

As previously mentioned, in normal individuals approximately 95% of serum copper is firmly bound to ceruloplasmin. This metalloenzyme is the only copper fraction known to fluctuate extensively in physiological and pathological conditions, including malignant diseases. The only exceptions are rare cases of multiple myeloma in which serum copper was shown to be bound to IgG immunoglobulin.[148,149] Simultaneous determinations of serum copper and ceruloplasmin levels in various conditions revealed a high degree of correlation between this metal and the main copper-bound protein, ceruloplasmin. It appears that this protein is responsible for the alterations in serum copper observed. Therefore, both tests seem to serve the same purpose, measuring the activity and the extent of malignant disease and the rate of neoplastic cell proliferation. In this review, serum copper and/or ceruloplasmin variations in malignant diseases in humans will be summarized chronologically as reported by various investigators. Since more or less constant serum copper and/or ceruloplasmin alterations have been observed in patients with various lymphoproliferative and myeloproliferative disorders, these malignancies will be reviewed first, followed by alteration of serum and/or ceruloplasmin concentrations in solid tumors, in which a less consistent serum copper or ceruloplasmin pattern has been reported. Experimental data from the literature will be discussed in connection with our animal related studies.

1. Serum Copper and/or Ceruloplasmin Levels in Myeloproliferative and Lymphoproliferative Diseases

There are a number of general references to the elevation of serum copper in lymphoproliferative and myeloproliferative disorders dating back to the early 1940s.[150] Lahey et al.,[58] in 1953, in their survey of blood copper in various pathological conditions found increased plasma copper levels in 9 of the 14 patients with Hodgkin's disease who were studied. Whole-blood copper values were elevated in 8 of the 14 patients. The highest values for whole-blood copper (230 and 256 μg/100 mℓ) and plasma copper (282 and 297 μg/100 mℓ) were found in two of the three boys studied, both of whom exhibited widespread visceral involvement with Hodgkin's disease. Red blood cell copper determinations were made on eight of their patients and all were within normal range. These authors have also studied blood copper in ten untreated patients with chronic myelogenous leukemia, eight cases with chronic lymphocytic leukemia, and four with lymphosarcoma. While the mean values for whole blood (119 μg/100 mℓ) and plasma copper (148 μg/100 mℓ) were higher than in the normal, in 12 patients (57%) the whole-blood copper values were within the normal range and in 10 (45%) the plasma copper levels were normal. The red blood cell copper was not significantly different from normal. The amount of copper in the average corpuscle was not significantly altered (88 μg) in these patients and hypercupremia was the result of an increase in the plasma copper. A favorable response to the administration of corticotropin (ACTH) in two patients with acute leukemia was accompanied by a decrease in their blood copper values and relapse was associated with a return to the previous high copper level. In another two patients with acute leukemia who were treated with ACTH, a decrease in the plasma copper occurred even though there was no demonstrable clinical improvement. Keiderling and Scharpf[151] reported in 1953 on the clinical significance of serum copper and iron determinations in 163 patients with neoplastic diseases. Clinical material consisted of 55 cases with reticuloendothelial malignancies (5 cases with polycythemia vera, 1 with erythroleukemia, 16 with chronic leukemias, 8 with acute leukemia, 9 with multiple myeloma, and 16 patients with Hodgkin's disease), and 98 patients with epithelial malignant neoplasms (15 of the respiratory tract, 24 of the gastrointestinal tract, 34 of the hepatobiliary system, 5 cases with carcinoma of the bladder, 14 patients with carcinoma of the endocrine glands, and 4 cases with metastatic malignant diseases of unknown primary). In addition, the authors studied 10 cases with mesenchymal malignancies and 60 normal patients were included as a control group. More or less increased serum copper levels were found in all patients with malignant diseases. The highest serum copper levels were observed in patients with carcinoma of the hepatobiliary systems, Hodgkin's disease and acute leukemia, and sarcoma in decreasing order followed by carcinoma of the lung, bladder, endocrine organs, gastrointestinal tract, and chronic leukemias. On the other hand, serum iron was depressed in all malignant diseases with the exception of the acute leukemia cases in which it was found to be increased. The mean serum copper value in 149 patients with various types of neoplastic disease was 178 μg/100 mℓ, as compared to 105 μg/100 mℓ in the control group. Elevated serum copper levels were found in 81.2% of the cases studied. The authors observed the normalization of serum copper and iron level following chemotherapy or irradiation therapy, resulting in control of the tumors. The observed changes in these metals in malignant diseases are considered nonspecific, reflecting tumor complications (necrosis, resorption of toxic cellular products) and perifocal tumor inflammation as it is seem in patients with acute or chronic infections.

Rechenberger[152] in 1957 reported serum iron and copper levels in 64 patients with chronic leukemia (15 of them with chronic lymphocytic leukemia), 28 cases with acute leukemia, 17 patients with Hodgkin's disease, 19 patients with bronchial carcinoma and 23 patients with gastrointestinal carcinomas. The normal range of serum copper and iron was from 68 to 136 μg/100 mℓ. The mean serum copper value in patients with chronic leukemias was 162 μg/100 mℓ (ranging from 148 to 206 μg%), acute leukemias 206 μg/100 mℓ (range 139 to 229 μg%), Hodgkin's disease 208 μg/100 mℓ (range 156 to 223 μg%), bronchial carcinoma 181 μg/100 mℓ (range 160 to 200 μg%), and gastrointestinal carcinoma patients 178 μg/100 mℓ (range 141 to 202 μg%). On the contrary, the serum iron level was decreased in all groups except for acute leukemia in which cases it was increased (156 μg%).

Koch et al.[153] in 1957 analyzed various trace elements in patients with malignant lymphoma. One of the trace elements studied in plasma and various tissues was copper. Their clinical material consisted of 34 patients with Hodgkin's disease, 14 cases with lymphosarcoma, 10 patients with reticulum cell sarcoma, 8 patients with chronic lymphocytic leukemia, four patients with acute myeloid leukemia, and one patient with chronic granulocytic leukemia. The mean serum copper value in Hodgkin's disease patients was 216 ± 68 μg/100 mℓ (range 75 to 428 μg/100 mℓ) vs. a control of 120 ± 23 μg/100 mℓ. In the lymphosarcoma group the mean serum copper values were 199 ± 29 μg/100 mℓ (range 163 to 250 μg/100 mℓ). Reticulum cell sarcoma patients had the mean value of 167 ± 32 μg/100 mℓ (range 123 to 217 μg%). Chronic lymphocytic leukemia cases had the mean serum copper values of 133 ± 50 μg/100 mℓ and a range of 147 to 294 μg/100 mℓ. Four patients with acute myeloid leukemia had a mean serum copper value of 197 ± 63 μg/100 mℓ (range 140 to 309 μg/100 mℓ) and one case with chronic myelocytic leukemia had a mean serum copper value of 176 μg/100 mℓ. The authors found the mean serum copper value to be elevated in every category of patients studied. The wide range of values observed (large standard deviation) was thought to be caused by the extent of lymphoma or leukemic involvement which varied considerably from patient to patient and by treatment (irradiation) response that produced a clinical remission. The authors found no correlation between the plasma copper level and hemoglobin or white cell counts in any of the disease categories studied. The copper content of tissue (bone, spleen, and liver) analyzed was normal or slightly increased.

Pagliardi and Giangrandi[154] in 1960 emphasized for the first time the clinical significance of blood copper in patients with Hodgkin's lymphoma studied throughout the course of the disease. Most cases had not been previously treated. Serum copper levels were determined by the colorimetric method. The mean normal serum copper levels for men was 126 ± 22 and for women 132 ± 20 μg/100 mℓ.

The range of the serum copper levels in the active phase of 23 Hodgkin's disease patients was 230 to 410 μg/100 mℓ (mean value of 307 μg/100 mℓ). Hypercupremia was apparently one of the most valuable signs of activity of the disorder and the degree of the hypercupremia was closely related with the evolutive stage. Remission induced by irradiation or chemotherapy was characterized by an early and considerable drop of the concentration of this metal in the serum and by relatively persistent normal values, whereas persistently high concentrations, even in the presence of an apparent clinical improvement, always represented an unfavorable sign and seemed to be an exact indication of the severity of the course of the disease. In this disease, the red blood cell copper shows no characteristic pattern according to these authors. These authors also observed the variations of temperature, WBC, and clinical findings to run approximately parallel to the changes of the serum copper. Improvement of the clinical picture was, in general, more marked in the cases in which the normalization of the sedimentation rate, temperature, and white blood count and the serum copper was greater. Serum copper often tended to return to the normal level more promptly than any other laboratory finding studied. The serum iron constantly showed normal or low normal values and its variation did not correlate with those of the other laboratory findings. In summary, these authors concluded that in Hodgkin's disease the serum copper level is constantly increased during relapse and its variations are intimately related with the course of the disease. A drop to normal values always indicated an imminent remission following treatment. This drop occurred early and preceded the normalization of the usual signs (sedimentation rate, temperature, white blood cell count). Likewise, persistent hypercupremia during or after treatment, or an increase of the serum copper level during a period of relative well-being, are always a faithful sign indicating an imminent relapse. Also, in such cases the variations of the serum copper level occurred before those of the other laboratory findings.

Herring et al.,[155] in 1960 reported trace metal analysis (magnesium, chromium, nickel, copper, and zinc) in human plasma and the red blood cells in 78 adult patients with various hematologic diseases. Clinical material consisted of 12 Hodgkin's disease patients (1 untreated, 11 treated), 16 lymphosarcoma cases (2 untreated, 14 treated), 8 multiple myeloma cases (5 untreated, 3 treated), 9 acute leukemia cases (3 untreated, 6 treated), 6 chronic lymphocytic (2 untreated, 4 treated), 5 chronic myelocytic leukemia (1 treated, 4 untreated), and 21 patients with nonmalignant hematologic disease. No information was given concerning the status of disease activity in the treated patients. Of 28 patients with Hodgkin's disease and lymphosarcoma, 19 showed moderate to marked elevations of the plasma copper. These two groups yielded average values which were 50% higher than normal. Only two patients with acute leukemia had extremely high plasma copper levels. One of these returned to normal following treatment of the patient with 6-mercaptopurine. No consistent abnormalities in the amount of copper in the red blood cells were observed.

In 1961, Cherry et al.[156] published their observations of plasma copper levels in 8 adult patients with acute leukemia observed during periods ranging from 6 to 191 days, and generally until death. The clinical material consisted of three cases of acute monocytic leukemia, two patients with chronic granulocytic leukemia in blastic crisis, one acute lymphocytic leukemia, and two cases with acute myelomonocytic leukemia, ages 23 to 54 years. There were four female and four male patients. The mean normal plasma copper levels were 127 μg% for males and 146 μg/100 mℓ for females. The mean pretreatment serum copper value for all eight cases was 273 μg% with a range of 171 to 612 μg/100 mℓ. With the exception of one patient with a fulminant course of disease, each patient showed at least one rise and one fall in the total plasma copper level during the observation period. In each of these seven cases the periods of leukocytosis were associated with corresponding periods of hypercupremia. Total plasma copper levels were also elevated during other periods when the leukocyte count was normal. In terms of the sequence of events, total plasma copper levels were found to fall following signs of remission and rise following signs of relapse. Based on the observed copper changes in the course of leukemic disease and possible purposeful role of ceruloplasmin as a host defensive mechanism, these authors investigated the clinical use of ceruloplasmin in several patients with cancer. A total of 66 g of ceruloplasmin was administered to four patients in various doses without evidence of significant side effects. Clinical trials of ceruloplasmin therapy resulted in the significant elevation of the serum copper level to as high as 900 μg/100 mℓ in one patient. No information was given concerning the benefits to the patients or altering their disease course with ceruloplasmin therapy.

Pagliardi et al.[157] in 1963 reported on serum copper, loosely bound plasma copper, ceruloplasmin, urinary copper, and their variations following intravenous injections of CaEDTA in patients with Hodgkin's disease. Clinical material consisted of 13 patients with Hodgkin's disease and 10 normal controls. Their results showed that the alterations of serum and ceruloplasmin bound copper correspond to the disease activity. The mean serum copper value in Hodgkin's disease patients was 246.3 μg/100 mℓ, loosely bound plasma copper was 12.18 μg/100 mℓ, and plasma oxidase activity was 0.329, and 24-hr urinary copper was 42.4 ± 19.26 μg. In the control group, the mean serum copper level was 114.5 μg%, nonceruloplasmin bound copper was 15.08 μg, and ceruloplasmin oxidase activity was 0.195, and 24-hr urinary copper was 30.5 ± 17.66 μg. Following the injection of CaEDTA, within 30 min slight initial fall (20%) in serum copper level was noted in normal and Hodgkin's disease patients which rapidly returned (within 3 hr) to normal pretreatment level. A rise in nonceruloplasmin-bound copper was noted after EDTA infusion accompanied by increased urinary copper the following day from 42 to 66.95 μg/24 hr. In summary, these authors stated that the results showed that the variation of the ceruloplasmin content represents the sensitive index of the clinical course of the disease, and that this is characteristic of Hodgkin's disease.

Another feature which was observed the first time is that in Hodgkin's disease, unlike previous observations in other pathological conditions, there is a double response to the administration of CaEDTA, on account of which it can be deduced that a different metabolic utilization of copper is possible. Preece et al.[158] reported in 1977 on a preliminary study with penicillamine in a small group of patients not responding to oncolytic therapy. It was shown that serum copper concentrations were, in fact, depressed by an amount greater than that accorded to the "free" (albumin-bound) copper. The average fall in total serum copper was 23% and each individual showed a decrease. In all cases the fall in serum copper was associated with a corresponding decrease in ceruloplasmin. The authors suggested that the interrelationship of the leukocyte system, copper, and metal-binding agents such as penicillamine and aspirin be investigated further.

Jensen et al.[159] reported in 1964 on the serum copper level changes in 29 patients with Hodgkin's disease, undergoing chemotherapy or irradiation. Most of these patients were followed by serial serum copper level determinations ranging from 4- to 16-week intervals. The favorable effects of the applied therapy and improvement of the clinical condition (remission) was always accompanied by a decrease in the serum copper. Absence of a fall in the serum copper during treatment seemed to definitely indicate resistance to therapy. An increase in the serum copper was, as a rule, accompanied by an aggravation of the patient's condition and by the activation of the disease (relapse). The authors noted that the serum copper changes during treatment seemed to be of great value in assessing the effect of chemotherapy or irradiation. Variations of the serum copper levels in normal subjects observed for more than 2 months were 14 μg/100 mℓ, whereas in Hodgkin's disease patients the serum copper changes were up to 165 μg/100 mℓ during one period of treatment. High serum copper levels were observed in most ill patients. Moderately increased or quite normal serum copper levels were found during the symptom-free intervals. Furthermore, good agreement between the serum copper changes and the erythrocyte sedimentation rate was often found. During the observation period the serum copper levels showed considerable variations and reflected much better clinical conditions (Hodgkin's disease activity) than the erythrocyte sedimentation rate.

Walberg et al.,[160] in 1966 reported erythrocyte magnesium, copper, and zinc in 47 patients with malignant diseases affecting the hemopoietic system. Among those patients, there were five patients with acute leukemia, seven cases with chronic lymphocytic leukemia, four lymphosarcoma patients, seven chronic myeloid leukemia cases, six multiple myeloma cases, and nine cases with carcinoma and secondary anemia. Increased levels of the erythrocyte copper were found in Hodgkin's disease and normal levels in the other malignant conditions, which was in accord with the observations of Pagliardi et al.[154] The results of their studies suggested that the amount of copper per cell is normal and only the mean level per cubic microgram of red blood cell is increased.

Goodman et al.[148] in 1967 reported significant hypercuperemia in a patient with multiple myeloma, with diffuse brownish-gold corneal pigmentation. The initial serum copper determination gave the result of 3350 μg/100 mℓ, a level far in excess of any previously reported. Another 11 IgG myeloma patients studied showed a mean serum copper of 137 μg/100 mℓ, with a range of 85 to 213 μg/100 mℓ. On the other hand, this particular patient consistently had serum copper levels ranging between 1780 to 3350 μg/100 mℓ. Ceruloplasmin levels were repeatedly shown to be normal or slightly increased; 24-hr urinary copper excretion was initially determined on four occasions and was found to be 18, 386, 73, and 346 μg/24 hr. D-Penicillamine therapy was administered in an oral dose of 1 g daily for 13 days and the 24-hr urinary copper excretion was measured. During the penicillamine therapy, no significant decrease in the serum copper or an increase in the urinary copper was observed. Serum protein electrophoresis, followed by the quantitation of copper on the individual protein fractions, showed that more than 90% of the serum copper migrated with the gamma globulin. Immunoglobulin determinations showed an IgG of 3240 mg/100 mℓ (normal — 620 to 1400 mg/100 mℓ), and IgA of 40 mg/100 mℓ (normal — 46 to 418 mg/100 mℓ) and an IgM of 20 mg/100 mℓ (normal — 70 to 384 mg/100 mℓ). The copper concentration of the protein in this patient was 35.3 μg/100 mℓ and the concentration of the control was 2.0 μg/100 mℓ, of the total serum copper (2320 μg/100 mℓ). The IgG myeloma protein showed normal immunological reactivity, no significant abnormality in the amino acid composition, and no gross increase in the copper binding capacity was noted.

Hobbs et al.[161] in 1967 reported the flame spectrophotometric measurement of serum copper and its use in the assessment of patients with malignant lymphoma. Their clinical material consisted of 12 patients with lymphosarcoma and 5 patients with Hodgkin's disease. In these cases, the variations of serum copper levels ran parallel to the clinical disease activity, regression of the tumor being accompanied by a fall in the serum copper level and clinical relapse of the lymphoma was accompanied by a rise in the serum copper level. Two cases with malignant lymphoma were studied in detail. The regression of massive tumors seen on chest X-ray examination was accompanied by a marked reduction in the serum copper level. Later, in both patients, a relapse of the disease was accompanied by increased serum copper levels. It is the authors' impression that the level of the serum copper may rise before actual clinical activity of the disease is apparent; conversely the fall in the serum copper level has been noted to precede clinical signs of disease regression. The authors also studied the relationship between the serum copper and ceruloplasmin level in 51 sera of patients with malignant lymphomas. The correlation coefficient was 0.96. The authors suggested that in view of serum copper alterations caused by estrogen therapy, female patients should be asked concerning the use of oral contraceptives.

Pisi et al.[162] in 1968 reported a behavior and diagnostic significance of serum ceruloplasmin concentration in 32 patients with Hodgkin's disease followed from 3 to 19 months. The ceruloplasmin was determined electrophoretically. These results showed increased ceruloplasmin concentration in all cases and a clear relationship between ceruloplasmin levels and the histopathological and clinical picture. Serum ceruloplasmin values were found to show a gradual increase from lymphocytic predominance to mixed cellularity and lymphocytic depletion of Hodgkin's disease. With reference to the clinical picture, it was clear that increased activity of the disease was always accompanied by an increased ceruloplasmin level and that remission was accompanied by decreased, though still pathological, values irrespective of the type of treatment. These increases are explained on the hypothesis of increased ceruloplasmin synthesis in response to enhanced metabolic needs on the part of the proliferating granulomatous tissue.

Warren et al.[163] in 1969 reported on the prolonged observations on variations of the serum copper in 64 patients with Hodgkin's disease who were followed for from 3 to 45 months. Serial serum copper determinations were done during this period of time. Their normal serum copper level was from 82 to 148 μg/100 mℓ. The authors divided their clinical material according to the time of the initial therapy given in regard to copper determinations and the patient's disease activity. The first group consisted of nine patients treated initially before the serum copper estimation, all of them were considered to have no evidence of active Hodgkin's disease. Only three patients from this group had serum copper levels constantly within normal range. The other six were all females and showed a rise above the normal serum copper levels. In four of these patients, elevated serum copper levels could be related to the use of estrogens, oral contraceptives, or pregnancy. In one patient increased serum copper levels were attributed to chronic inflammation (pericardial effusion due to previous radiotherapy or still active Hodgkin's disease); in another patient from this group, with persistently high serum copper levels, no obvious explanation was found and the authors suspected that this patient will eventually develop a clinical relapse. The second group consisted of 18 patients treated initially after the copper investigation started, all of whom had no evidence of Hodgkin's disease activity since the initial treatment. In the 15 patients from the second group, increased pretreatment serum copper levels fell to normal range. The time interval for normalization of elevated serum copper levels ranged from a couple of weeks to a couple of months. In three patients of this group with stage I-A active Hodgkin's disease, the serum copper level has never been elevated above the normal range. The third group was comprised of 27 patients who have died of Hodgkin's disease and who had persistently elevated serum copper levels. The fourth group consisted of 10 patients who had active Hodgkin's disease or have recovered recently from active disease, with fluctuations in the serum copper level. Three of these patients had persistently normal serum copper levels in spite of recurrent disease. A linear correlation between the serum copper level and ceruloplasmin was also found, suggesting that the variations of the serum copper level are due to the alteration in the ceruloplasmin concentrations and that the serum nonceruloplasmin-bound copper remained constant throughout the period of investigation. Based on these studies, the authors concluded that serial examinations of the serum copper level can be extremely valuable in the assessment of the response to treatment and of the prognosis of the Hodgkin's disease patient.

Tura et al.[164] in 1969 reported on serum copper and ceruloplasmin concentrations in 25 patients with Hodgkin's disease and in 45 patients with lymphosarcoma and reticulum cell sarcoma, with particular attention to their variations in regard to the state of disease activity. Each series of patients was divided in two groups: (1) active and (2) partial and complete remission. Patients with active Hodgkin's disease had increased serum copper levels of 218 ± 33 μg/100 mℓ and those in partial or complete remission (combined) had serum copper levels of 169 μg/100 mℓ, compared to 122 ± 15 μg/100 mℓ found in the control group. Active phase of non-Hodgkin's lymphoma patients was also associated with elevated serum copper levels of 204 μg/100 mℓ and partial or complete remission (combined) of these patients was accompanied by decreases in serum copper levels of 159 ± 21 μg/100 mℓ vs. normal of 122 ± 15 μg%. The elevated serum ceruloplasmin concentrations were found in active phase of Hodgkin's disease (81 ± 14 mg/100 mℓ) and decreased (59 ± 14 mg/100 mℓ) in a group of partial and complete remission, vs. the control group of 49 ± 11 mg/100 mℓ. Similar ceruloplasmin concentrations according to disease activity were found in non-Hodgkin's lymphoma. In both series the average values of serum copper and ceruloplasmin were indisputably higher than in normal controls. Also, the average serum copper and ceruloplasmin level was indisputably increased in lymphoma in active phase (second group) by comparison with lymphoma in remission (first group). The confrontation of the respective confidence limits has shown that the differences found between the average values of the two series and the normal average, as well as those between first and second group of each series, are significant and represent different populations. Higher serum copper and ceruloplasmin concentrations in lymphoma patients responding to therapy (partial or complete remission combined) were attributed to partial control of disease activity. Based on their findings, these authors believe that the increase of serum copper and ceruloplasmin in lymphoma diseases has been confirmed and the validity of the relation between them and the phase of disease activity has been demonstrated.

Ilicin[165] in 1971 reported serum copper and magnesium levels in 22 patients with leukemia and 15 patients with malignant lymphoma. A group of 17 healthy men were used as controls. Serum copper levels in untreated leukemia patients were 334 ± 21 μg/100 mℓ and 242 ± 41.2 during treatment vs. 102 ± 5 μg/100 mℓ in the control group. In lymphoma patients, pretreatment serum copper levels were 298 ± 37 μg%, and 168 ± 33.7 μg during therapy. The falling serum copper level during treatment was accompanied by clinical and laboratory evidence of improvement. In both patient group, the serum copper was significantly lower and the serum magnesium was significantly higher during treatment than before therapy. The author believed that the determination of the serum copper levels in leukemia and malignant lymphoma is a valuable means of monitoring the efficacy of therapy.

Alexander et al.[166] in 1972 reported the concentration of copper and zinc in the plasma of leukemic children before and after treatment and compared these with healthy controls. The plasma copper concentrations were higher and the plasma zinc concentrations were lower for the untreated leukemic children than for the other two groups. The concentration of both metals was altered after treatment, the copper was lowered and the zinc increased to values approaching the normal range. The plasma copper:zinc ratio discriminated well between the three groups of children and would be valuable both in the diagnosis and response to treatment of leukemia in childhood. There was no correlation between this ratio and the total white cell or peripheral blast cell count. However, the authors concluded that the ratio was proportional to the extent of the leukemic process.

Mortazavi et al.[167] reported in 1972 the value of serum copper measurement in 42 patients with lymphomas (Hodgkin's disease — 19 cases, reticulum cell sarcoma — 9 patients, lymphosarcoma — 14 cases, 5 patients with multiple myeloma, and 3 with lymphoepithelial tumors). Increased copper levels were found in 23 of 24 lymphoma patients with generalized disease. On the contrary, only 6 lymphoma patients with localized disease had elevated serum copper levels, whereas 12 others had normal copper. Elevation of the serum copper levels were also found in multiple myeloma and patients with lymphoepithelial tumors (Schminke tumor). A group of 11 patients with Hodgkin's disease were closely followed over a period of 3 to 15 months with serial copper determinations. In all patients studied, the serum copper level reflected the disease activity course. Favorable response to therapy was associated with the lowering of the serum copper level. Those patients who did not clinically respond to treatment did not show, as a rule, a lowering the serum copper level; their condition deteriorated and they died, or they improved with a change of therapy. A significant finding in six patients with localized lymphoma and an increased serum copper level was that three developed generalized disease within 6 months. The authors believe that the serum copper measurements can be used as a prognostic and therapeutic index and that this is a useful auxilliary test in the management of patients with lymphomas.

Jelliffe[168] emphasized that his continued observation of the serum copper level changes over the last 3 years confirmed the views expressed in the original publication in 1969.[163] In general, the serum copper level mirrors the control of Hodgkin's disease in nearly all patients. Control of disease was accompanied by a steadily falling copper level and the lack of control by a fluctuating and usually high serum copper level. When control of disease is followed by relapse, this is usually accompanied by an increase of the serum copper from a normal to a high level and often this rise precedes any clinical signs of disease. This elevation of the serum copper level has been observed up to 6 months before the disease was clinically or radiographically demonstrable. Serial serum copper level determinations are of considerable value in long-term follow-up of disease activity.

Delves et al.[169] reported in 1973 copper and zinc concentrations in the plasma of leukemic children. The first group consisted of 14 children with untreated newly diagnosed acute lymphocytic leukemia. The second group consisted of 12 of these 14 children, together with 7 other children with acute lymphocytic leukemia in complete remission; 13 healthy children used as a control group had a mean serum copper value of 126 ± 21.2 μg/100 mℓ. The pretreatment mean plasma copper value in 14 children with acute lymphocytic leukemia was 256.8 ± 73.8 μg/100 mℓ, whereas the mean concentration of plasma copper determined on 42 samples of acute childhood leukemia in remission was 105.9 ± 31.2 μg/100 mℓ. Following chemotherapy the plasma copper concentrations were reduced to normal values in every child studied. The plasma copper:zinc ratio discriminated well between the untreated leukemic children and those in remission or normal control. None of the children from the normal group had a copper:zinc ratio greater than 1.80 (mean +2 SD), whereas 11 of the 14 untreated leukemic children had ratios in excess of 1.80 (range — 1.65 to 9.10). Of the plasma samples taken from the leukemic children in remission 74% had copper:zinc ratios of below 1.80. The correlation between the stage of disease and the copper:zinc ratio was significant at the 5% level. The relationship is further confirmed by the observation that serum copper rises with the percentage of blast cells in the bone marrow. The plasma copper:zinc ratio was not significantly correlated with the total leukocyte or the blast cell count in the peripheral blood, which is contrary to the findings of Cherry et al.[156] However, the significant correlations between the copper:zinc ratio and the extent of the leukemic process suggested that the changes in the copper and zinc are related to the total leukemic cell mass and not to the peripheral blood leukocyte count alone. The authors concluded that the determination of the concentrations of copper and zinc in the plasma of leukemic children and the calculation of copper:zinc ratio may prove to be useful parameters of the disease activity and the response to treatment.

Ray et al.[170] in 1973 reported on the value of laboratory indicators in Hodgkin's disease. The erythrocyte sedimentation rate (ESR), serum copper, ceruloplasmin, leukocyte alkaline phosphatase (LAP), muramidaze, magnesium, and iron were measured in 100 consecutive untreated patients with Hodgkin's disease and 75 concurrent controls. A highly significant difference was found between the mean values of the ESR, copper, and ceruloplasmin for the study group and controls. The ESR and serum copper showed the best correlation with disease activity, being elevated in 87 and 80% of the patients, respectively. In general, the ceruloplasmin and copper values parallel each other.

Statistical analysis supported the findings of Warren et al.,[163] that the increase in the total serum copper was due to the increase in ceruloplasmin concentration and that the serum nonceruloplasmin copper remained relatively constant. Of all of the tests performed, the three most reliable indicators of disease activity appear to be the ESR and the serum ceruloplasmin and copper levels. The authors considered the ESR and serum copper as being complimentary to each other, since in 50% of cases in which one test did not accurately reflect the patient's status, the other did. Since a linear relationship between the ceruloplasmin and copper levels was found, there appeared to be no reason to perform both tests. These authors have chosen to eliminate the ceruloplasmin determination, since the copper assay by atomic absorption is inherently more accurate than the test for ceruloplasmin which depends upon the enzymatic process. The frequency with which abnormal values were obtained with the ceruloplasmin, copper, and ESR did not differ significantly ($p > 0.05$) from each other. Multiple regression analysis revealed no significant relationship between the frequency of abnormal values and sex, age, histology, and/or specific sites of involvement. There was an increased frequency of abnormal values with the extent of disease, being higher in patients with more advanced disease. All patients with stage IV Hodgkin's disease had abnormal serum copper, ceruloplasmin, and sedimentation rates. Based on this data, the authors suggested that the ESR and serum copper should be considered as being complementary examinations which may be of value in monitoring the patient's response to treatment and possible early detection of recurrent disease.

Masi et al.[171] in 1975 studied serum ceruloplasmin levels in 35 children with acute lymphoblastic leukemia and 12 with Hodgkin's disease. Ceruloplasmin levels were determined immunologically. The mean ceruloplasmin values were markedly increased during active phase of disease compared to those in remission. Based on their findings and literature data, these authors concluded that activity of disease is reflected in serum copper and ceruloplasmin concentrations and stressed the usefulness of these blood indices in clinical oncological practice.

Cappelaere et al.[172] in 1975 reported plasma copper and zinc levels in the course of Hodgkin's disease patients. These authors studied 92 patients with Hodgkin's disease and 85 patients with other malignancies, in addition to 103 noncancerous subjects. Serum copper levels in control group were 87 ± 16 μg/100 mℓ for males and 120 ± 32 μg% for females. Both groups combined had a mean plasma copper level of 109 ± 31 μg/100 mℓ. Serum zinc levels were 92 ± 36 μg/100 mℓ for females, both groups combined had plasma zinc levels of 149 ± 69 μg/100 mℓ.

Patients with Hodgkin's disease in the pretreatment period had elevated serum copper levels of 174 ± 43 μg/100 mℓ, which fell to 145 μg/100 mℓ during therapy, and values of 137 ± 38 μg/100 mℓ were found after completion of treatment. The zinc levels of 149 μg/100 mℓ prior to therapy decreased to 127 μg/100 mℓ following completion of treatment. A group of 24 patients were treated with combination chemotherapy and 42 with irradiation. Patients in relapse had reelevation of serum copper levels of 216 ± 22 μg/100 mℓ and zinc levels of 115 ± 82 μg/100 mℓ, which decreased under therapy to 152 and 92 ± 36 μg/100 mℓ, respectively. Patients with localized (stage I and II) disease had a mean serum copper level of 167 ± 39 μg/100 mℓ, whereas those with generalized disease (stage III and IV) had higher serum copper levels of 196 ± 38 μg/100 mℓ. Also patients with systemic symptoms had higher serum copper levels (188 ± 42 μg/100 mℓ) as compared to those without symptoms (169 ± 37 μg/100 mℓ). Among the histological types, the highest serum copper levels were found in the mixed cellularity and the lowest in the lymphocytic predominance group followed by nodular sclerosis and lymphocytic depletion. Increased serum copper levels over 150 μg/100 mℓ were observed in 78% of patients with active disease, whereas elevation of the sedimentation rate of over 30 mm was found in 64% of cases. Other biological parameters measured such as fibrinogen, alpha 2 globulin, and alkaline phosphatase were elevated in 48, 57, and 65%, respectively, of patients with active Hodgkin's disease. In six patients with active stage I or II Hodgkin's disease, serum copper levels below 150 μg/100 mℓ were observed. Increased serum copper levels over 150 μg/100 mℓ were observed in 10 out of 45 patients during remission, in 6 cases it was explained either by oncoming relapse, estrogen therapy, or infection. In two cases, the cause of increased serum copper levels was not found. In the same group of 42 patients, an increased sedimentation rate of over 30 mm was found in 5 cases.

There were 12 patients with reticulum cell sarcoma, 47 with solid tumors, and 26 with metastatic carcinoma who showed serum copper levels of 142, 143, and 167 μg/100 mℓ and zinc levels of 146, 185, and 167 μg/100 mℓ, respectively.

In conclusion, the authors stated that increased serum copper levels and, to less degree, zinc levels are found in active disease, decreasing during therapy (radiation or chemotherapy) and normal in remission. The relapse of disease was accompanied by reelevation of the serum copper levels. They believe that serum copper levels are more reliable in following disease activity than any other biological parameter studied, including sedimentation rate, particularly in male patients with Hodgkin's disease.

Lewis et al.[149] in 1976 reported a second case of marked hypercupremia associated with multiple myeloma. This patient also presented with a golden-brownish corneal pigmentation. The initial serum copper level was 1250 μg/100 mℓ, ranging during a 12-month period of observation from 978 to 1300 μg/100 mℓ. The serum ceruloplasmin concentration in this patient averaged from 37.0 to 41.2 mg/100 mℓ. Immunoelectrophoresis of the serum protein documented a marked IgG ranging from 2350 to 3200 mg/100 mℓ, with a depression of the IgA (35 to 37 mg%) and IgM (38 to 42 mg/100 mℓ). A bone marrow survey was consistent with multiple myeloma. Cerebrospinal fluid, urine, saliva, hair, liver, and bone marrow biopsy specimens for copper were found to be within normal range. Heavy-metal screening (zinc, nickel, silver, and molybdenum) was negative. Electrophoresis of the serum protein, following the administration of ^{64}Cu and gel filtration of the untreated serum demonstrated that most of the copper was associated with the gamma globulin fraction. These techniques, together with membrane filtration, demonstrated that the hypercupremia was not a consequence of elevated free ionic copper, albumin, or ceruloplasmin-bound copper. The immunologic studies showed that the abnormal protein is a monoclonal IgG with lamda chains. Penicillamine therapy resulted in the loss of approximately 15 mg (44%) of serum copper, but had no effect on the level of serum myeloma protein. Despite the large loss of copper, the serum copper levels were ten times normal levels after treatment. The cause-effect relationship between hypercupremia and the myeloma was not clear.

Thorling and Thorling[173] in 1976 reported the clinical usefulness of serum copper determinations in Hodgkin's disease observed over a 10-year period. This was an extensive retrospective study of 241 patients with Hodgkin's disease followed with serial copper determinations from 1963 through 1973. The clinical significance of serum copper determinations in evaluating the course of Hodgkin's disease and its prognostic value was reported in their pilot study which was published in 1964.[159] In the present paper,[173] the authors extended their observations and obtained new ones concerning correlation between the initial serum copper level and the stage of the disease, as well as the histological classifications of Hodgkin's lymphoma. Clinical material consisted of 141 male and 100 female patients. The mean age group for men was 40.3 years and for women 37.6 years. Luke's classification was used for the histologic classification of Hodgkin's lymphoma. In the three groups: mixed cellularity, lymphocytic depletion, and unclassifiable Hodgkin's disease, there was almost identical representation of males and females. On the other hand, the group lymphocytic predominance comprised 33% males and only 16% females, whereas in the group of nodular sclerosing there were 17% males and 41% females. Complete remission of the disease in this study was defined as the disappearance of all enlarged nodes or organs, normalization of the sedimentation rate, and the disappearance of all subjective symptoms. The patients were treated with irradiation and chemotherapy. Serum copper levels were determined by the colorimetric method and their normal values were reported in the pilot study.[159] Three consecutive serum copper level determinations were done in the pretreatment period and three values in the days immediately after therapy, whenever it was possible, and the averages of these figures were used. The correlation between serum copper levels and the histologic classification in the pretreatment period was observed. These correlations were more obvious in men and less consistent in women. In men, there was an increase in the mean serum copper level from the group with lymphocytic predominance (192 μg/100 mℓ) to those classified as mixed cellularity (213 μg/100 mℓ — $p < 0.02$). The group with lymphocytic depletion had the highest average of serum copper levels in both sexes (251 μg/100 mℓ in women and 243 μg/100 mℓ in men). Inconsistent serum copper level patterns (being normal in the active phase of disease) were noted in some cases of Hodgkin's disease, particularly in the lymphocytic predominant group and in patients with stage I disease, probably reflecting clinical behavior and low tumor volume. The alterations of the serum copper level according to disease activity (increased in the active phase of disease and normal in remission) could be observed several times in the same patients during their disease course and according to these authors is now generally accepted as a useful tool in the surveillance of the efficacy of therapy and a reliable and often very early indicator of relapse.

In their reviews the authors looked in particular at the events on the first patient's admission: the effect of the treatment (successful or not) and the alteration in the serum copper level at first relapse. There was a fairly good correlation between the average decrease in the serum copper level with complete and partial remission as the result of treatment response. The serum copper level fall was highest in a case of complete remission. In 55 of 56 male patients achieving complete remission, serum copper levels were normalized. In partial remission, only in one third to one fourth of the patients did the serum copper level fall to the normal range. In patients who did not respond to treatment, the serum copper level remained high and fell to normal limits in two terminally ill patients as the result of liver failure. Of the 141 patients, an increased serum copper level at first relapse was observed in 66 and in 9 more cases of disease progression where the primary treatment failed to induce a complete remission.

In five patients, an increase in the serum copper level was either missing or negligible at first relapse. However, the authors stress that variations in the serum copper level within the normal limits were sometimes observed to be associated with relapse, yet without serum copper level increasing to decidedly pathologic values. At the conclusion of the study 36 patients remained in remission with normal serum copper levels and in the remaining patients either sudden death or deaths from unrelated causes excluded the evaluation as to a possible increase at first relapse. In the 100 women, the results are similar. Elevation of the serum copper level was observed in 40 patients at first relapse and in 5 at progression of disease; 32 patients were still in remission and in the rest of the cases sudden death or deaths from other reasons prevented the observation of an increased serum copper level at relapse. The authors have demonstrated a good correlation between the serum copper level and ceruloplasmin concentrations in patients with Hodgkin's disease and stated that there is no doubt that these changes in the serum copper are mainly reflecting changes in the ceruloplasmin level.

As far as the cause of the serum copper level increase in Hodgkin's disease, the authors speculated that some product produced by the tumor, possibly cross-reacting in some respects with estrogens, or, less likely, stimulating estrogen production or interfering with its metabolism might be considered (e.g., occurrence in Hodgkin's disease of gynecomastia and pigmentation). Based on their studies, these authors concluded that the activity of Hodgkin's disease is generally reflected in the serum copper level, increasing with progression and decreasing with improvement. Also, the serum copper level during the pretreatment period was statistically significantly correlated to the stage of disease and consequently probably to the volume of diseased tissue. The observations of the serum copper level in stage I differ from the observation in stage II and stage III, with $p < 0.01$. Stage II differs from stage III with $p < 0.01$. Also, patients with systemic symptoms (night sweats, fever, and body weight loss) had higher serum copper levels than those without. There was also a statistically significant correlation between the serum copper levels and the histological classification of Hodgkin's disease in men, whereas this correlation was blurred by the effects of estrogen and oral contraceptive medications in women. The serum copper level was regularly reduced to within normal range in patients achieving complete remission and therefore the authors proposed that this parameter be included in the criteria for complete remission.

In 1977, Kesava[174] et al. reported the serum copper assay as a biochemical marker to assess the response to therapy in Hodgkin's disease. Serum copper levels were determined in 100 patients with Hodgkin's disease to assess the relationship between this parameter and disease activity. Over 70% of the patients showed serum copper levels (150 to 420 μg/100 mℓ of serum) that were higher than normal. The remaining patients had serum copper levels in the normal range of 85 to 150 μg/100 mℓ and were found to be asymptomatic or under active treatment. A comparison of five patients with normal serum copper levels (102 to 127 μg/100 mℓ) with five patients with high serum copper levels (282 to 318 μg/100 mℓ) revealed that those with normal levels favorably responded to chemotherapy. Patients with high serum copper levels either had no treatment or did not respond to treatment. In four patients who responded to chemotherapy, the serum copper levels before treatment were 178, 269, 230, and 305 μg/100 mℓ as compared to posttreatment values of 153, 169, 169, and 114 μg/100 mℓ, respectively. The above results suggested that serum copper is a useful biochemical marker for evaluating the state of disease activity in Hodgkin's disease and for the efficacy of chemotherapy. El-Haddad et al.[175] in 1977 reported the value of serum copper measurement in acute leukemia in childhood. Serum copper level determinations were carried out in 16 normal children and on 16 with acute leukemia. A significant increase in the serum copper level was observed in cases of leukemia than in normal controls. A fall in the serum copper levels occurred in cases which responded to quadruple chemotherapy, while those which failed to respond showed persistently high serum copper levels.

Bucher and Jones[176] in the same year (1977) reported serum copper-zinc ratios in patients with malignant lymphoma. The determination of serum copper, zinc, and magnesium levels were made on 150 serial samples from 92 adult patients with malignant lymphoma. The results were correlated with histology and disease activity. Serum magnesium levels did not relate well to either. Serum copper levels were significantly higher ($p > 0.02$) in patients who had Hodgkin's disease (mean 184 μg/100 mℓ) compared with patients who had diffuse lymphoma (mean 156 μg/100 mℓ) or nodular lymphoma (mean 166 μg/100 mℓ). Mean zinc levels were similar for all groups. Mean copper levels were also significantly higher in untreated or relapsing patients compared with those in remission ($p > 0.01$) whereas the highest mean zinc levels were observed in patients in remission ($p > 0.05$). When the copper-zinc ratio was analyzed, the mean value was only slightly higher for patients with Hodgkin's disease (1.7 compared with 1.5 for diffuse lymphoma and nodular lymphoma) but correlated well with disease activity ($p = 0.001$). For example, untreated or relapsing patients with Hodgkin's disease had mean copper:zinc ratios of 2.1 and 2.2, respectively, whereas those with partial responses and those in complete remission had mean ratios of 1.7 and 1.5, respectively. These authors concluded that the copper:zinc ratio can be a useful monitor of disease activity in patients who have lymphomas, particularly Hodgkin's disease.

Rawat and Vijayvargiya[177] in 1977 reported the usefulness of serum copper level determinations in 50 patients with various types of lymphomas. They studied 11 patients with follicular lymphoma, 9 with lymphosarcoma, 19 with Hodgkin's disease, and 11 with reticulum cell sarcoma. The majority of cases (37 patients) was treated with irradiation therapy alone. In the remaining 13 patients, radiotherapy was combined with chemotherapy. Most of the patients treated were in stages III and IV. In all of the patients with active lymphoma the serum copper levels were invariably found to be increased. The lowest serum copper levels were found in stage I (mean 152 ± 14 μg/100 mℓ) whereas the cases with stage IV showed a mean level of 408 μg/100 mℓ of serum copper. The radiotherapy of one region (upper torso using Mantle's technique) resulted in a significant reduction in the serum copper level. The second irradiation to another region (lower torso-inverted Y technique) further decreased the copper level in patients of all stages. Some of the cases which were followed for 3 months showed a further decline in the serum copper levels. The authors concluded that the serum copper level can be a satisfactory guide for determining the prognosis and efficacy of treatment in cases of lymphomas and could serve as a useful auxilliary test in the management of patients with lymphoproliferative malignancies.

Davidoff et al.[178] in 1977 studied serum copper and ceruloplasmin levels as an adjunct to the diagnosis of Hodgkin's and non-Hodgkin's lymphoma. The authors proposed a new diagnostic index to add in the differential diagnosis of patients suspected of having lymphoma. This index, according to these authors, takes advantage of the vastly different regression plots of total serum copper vs. activated ceruloplasmin concentrations in lymphoma patients (12), smokers (16), oral contraceptive users (8), and normal individuals who do not smoke or use oral contraceptives (36). The data were coded and keypunched. Linear regression studies developed the following four correlation coefficients for the different strata — normal: $r = 0.48$; smokers: $r = 0.83$; users of oral contraceptives: $r = 0.91$; and lymphoma patients: $r = 0.83$. Classical discriminant analysis developed an index based on these different linear functions that showed a false-negative rate of 8% and a false-positive rate of 6%. This seems to be a more sensitive technique than use of copper levels alone. Ceruloplasmin concentrations seemed to anticipate the changes in serum copper levels in three treatment patients studied longitudinally, suggesting that ceruloplasmin may be a chief regulating factor in the elevation of serum copper during active lymphoma.

Foster et al.[179] in 1977 reported the measurement of electron spin resonance as a useful and simple technique in the management of Hodgkin's disease. The technique used for obtaining electron spin resonance spectra from the blood was described previously.[127] They measured the levels of ceruloplasmin and iron transferrin in whole blood from patients with Hodgkin's disease. In addition, erythrocyte sedimentation rate, neutrophil alkaline phosphatase, hemoglobin, packed cell volume, white blood count, and platelet count were determined. The study was made to assess thc value of blood ceruloplasmin and iron transferrin levels in the management of patients with Hodgkin's disease as compared with the standard clinical and hematological methods and, in particular, to assess the use of electron spin resonance spectroscopy to perform these tests. The study group comprised 50 patients with histologically proven Hodgkin's disease. There were 30 males and 20 females ranging in age from 16 to 75 years and the study was made over a period of 20 months. The patients were unselected with the exception of females taking oral contraceptives known to disturb copper and ceruloplasmin levels. A total of 221 samples were obtained, one blood sample being obtained at each visit to the outpatients' clinic.

As far as disease activity was concerned, patients were divided into three categories: (1) inactive, (2) intermediate disease activity in which there was some suspicion but no definite evidence of disease activity found, and (3) active group. Those patients with clinically active disease show higher ceruloplasmin levels and lower iron transferrin levels than those with inactive disease. The results indicated that these tests are good indicators of the state of disease activity and that serial measurement of these parameters may help in early prediction of clinical relapse and in monitoring response to treatment. The combined information from iron transferrin and ceruloplasmin levels appears to be more predictive than that from the erythrocyte sedimentation rate and neutrophil alkaline phosphatase score. The authors concluded that electron spin resonance is a useful technique in the management of patients with Hodgkin's disease.

In the same year, Foster et al.[180] reported the measurements of blood ceruloplasmin and iron transferrin by electron spin resonance in 32 patients with non-Hodgkin's lymphoma.

Ceruloplasmin and iron transferrin levels were measured in the blood of patients with non-Hodgkin's lymphoma in different stages of disease activity and compared with the erythrocyte sedimentation rate and neutrophil alkaline phosphatase level in the same samples. It was found that both ceruloplasmin level and sedimentation rate showed a slight increase in mean level in patients with active disease as compared with those in remission, particularly in the group of patients with poorly or undifferentiated diffuse malignant lymphoma disease. No difference was observed in levels of iron transferrin or neutrophil alkaline phosphatase. Both the ceruloplasmin and sedimentation rate showed occasional abnormal values in patients with remission, but in most cases where both were elevated the patients subsequently entered a more active phase of the disease. These data as well as other observations of these authors concerning serum ceruloplasmin and transferrin determination in malignant diseases, in addition to a literature survey, were well summarized in the authors' review article in 1981.[127]

In conclusion of that review article, in the section of lymphoma, the authors stated that normal ceruloplasmin levels determined by electron spin resonance were found in polycythemia rubra vera and slightly increased in nonleukemic myeloproliferative disorders. Ceruloplasmin levels were found to be increased in chronic granulocytic leukemia and even more so in acute myelogenous leukemia in adults. A slight increase in electron spin resonance detectable ceruloplasmin levels were found in some other solid tumors, especially carcinoma of the breast, but there was no increase in patients with squamous cell carcinoma of the head and neck or malignant melanoma. Patients with infections or following surgical procedures had higher ceruloplasmin concentrations than did those with malignant disease alone. Two weeks after surgery, the patients' elevated ceruloplasmin concentration fell to preoperative levels. The authors concluded that the measurement of serum copper or ceruloplasmin should be a part of patient management in Hodgkin's disease. Also, it would seem that this test should normally be supported by either measurement of the erythrocyte sedimentation rate or of the iron transferrin level of the blood.

Pizzolo et al.[181] in 1978 reported the diagnostic value of serum copper levels and other hematochemical parameters in malignant diseases. In order to explore the diagnostic value of serum copper compared to other hematochemical parameters frequently abnormal in malignancies, serum copper, serum alpha-2 globulin, plasmatic fibrinogen, the erythrocyte sedimentation rate, and serum iron were evaluated in 267 patients affected with the following diseases: Hodgkin's lymphoma, non-Hodgkin's lymphoma, acute leukemias, chronic myeloid leukemia, chronic lymphocytic leukemia, myeloma, and breast carcinoma. In 74 untreated patients with Hodgkin's disease, the mean serum copper was 186 μg/100 mℓ and 131 ± 37 μg% in remission. The mean serum copper values in 36 patients with active non-Hodgkin's lymphoma (86 determinations) were 174 ± 50 μg/100 mℓ and 139 ± 40 μg/100 mℓ in the remission group. The difference was significant ($p > 0.005$). In these diseases, serum copper measurements were more valuable than any other considered parameters. The erythrocyte sedimentation rate, which in the active stage of disease had the highest frequency of abnormal values, is also more often increased in remission. The mean values of 96 serum copper determinations on 38 adult patients with acute leukemia were 180 ± 66 μg/100 mℓ in the active phase of disease and 116 ± 40 μg/100 mℓ in the remission group ($p > 0.001$). Serum copper levels and erythrocyte sedimentation rates were the parameters more frequently abnormal in the active phase. Their incidence was greatly reduced in remission: 15% of acute leukemia patients in remission had abnormal serum copper levels and 35% of cases had elevated erythrocyte sedimentation rates. In chronic myeloid leukemia the mean serum copper value (44 determinations on 23 patients) was 148 ± 56 μg/100 mℓ; in chronic lymphocytic leukemia (47 determinations on 35 patients) serum copper levels were 182 ± 60 μg/100 mℓ; in multiple myeloma (18 determinations in 13 patients) serum copper levels were 167 ± 79 μg%. In chronic lymphocytic leukemia the serum copper level was the only parameter abnormal in a high percentage of cases; in almost all patients with multiple myeloma the erythrocyte sedimentation rate was abnormal while the serum copper level was increased in 50% of patients. The authors also studied 48 patients with breast carcinoma (115 serum copper level determinations were performed). In untreated patients or in relapse with carcinoma of the breast the serum copper level was 186 ± 68 μg%, whereas in the remission group the mean value was 132 ± 33 μg/100 mℓ. The statistical difference between the two groups was highly significant ($p > 0.001$).

The best correlation between copper alterations and disease activity was found in cases of Hodgkin's lymphoma, non-Hodgkin's lymphoma, acute leukemia, and carcinoma of the breast. In these diseases, when the considered parameters were compared, serum copper and erythrocyte sedimentation rate showed a similar pattern, i.e., a high frequency of abnormalities in active disease. It is concluded that serum copper represents a good complement to some other nonspecific parameters in evaluating the activity and extent of malignant disease and response to treatment, particularly in patients with Hodgkin's disease, non-Hodgkin's lymphoma, acute leukemia, and breast carcinoma.

Legutko[182] in 1978 reported on serum copper investigations in 57 children with acute lymphoblastic leukemia. It was found that the changes in copper were, to a greater extent, related to the clinical course of acute lymphocytic leukemia than to age of the children examined. The highest mean serum copper level was obtained in untreated patients (261.2 μg/100 mℓ). Normalization of the bone marrow during treatment was accompanied by a decrease of the serum copper level. The significant increase of the serum copper level in comparison with the serum copper level in remission (129.8 μg/100 mℓ) occurred in extramedullary localizations (163.8 μg/100 mℓ). Based on his observations, the author suggested that serum copper level determinations may be useful as an auxiliary test in the clinical evaluation of the disease activity and the efficacy of therapy and in predicting relapse of acute lymphocytic leukemia.

Williams et al.[183] in 1978 reported their observations concerning the value of the serum copper level and erythrocyte sedimentation rates as indicators of Hodgkin's disease activity in 29 children. Clinical material consisted of 12 boys and 17 girls ranging in age from 8 to 18 years (median 15 years). Serum copper level determinations on 128 healthy black children ranging in age from 6 to 15 years were used as controls. The mean serum copper value was 143 μg/100 mℓ and ranged from 73 to 213 μg/100 mℓ. The serum copper level varied considerably among patients and within the same patients followed from 2-week to 6-month intervals. The serum copper levels were frequently increased at diagnosis and relapse and there was a statistically significant difference ($p = 0.04$) between the overall mean serum copper level of therapy (136 μg/100 mℓ) and at relapse (166 μg/100 mℓ). However, serum copper levels above the upper normal limits for adults have occurred among those patients without active disease. Of the 10 patients at relapse, 7 had serum copper levels greater than those for healthy children of similar age.

Also, 9 of 19 patients in remission had elevated age-corrected serum copper levels. Although 9 of the 10 patients at relapse had an increase in the serum copper level over the preceding value, almost one fifth of the patients in remission had increases in consecutive serum copper levels greater than the average increase of patients who relapsed. Based on their observations, the authors concluded that sedimentation rates and serum copper levels are not useful as indicators of disease activity in children with Hodgkin's disease.

Davidoff et al.[8] in 1979 reported the effect of smoking habits on serum and erythrocytic copper and ceruloplasmin concentrations. Their clinical material consisted of 40 healthy nonsmoking subjects, 25 smokers, 14 oral contraceptive users, 4 nonsmoking pretreatment lymphoma patients, and 8 lymphoma patients during treatment. Significant elevation of the serum copper level and ceruloplasmin concentrations were observed in lymphoma patients and oral contraceptive users. The mean serum copper value in healthy nonsmoking controls was 99, in smokers 116, oral contraceptive users 175, pretreatment lymphoma patients 197, and in treated patients 143 μg/100 mℓ. The mean serum ceruloplasmin values in nonsmoking controls was 30.3, smokers 41.1, oral contraceptive users 52.5, pretreatment lymphoma patients 52.5, and 46.9 μg/100 mℓ in treated lymphoma patients. Significant correlation coefficient of the linear relationship between serum copper and ceruloplasmin were observed for every group studied except for smokers. In addition, all groups had elevated red cell copper compared with the healthy subjects. Also, significantly increased serum copper ($p < 0.0053$) and ceruloplasmin ($p < 0.0001$) levels were found in smoking relative to non-smoking subjects. No correlation was observed between the duration of smoking (pack-years smoked) or frequency of usage (packs per day) and elevation of the serum copper, erythrocytic copper, or ceruloplasmin concentration.

Subrahmaniyam et al.[184] in 1979, published their observations in serum copper in patients with non-Hodgkin's lymphoma. Serum copper levels were determined in 22 healthy adults (18 men and 4 women) and 15 patients with non-Hodgkin's lymphoma in order to study the significance of the serum copper level in the management of non-Hodgkin's lymphoma. The mean serum copper level for control subjects was 84.45 μg/100 mℓ with a range of 49 to 150 μg/100 mℓ; female control subjects had higher serum copper levels (mean 91.83 μg/100 mℓ) as compared to males (mean 83.53 μg/100 mℓ). The mean pretreatment serum copper level for the non-Hodgkin's lymphoma patients was 117.9 μg/100 mℓ with a range of 78 to 200 μg/100 mℓ, significantly higher than the serum copper level for the control subjects. The mean serum copper level in 10 patients with stage I and II disease was compared to the obtained value in 5 patients with stage IV disease and the difference was not significant. The mean serum copper level in 10 patients with diffuse lymphoma was compared to the values obtained in 5 patients with nodular lymphoma and the difference was not significant. The mean serum copper level in patients before treatment was 117.9 μg/100 mℓ whereas the value after treatment was 48.3 μg/100 mℓ, a significant fall in the serum copper level as a response to therapy. The fall in the serum copper level after treatment correlated well with the clinical observation of the regression of tumor and the return of the patient's general condition to normal.

In 1979, Sirsat[185] published serum ceruloplasmin levels in Hodgkin's disease and malignant lymphomas in correlation with serum copper levels. Serum ceruloplasmin and copper levels were determined in 25 healthy male subjects and 54 patients with Hodgkin's disease and malignant lymphoma. The mean ceruloplasmin level (mg%) for the healthy subjects was 20.716 ± 0.598, while for the patients the mean level was 58.874 ± 2.268. The mean serum copper levels (μg%) for healthy subjects was 105.112 ± 2.646, while for the patients it was 272.680 ± 10.111. The increase in both substances in Hodgkin's disease and malignant lymphoma was statistically significant ($p > 0.01$). A linear correlation between the total serum ceruloplasmin and copper levels was found throughout the investigation. The author suggested that the more easily determined serum ceruloplasmin level may therefore be used as confidently as serum copper levels in monitoring the course of Hodgkin's disease and malignant lymphoma during treatment.

Mitta and Tan[186] in 1979 reported on serum copper levels in 72 children with Hodgkin's disease, ages 14 to 16 years. At diagnosis 12 of 72* patients (stage I:0/15; Stage IIA:3/26; Stage III: 6/17 — 3A, 3B; and stage IVB: 3/14) had elevated serum copper levels (range 234 to 326 μg/100 mℓ). All of these serum copper levels returned to normal within 1 to 6 months after treatment, and remained normal. In 14/72 patients there were 20 relapses, only 1/20 relapses had elevated serum copper levels. In 16/20 relapses the serum copper level remained normal. In the other three, serum copper levels were not done at relapse but were normal when studied 2 to 9 months after relapse. Two patients died from disease who had normal levels at the beginning of relapse, but became elevated during the terminal stage 10

* A = Patients without systemic symptoms and B = patients with systemic symptoms.

to 15 months later. Of 58 patients who were free from disease, 12 had elevated serum copper levels 6 to 35 months after diagnosis, and remained in remission. Elevated serum copper levels during remission were in the range of 248 to 431 in 9/12 patients, three patients had mildly elevated serum copper levels (236, 235, and 237). Correlation with other laboratory parameters in these 12 patients showed that serum iron was normal in 11 patients. Erythrocyte sedimentation rate was slightly elevated in five (13 to 35 mm/hr), normal in two, and not done in five. All 12 patients were afebrile and their mean Hb was 12 g/dℓ at the time that the serum copper levels were done. This review showed that serum copper levels are elevated in more advanced stages at diagnosis; but it is not a reliable measure of follow-up for recurrence.

Asbjörnsen[187] in 1979 compared the serum copper to the erythrocyte sedimentation rate as an indicator of disease activity in Hodgkin's lymphoma. Serum copper and erythrocyte sedimentation rates were recorded in 54 patients with active Hodgkin's disease and in 186 occasions in 78 patients during stable complete remission. Relative high and age-dependent normal limits for erythrocyte sedimentation rates were used. Each of the tests was elevated in 70% of the patients with active disease. During remission, the serum copper was elevated in 14% and the erythrocyte sedimentation rate in 15.6% of the determinations. Thus, the two tests are considered not far from equal in their ability to discriminate between the presence and absence of specific disease activity in Hodgkin's lymphoma. Simultaneous elevation during remission occurred in less than 5% of the recordings, as compared to 61% during active disease. The author concluded that the serum copper level may be of value as a supplement to the erythrocyte sedimentation rate.

Horn et al.[188] in 1979 reported electron spin resonance studies on properties of ceruloplasmin and transferrin in 278 normal donors and 97 cancer patients. The average Cu^{++}-ceruloplasmin electron spin resonance signal intensity was significantly different for the control groups of males, females who were not taking estrogen medication, and females who were taking estrogens. The mean electron spin resonance signal intensities of Cu^{++}-ceruloplasmin from cancer patients who were separated into the same groups as the control data were approximately twice as great as the mean control levels. The total serum copper levels were correlated with electron spin resonance of Cu^{++}-ceruloplasmin and indicated that the ratio of Cu^{++}/Cu^{+} in ceruloplasmin is higher in serum from cancer patients than from the controls.

Hamberg[189] in 1979 reported electron paramagnetic resonance and X-ray fluorescence determinations of serum copper in 47 patients with Hodgkin's disease, 59 patients with leukemia, 57 patients with lung cancer, and 69 controls. Samples from 15 patients with infections and 9 patients with myocardial infections were also studied. Electron paramagnetic resonance signals and total copper levels were both found to be elevated on the average in the serum of patients with cancer, as compared with normal subjects. When the cancer groups were divided into active and complete remission subgroups, all of the mean values in the active disease group were significantly increased above normal mean values, while for the complete remission groups in leukemia and lung cancer, copper levels could not be distinguished from normal levels. Complete remission values in Hodgkin's disease, however, while statistically lower than active disease value, remained increased above normal levels.

Gobbi et al.[190] in 1979 reported a comparison between plasma iron and copper with other laboratory indicators in the evaluation of disease activity in 55 patients with Hodgkin's disease. The laboratory indexes studied include plasma iron, total iron binding capacity, plasma copper, erythrocyte sedimentation rate, $alpha_2$-globulins, and fibrinogen. The authors' chief clinical aim was to investigate the discriminatory ability of such laboratory tests between the crucial conditions of evidence of disease — being it palpable or concealed — and of undoubtable complete remission. There were 30 males and 25 females, age 13 to 63 years; 35 patients were untreated and 10 were in relapse after remissions lasting from 4 months to 11 years without maintenance therapies. The last 10 patients have been in complete uninterrupted remission from the beginning to the end of this study. Out of 55 patients, 29 underwent pathological staging. Plasma iron, total iron binding capacity, and plasma copper levels were determined by means of atomic absorption spectrophotometry. A total of 145 plasma iron and plasma copper determinations were made: in the 45 patients who had active disease (untreated or in relapse) they were performed at least once before therapy; they were repeated more than once during the treatment in 27 patients and also after it in 25 of them. Data concerning erythrocyte sedimentation rate, $alpha_2$-globulinemia, and fibrinogenemia with iron and copper were available in 88 observations of patients: 57 of them were untreated or in relapse and 31 were in complete remission.

In patients with active Hodgkin's disease, the mean plasma iron levels were significantly lower (67.6 μg%) and the mean plasma copper levels were significantly higher (170.9 μg%) than in patients in complete remission (Fe 111.5 and Cu 119.1 μg%). The total iron binding capacity displayed nearly unchanged mean values at the onset, in remission, and in relapse of Hodgkin's disease. Patients' sex, histological type, clinical stage, and general symptoms did not affect plasma iron or total iron binding capacity levels significantly. Only stages I and II seemed to support plasma iron values somewhat higher than stages III and IV, but conventional significance levels were not obtained. The plasma copper levels confirmed their well-known positive correlation with the clinical stage of Hodgkin's disease, whereas they proved higher in less severe histological types than in more severe ones. As acute phase indicators, plasma iron and plasma copper individually showed a clearly higher diagnostic value than erythrocyte sedimentation rate, $alpha_2$ globulinemia, and fibrinogenemia. The allocation of patients into the clinical conditions of evidence of disease and remission made by the combined plasma iron and plasma copper levels is the greatest (exactly classified patients: 75%) when compared with the diagnostic effectiveness shown by all five indexes together. The discrimination error made by combined erythrocyte sedimentation rate, $alpha_2$ globulin, and fibrinogen is 20% higher than by associated plasma iron and plasma copper, which appeared the relatively most useful tests in monitoring patient's response to treatment and in early detection of relapse. The relations of such elements with the staging parameters and the clinical course of Hodgkin's disease suggest that it may also be of great interest to investigate their prognostic value. The findings of higher serum copper levels in patients with less aggressive types (lymphocytic predominance and nodular sclerosis) as compared to mixed cellularity and lymphocytic depletion is unusual and will be alluded to later.

DeBellis et al.[191] investigated, in 1979, the metabolic changes in the red blood cells in malignant lymphoma. The serum copper levels which were concomitantly related to red blood cell free copper were significantly increased in some malignant lymphomas in the face of disease activity. This resulted in a profound inhibition of red cell key glycolytic enzymes, hexokinase being the most sensitive. A total of 15 patients (8 with Hodgkin's disease and 7 with non-Hodgkin's lymphoma) were studied for serum and red cell copper concentrations and hexokinase activity. The mean red cell life span was determined using ^{51}Cr-labeled red cells. The resulting data showed that in active disease, an increase in the serum copper level was associated with a decrease in the hexokinase activity and a shortened red cell survival. In these cases, there was no evidence of an autoimmune phenomenon or of direct bone marrow involvement by the disease. The authors suggested that the increase in the copper levels resulted in a shortened red cell life span through a copper-induced inhibition of red cell hexokinase.

Wolf et al.[192] in 1979 reported an evaluation of copper oxidase (ceruloplasmin) and related tests in Hodgkin's disease. A panel of laboratory tests was undertaken in 100 consecutive patients with Hodgkin's disease, proven by biopsy, who were part of a controlled, randomized clinical study to evaluate treatment by radical radiotherapy and/or chemotherapy. There were 50 control patients studied. Single pretreatment values were obtained 2 to 24 months posttreatment. Pretreatment values demonstrated that electron spin resonance was elevated in 86 of the cases, copper oxidase (ceruloplasmin) in 78%, and serum copper in 80%. Leukocyte alkaline phosphatase was elevated in only 20%, muramidase in 4%, and magnesium in 2% of the cases. Hypoferremia occurred in 57% of patients. A distinct relationship existed between the frequency of elevated electron spin resonance and serum copper oxidase and the extent of disease. Serial determinations may be valuable in the assessment of the patient's response to treatment. Increased values of copper oxidase in the absence of infections or estrogens, according to these authors, may suggest the possibility of occult active Hodgkin's disease.

Shah-Reddy et al.[31] in 1980 reported on serum copper levels in 34 patients with adult non-Hodgkin's lymphoma at different phases of the disease activity. Clinical material consisted of 22 patients with poorly differentiated lymphocytic lymphoma (10 nodular and 12 diffuse), 6 had histiocytic lymphoma disease and 3 had diffuse well-differentiated lymphocytic lymphoma. The patients were treated either with standard chemotherapy or extended field irradiation. The range of the serum copper levels for 17 patients considered to be in complete remission by all criteria were 97 to 134 μg/100 mℓ (mean value 114.76) whereas 17 patients with active disease, i.e., nonresponders or those in relapse had serum copper levels ranging from 157 to 279 μg/100 mℓ. Three patients who were judged to be completely free of demonstrable disease had high serum copper levels and on reevaluation were found to be in early relapse. There was no overlap of the serum copper levels among patients with active disease and controls or those in remission. Using the *t*-test and two sample Rank tests, the probability was less than 0.005 that the active disease values were part of the population of normal controls. The authors concluded that the serum copper level is of definite value in non-Hodgkin's lymphoma to monitor the status of disease activity, including detection of early relapse. Green et al.[193] in 1980 reported electron spin resonance studies on ceruloplasmin and iron transferrin in 41 patients with chronic lymphocytic leukemia. The ceruloplasmin levels were found to be above normal range in all stages of the disease, increasing with increased clinical activity and were higher in progressive than in inactive disease. The whole-blood iron transferrin levels were more variable and were significantly elevated only in patients with marrow failure.

Larsen et al.[194] in 1980 reported an inhibitory effect of ceruloplasmin on lymphocyte response in vitro. The authors studied 187 sera of patients representing different stages of Hodgkin's disease. Elevated serum copper and ceruloplasmin concentrations were found. The ratio between serum copper and ceruloplasmin levels was found to be relatively constant indicating that about 90% of the total serum copper is bound to ceruloplasmin in most stages of the disease and that this is not significantly different from normal.

Since patients with Hodgkin's disease usually have decreased immunoreactivity, it was investigated whether the high ceruloplasmin concentrations found in sera from these patients would be immunosuppressive. In this study, ceruloplasmin was found to markedly inhibit the in vitro proliferative response of normal human peripheral blood lymphocytes stimulated with phytohemagglutinin, concanavalin A, and purified protein derivative, and to inhibit the mixed lymphocyte culture response. Generation of cytotoxic killer cells during mixed lymphocyte cultures was likewise inhibited whereas killing by preformed cytotoxic T cells was not influenced. It was concluded that increased ceruloplasmin which was found in 36% of the Hodgkin's sera may act immunosuppressive, but the extent of such effects in vitro can only be suggestively evaluated by means of measurements of the lymphocyte response in vitro.

Kaplan[195] in his excellent book of Hodgkin's disease published in 1980 reported serial determinations of the serum copper level in 50 patients with Hodgkin's disease of all stages treated and subsequently followed at Stanford University Medical Center from 1971 through 1975. This group included 37 males and 13 females (3 of whom started taking contraceptive pills some months after completion of treatment). The initial serum copper value was elevated in 45 (90%) of these 50 patients. The data were suggestive of a trend toward an increasing proportion of elevated values among patients with advanced stage of disease and among those with constitutional symptoms and extra lymphatic lesions confirming correlations noted in previous studies by Hrgovcic et al.[29] and by Thorling and Thorling.[173] Serial determinations of the serum copper level were particularly useful in the 6- to 12-month period immediately following completion of radiotherapy and/or combination chemotherapy, since the erythrocyte sedimentation rate tends to rise and to remain elevated for many months as a consequence of treatment and is of little help during this interval as a prognostic indicator. Kaplan stated that the serum copper falls to normal levels and remains normal in successfully treated patients, whereas failure of the serum copper values to return to normal following treatment, or a transient fall followed by a renewed elevation during the first 12 months after treatment usually denotes impending relapse.

Cohen et al.[196] in 1981 reported serum copper levels in 125 patients with non-Hodgkin's lymphoma. The mean serum copper values observed by these authors in patients with active disease was significantly higher than in patients in complete remission and in the controls ($p < 0.001$). The serum copper levels of patients in remission were significantly higher than those of the controls ($p < 0.001$). In patients with active disease, there was no significant difference in the serum copper levels either between lymphocytic (number = 71) and histiocytic lymphoma (number = 31), or between diffuse (number = 87) and nodular lymphoma (number = 24). There were 26 untreated patients with stage I or II who had similar values to those of 23 patients with stage III or IV. Thirty-two of 80 patients in complete remission had at least one elevated serum copper level greater than 160; 6 of these have relapsed as compared to 7 of 48 with normal values. This study did not show a return of the serum copper level to normal values in complete responders nor the difference between limited and extensive disease. The authors concluded, based on their experience, that the serum copper level seems to be useful in monitoring the disease activity for the individual patient with non-Hodgkin's lymphoma although its value in detecting an earlier relapse has yet to be established.

2. *Tissue Copper Content in Malignant Diseases*

Several investigators studied various trace element tissue content in patients with malignant diseases. The different analytical technique, tissue sampling, and histopathological procedures may explain the differences in concentrations of certain metals in cancerous and noncancerous tissues observed. Therefore, these investigations were far from complete and studies consisting of many paired samples for each type of cancer with uniform standard tissue classification and analytical procedures are needed before a definite conclusion can be met.

Sandberg et al.[150] in 1942 reported the results of a study on the copper and iron content of the liver and the spleen in 146 cases of various diseases accompanied by secondary anemia. The primary diseases comprised 38 cases with cancer of the gastrointestinal tract, 13 patients with genitourinary tract carcinoma, 12 cases with carcinoma of the female pelvic organs, 14 patients with breast carcinoma, and 11 patients with carcinoma of the lung. A lymphoblastoma group consisted of 16 cases and 41 cases with noncancerous disease were included. The retention of iron was more prevalent and usually greater than that of copper. The liver appeared to maintain close guard over copper storage. The abnormal copper values were found in only 16% of cases studied. In cancer accompanied by anemia the marked increase in both copper and iron storage was out of all proportion to the anemia. In some cases of cancer, such increases took place even in the absence of anemia, with cancer as the only apparent causative factor. The excessive copper and iron retention in the liver and the spleen encountered in cases of cancer was significantly higher in cases with extensive metastatic disease.

It was interesting to note that patients with lymphoreticular malignant diseases had frequently increased copper and iron content and some of the highest values in the entire series were encountered in this group. Although the large concentrations of these metals in the liver and spleen of these patients were so striking, no clear explanation was found.

Olson et al.[197] analyzed by spectrographic method five trace elements (Fe, Zn, Cu, Mo, and Mn) in the liver of 29 autopsy cases who died of various nonmalignant diseases, 19 patients with cancer metastatic to the liver, and 9 patients without liver metastases. The mean values for copper content were not notably altered. Most fell well within the wide range found in noncancerous livers. One exception was the value of 14.0 μg/g found in the liver of a patient with cancer of the breast (290% of the noncancerous mean). Lesser increases of 47% were found in two cases of cancer of the hypopharynx and kidney, and moderately low values were found in cancer of the breast and stomach. The mean zinc value was 70% higher in patients with liver metastases compared to noncancerous liver. The authors hoped that their findings would give additional impetus to the study of zinc metabolism in patients with cancer.

Mulay et al.[198] in 1971 reported 22 trace metal analyses of cancerous and noncancerous human tissue determined by emission spectrography. One of the elements studied was copper. The following types of cancer were studied: eight patients with cancer of the breast, six cases with cancer of the lung, and six cases with adenocarcinoma of the colon. The results showed that the copper, magnesium, manganese, and zinc content of cancerous breast tissue was significantly higher than that of the noncancerous tissue. In carcinoma of the lung, the iron content of the cancerous tissue was significantly lower and the zinc content significantly higher than contents in the noncancerous bronchial tissue. The patients with adenocarcinoma of the colon showed the concentration of tin was considerably lower in cancerous than in noncancerous tissue. The significant differences in the concentrations of selective metals were noted between cancerous tissues. It was also apparent that these differences pertain to a specific type of cancer.

Morgan[199] reported in 1971 hepatic copper, manganese, and cadmium contents of 104 autopsy cases of which 42 died of carcinoma of the lung. The copper content of liver was not significantly different in any of the groups studied.

Schwartz et al.[200] in 1971 reported trace elements in normal and malignant human breast tissue. These authors studied normal and malignant breast tissue from nine patients undergoing radical mastectomy. The tissue was analyzed for concentration of trace metals by neutron activation analysis. The levels of potassium, phosphorus, copper, magnesium, and zinc were substantially higher in the malignant tissue. The significance of this difference remains a matter of speculation according to these investigators.

Janes et al.[201] in 1972 reported trace metals in human osteogenic sarcoma. One of the trace elements studied was copper, which was found to be decreased. The significance of these findings was not known, although from the authors' experimental work, a possible relationship between certain trace elements and the production of osteogenic sarcoma was suggested.

Santoliquido et al.[202] reported trace metal analysis in 1976 in cancer of the breast — 20 sample pairs of cancerous and noncancerous tissue from the human breast were analyzed for trace metals. The iron, zinc, copper, lead, calcium, manganese, chromium, silver, strontium, and aluminum contents were determined. These authors found magnesium and zinc levels to be significantly higher in the cancerous than in the noncancerous tissue.

Karcioglu et al.[203] in 1978 reported trace element concentrations in five cases with renal cell carcinoma. Cadmium, zinc, and copper levels and zinc:copper, zinc:bromine, iron:zinc, iron:copper, and iron:bromine ratios were measured in neoplastic and normal kidney samples from humans by the proton-induced X-ray emission analysis. Decreased zinc:copper ratios were found in all neoplastic tissues, but this was not observed in other element ratios. The authors believe that some of the trace elements may very well play important roles in the pathogenesis of neoplasia and it is quite likely that research in this field will produce a better understanding of the etiology of tumors and valuable practical information that can be applied to clinical medicine.

In 1980, Roguljic[204] et al. reported iron, copper, and zinc tissue levels in 53 patients of untreated malignant lymphoma. There were 14 cases of Hodgkin's disease, 17 patients with lymphocytic lymphoma, and 22 patients with histocytic lymphoma. Of 53 cases, 26 had localized (stage I and II disease) and 27 cases had generalized disease; 23 subjects with clinically suspected hepatic disorders underwent liver biopsies, and since no histological abnormalities were found these were used as a control group. Normal serum copper levels were determined on a control group of 45 men and 15 women. No significant sex difference in the serum copper was found. Normal serum copper values for both sexes was 107 ± 19 μg/100 mℓ. The studies of iron, copper and zinc content in the liver tissue of Hodgkin's disease patients showed no significant difference in the levels of these metals as compared to the control group ($p > 0.05$). None of these Hodgkin's disease cases had liver involved with basic disease. On the contrary, serum copper levels in these patients were statistically significantly higher ($p > 0.05$), confirming the reports of Hrgovcic et al.[28,29]

Normal liver copper content was also found in a group of lymphocytic lymphoma patients without hepatic involvement by lymphoma. In cases in which liver invasion by lymphomatous disease was found there were significantly lower copper levels in spite of hypercupremia which was present in all lymphocytic lymphoma cases studied. Copper-liver content was not significantly higher in the histiocytic lymphoma group when compared to the normal. The only correlation in the redistribution of copper between the liver tissue and sera was noted in patients with lymphocytic lymphoma with liver involvement. The authors concluded that it would be very difficult to draw any final conclusion based on this study because the patient population was too small.

D. Serum Copper and/or Ceruloplasmin Alterations in Solid Tumors

Pirrie[205] reported in 1952 the serum copper and iron levels in 19 patients with malignant diseases. The mean values of 105 ± 5.74 and 120 ± 6.72 μg% were found for serum copper in 20 healthy adult males and 20 healthy adult females, respectively. The sex difference was not significant. The mean serum copper levels in patients with malignant diseases were 223 ± 14.18 μg/100 mℓ and the mean serum iron was 67 ± 5.96 μ%. Serum copper levels in these patients were significantly higher than normal and serum iron were significantly lower. A statistically significant inverse correlation between the degree of hypercupremia and hypoferremia in patients with neoplastic disease was observed. There was no significant correlation found between serum copper levels and hemoglobin concentrations. The authors speculated that a reduced demand for copper, an essential catalyst in hemoglobin synthesis, may be the explanation of the observed increased serum copper level.

Graf[206] in 1965 reported the significance of ceruloplasmin levels in various stages of colon carcinoma. Serum ceruloplasmin concentrations were measured in 43 patients with colon carcinoma before and during therapy (irradiation or chemotherapy). There were six patients with stage I, 14 cases with stage II, and six patients with stage III classification. In localized colon carcinoma, ceruloplasmin levels were found to be 35.2 mg/100 mℓ in stage II and 49 mg/100 mℓ in stage III, as compared to 33 mg/100 mℓ in the control group. Likewise, the serum copper levels reflected the extent of disease being 154 μg/100 mℓ in stage I and 184 μg/100 mℓ in stage III of colon carcinoma. Various complications such as proctitis, pyelitis, or hydronephrosis caused a further increase in the ceruloplasmin concentrations during irradiation.

According to this author, ceruloplasmin determinations during and after treatment can be of prognostic value. The increase of the ceruloplasmin during therapy and the slow decrease after treatment is a favorable sign of improvement. A decrease in the serum ceruloplasmin levels with treatment is a bad prognostic sign suggesting progression of the malignant condition. Reevaluation of the ceruloplasmin concentrations after therapy was indicative of a relapse or inflammation. It was not possible, on the basis of the ceruloplasmin concentration alone, to distinguish between inflammation and/or relapse of carcinomatous disease.

DeJorge et al.[207] in 1965 reported serum copper levels and tissue copper contents in ten patients with squamous cell carcinoma of the larynx. They found serum copper levels of 273.5 ± 36.8 mg/100 mℓ, which were significantly increased ($p > 0.01$) compared to normal subjects (108.1 ± 9.7 mg/100 mℓ). However, the copper oxidase activity (35.7 ± 4.6 mg/100 mℓ) did not differ ($p < 0.02$) from the normal values (33.4 ± 3.1 mg/100 mℓ). The copper contents of the laryngeal carcinoma tissue was also significantly higher ($p > 0.001$) than the copper contents of the contralateral normal larynx. The mean of the difference between the two tissues was +4664.6 μg/100 g of dry tissue.

The same authors[108] in 1965 studied copper, copper oxidase, magnesium, sulfur, calcium, and phosphorus in the serum and mammary tissue in 23 women with carcinoma of the breast. The mean serum copper values were 228.8 μg/100 mℓ ± 33.7%, compared to the control group of 108.1 μg/100 mℓ ± 9.7 μg%. Copper oxidase, calcium, and phosphorus were found to be within normal range. The mean copper values in mammary carcinomatous tissue were 761.2 μg/100 g of dry tissue ±78.3 μg%, as compared to the normal value of 64.9 μg/100 g of dry tissue ±11.1. The authors concluded that a high copper content in malignant breast tissue, without parallel increase of the phosphorus content, suggests that one of the first tissue disorders in carcinoma of the breast is the accumulation of copper.

Wu[209] in 1965 studied the serum iron and copper in hepatobiliary disease. This study comprised 22 cases with primary carcinoma of the liver, 25 cases with liver abscess, 10 cases with cholecystitis, 14 cases with infectious hepatitis, 4 cases with chronic hepatitis, 10 cases with liver cirrhosis, and 1 case with Banti's syndrome, making a total of 86 cases. The serum iron level was diminished in primary carcinoma of the liver, liver abscess, and cholecystitis, and was most marked in carcinoma of the liver. A remarkable increase in the serum iron level was observed in infectious hepatitis over 200 μg/100 mℓ. The serum copper level was increased in primary carcinoma, liver abscess, and cholecystitis, but most markedly in the primary carcinoma of the liver. In primary carcinoma of the liver 73% of cases showed a copper level of over 200 μg/100 mℓ, however, only 20% of cases with liver abscess and 50% of cases with cholecystitis showed elevated serum copper levels. A mild to moderate degree of an increase in the serum copper level, with a marked increase in the iron level in infectious hepatitis, was also found. In portal cirrhosis no definite change was noted in the serum copper level, but it moderately increased in the postnecrotic type of cirrhosis. The iron:copper ratio diminished in primary carcinoma and liver abscess, and was most marked in the first one. Among 41 patients of the primary carcinoma group, the ratio was less than 0.3, while in liver abscess or cholecystitis such a low value was not seen. Accordingly, where the serum iron:copper ratio showed less than 0.3, primary carcinoma of the liver was likely to be the diagnosis in 41% of cases. Moreover, when the serum copper level was exceeding 200 μg/100 mℓ the diagnosis of liver malignancy was further more likely. When the iron:copper ratio was over 1.3 and the iron level exceeded 200 μg/100 mℓ it was proper to establish the diagnosis of infectious hepatitis with certainty. According to this author, the above-mentioned criteria are helpful in the differential diagnosis among primary carcinoma of the liver, liver abscess, infectious hepatitis, and other hepatobiliary tract diseases.

Zamello[210] in 1966 reported serum copper and iron levels in 55 patients with reproductive organ malignancies. The serum copper levels in the studied group was 163 μg% as compared with 122 μg% in the control group. The serum iron level was 95 μg% for the patients and 119 μg% for the control.

Lyko[60] reported in 1967 serum ceruloplasmin levels in 83 healthy persons and 25 patients with various pathological conditions (6 cases with Wilson's disease, 13 patients with schizophrenia, 4 cases with female pelvic organ malignancies, 1 with photodermatosis, and 1 patient with chronic alcoholism). Low ceruloplasmin concentrations were observed in neonates and in adult patients suffering from Wilson's disease. Normal levels of ceruloplasmin were found in healthy adults, ceruloplasmin activity in women being somewhat higher than men. In patients with schizophrenia ceruloplasmin concentrations were found to be within normal range. High activity of enzyme was noted in the serum of pregnant women and in patients with gynecological tumors. Correlation between ceruloplasmin activity and serum copper levels was studied in 44 persons, including 28 healthy subjects and 16 patients. Under physiological conditions, the coefficient of correlation was fairly high and statistically significant. On the contrary, in the pathological conditions that were studied, the coefficient of correlation was low and only slightly or not significant.

Maracek and Hoenigova[211] in 1969 reported serum copper levels and ceruloplasmin concentrations in 80 normal subjects and 100 patients with malignant disease. The serum ceruloplasmin concentrations were measured by the serum oxidase activity and by starch electrophoresis. Serum copper was determined by the colorimetric method. The mean ceruloplasmin concentration in normal subjects was 32.8 mg% and in patients with malignant disease was 52.6 mg/100 mℓ. The mean serum copper level in control group was 103 μg% and in the patients with neoplastic diseases was 169 μg%. The statistically significant difference in serum ceruloplasmin concentrations was found between localized and the generalized disease. The mean ceruloplasmin level in localized disease was 43.9 mg% and 60.8 mg% in generalized disease.

Wiederanders and Evans[212] in 1969 reported the copper concentration of hyperplastic and cancerous prostates. The copper concentrations of 16 patients with benign prostatic hypertrophy were compared to 15 cases with a cancerous prostate gland and were found to be essentially the same. The adenomatous prostatic tissue had a little higher mean serum copper concentration (1.89 ± 0.63 μg/g of wet tissue) than cancerous prostate (1.64 ± 0.56 μg/g of wet tissue). No statistically significant difference was found between the two groups studied.

O'Leary and Feldman[213] in 1970 studied the serum copper alterations in 96 women with genital cancer (cervix, ovary, endometrium, vagina, and vulva). The control group had a mean serum copper level of 117 μg% with a range of 78 to 147 μg/100 mℓ. Women with genital cancer had levels of 162 to 219 μg/100 mℓ. After successful treatment, the serum copper level returned to near normal values: 122 to 145 μg%. Women who subsequently died from their tumors still had elevated copper levels ranging from 133 to 227 μg/100 mℓ. As the stage of the disease increases so does the serum copper level. In stage I of the disease the serum copper level was 162 μg% and in stage IV it was 219 μg%. The level of serum copper level increases with severity of disease and decreases after favorable treatment response. In general, the response to therapy was mirrored in the serum copper level. These investigators confirmed the findings of previous investigators who found variations in the serum copper level with the stage of the tumor and the form of therapy.

In summary, these authors suggested the serum copper level as a prognostic tool in the management of female genital cancer. Significant elevation of the serum copper level decreased with successful treatment of the malignant condition.

Bélanger[214] in 1971 reported the ceruloplasmin concentrations in patients with prostatic disease. His clinical material consisted of 15 cases with localized carcinoma of the prostate and 12 patients with generalized prostatic malignancy. In addition, ceruloplasmin levels were determined in 37 cases with benign prostatic hypertrophy and 10 patients with acute and chronic prostatitis. Also studied were 20 cases with nonneoplastic urinary disorders, 20 cases with different types of localized malignancies, and 15 patients with metastatic disease. The mean ceruloplasmin concentration in 23 normal subjects aged 20 to 45 years was 26

± 3.1 mg/100 mℓ (range 18 to 32 mg%). In patients with prostatic carcinoma a mean ceruloplasmin level of 45.3 ± 9.6 mg% was found, and in those on estrogen therapy ceruloplasmin concentrations were over 50 mg%. Increased ceruloplasmin levels were also found in patients with disseminated malignancies of different origin, and normal in localized disease. Control cases of benign hypertrophic prostate, as well as any other type of non-neoplastic disease of the urinary tract, have shown no appreciable increase in serum ceruloplasmin concentrations.

Tani and Kokkola[215] in 1972 studied the serum iron, copper, and iron-binding capacity in 43 untreated patients with lung carcinoma. The mean serum copper level in the whole series was approximately the same as that of the control group. However, an internal trend established in the series was a gradual increase in the serum copper level in accordance with the size of the tumor concerned. A highly significant positive correlation was found between the serum copper level and the erythrocyte sedimentation rate and a highly significant negative correlation between the serum copper, hemoglobin, and also between the serum copper and the serum iron. The authors interpretation of this was that the carcinoma had caused disorders in the iron and copper metabolism giving rise to these correlations. In the control group the serum copper did not correlate with the erythrocyte sedimentation rate, hemoglobin, serum iron or total iron-binding capacity.

In 1972, Schapira[216] observed green plasma in patients with breast and female genital organs. Clinical material consisted of six patients with breast carcinoma, six with carcinoma of the cervix, three cases with cancer of the vulva and ten with other forms of malignant diseases. In addition, and for comparison purposes, serum copper levels were determined in six patients with normal pregnancies, six cases taking oral contraceptives, six patients who discontinued taking oral contraceptives, and six ''control'' patients on no contraceptive medication. An intense green color of plasma samples was noted in patients in the third trimester of pregnancy and in oral contraceptive users. The moderate intensity of green plasma sample color was seen in patients with malignant diseases and normal appearance of plasma, in patients who discontinued taking oral contraceptives, and the control group.

This author concluded that the detection of ceruloplasmin in various stages of carcinoma of the breast and female genital organs, before and after the operation as well as during the follow-up period, could supply valuable information and also be useful in assessing the prognosis.

Hughes[217] in 1972 reported serum transferrin and ceruloplasmin concentrations in 1032 patients with various types of malignant diseases. Ceruloplasmin concentrations were significantly increased in most groups of patients with malignant diseases studied (91 ± 29 mg in males and 96 ± 30 mg/100 mℓ in females) compared to the mean ceruloplasmin concentration in the control group (71 ± 17 mg in males and 84 ± 22 mg/100 mℓ in females). There was no correlation observed between ceruloplasmin and transferrin concentrations in 134 patients with different types of cancer studied.

Dines et al.[218] in 1974 studied the zinc, copper, and iron content in 300 pleural effusions involving 114 patients with benign disease and 186 cases with malignant disease. The pleural fluid levels of these heavy metals were not helpful in differentiating benign from malignant effusions. There was, however, positive regression of the serum copper on pleural fluid copper, with the high slope being associated with the lymphoma group and the next highest slope being associated with the malignant disease group. In the benign disease group, pleural fluid copper was positively and significantly correlated with pleural fluid protein.

Kolaric et al.[219] in 1975 studied serum copper levels in 125 patients with solid tumors. Clinical material consisted of 34 patients with lung carcinoma, 35 cases with carcinoma of the stomach, 31 patients with breast carcinoma, and 25 melanoma cases. Two thirds of the examined patients had regional or distant metastases. Blood samples were taken before therapy and every 3 weeks during treatment or follow-up. The number of analyzed blood samples per patient ranged from 4 to 10 with a total of 750 determinations. Of the 34 patients with untreated lung carcinoma, 28 (82%) had a definitely increased serum copper level. Of the 28 patients with increased serum copper levels, 15 had regional or distant metastasis, while 13 had localized disease. In six of seven patients who underwent irradiation with complete regression of disease, pretreatment serum copper levels fell to normal range. Of the 35 patients with untreated carcinoma of the stomach, only 14 (40%) showed increased serum copper levels. In five patients who had operable tumors and had undergone complete resection, no significance in the serum copper levels was noted before or after surgery. In 31 patients with untreated carcinoma of the breast, increased serum copper levels were noted in only 8 patients (26%). In 25 patients with metastatic melanoma, the serum copper levels were increased in 12 (48% of cases).

In summary, these authors' data showed that the serum copper was extremely high in 82% of the patients with bronchial carcinoma, while in the other examined groups no significant changes were observed. According to these investigations the authors concluded that serum copper could be a diagnostic factor in patients with bronchial carcinoma.

Monari et al.[220] in 1976 reported the ceruloplasmin concentration in the serum and in cerebrospinal fluid in normal patients, patients with neurological diseases, and patients with neoplasms. The values found in normal patients agreed with those previously reported. There was a good correlation between the cerebrospinal fluid ceruloplasmin and proteinorachia. In intracranial tumors, there was a significant increase in the serum ceruloplasmin concentrations. This may indicate an existence of cerebral metastasis. There was a largely diminished serum ceruloplasmin in Wilson's disease. In the acute stage of pontine hemorrhage, both serum and cerebrospinal fluid ceruloplasmin concentrations were slightly elevated. Carcinomas without involvement of the CNS had increased ceruloplasmin levels.

Fisher et al.[221] in 1976 reported serum copper and zinc levels in 19 patients with sarcomas, 12 of which were osteogenic sarcomas at various stages of disease. Their data indicated that the activity and extent of disease were associated with the degree of elevation of the serum copper levels. The highest serum copper levels were found in patients with the most active and extensive malignancy and the poorest prognosis. Patients with primary or metastatic osteosarcoma had increased serum copper levels, whereas amputated osteosarcoma patients clinically considered to be tumor-free had nearly normal serum copper concentrations. On the contrary, serum zinc levels were lower in patients with metastases than in those with primary tumors only. According to these authors, it appears that the determination of the serum copper level and the serum zinc level and the ratio of serum copper and serum zinc levels may be of value in the prognosis and evaluation of therapy and may be useful in discriminating between patients with primary and metastatic osteosarcoma.

In the same year, Shifrine and Fisher[222] investigated and reported serum ceruloplasmin concentrations in 22 normal individuals, 7 women receiving chemical contraceptives, and 7 patients with osteogenic sarcoma. They found the ratios of the serum copper level:ceruloplasmin concentration to be constant in all of the groups studied indicating that the ratio of bound to unbound copper in serum is maintained in normal persons, oral contraceptive users, and in patients with osteosarcoma. The serum copper levels and ceruloplasmin concentrations were found to be markedly elevated ($p > 0.001$) both in women taking oral contraceptives and in osteosarcoma patients. However, the serum copper level:ceruloplasmin ratio in each group was not different from normal. The authors concluded that the determination of serum ceruloplasmin in osteosarcoma patients may be of value in the differential diagnosis of the disease and in the evaluation of the efficacy of therapy.

Scanni et al.[223] in 1977 studied serum copper and ceruloplasmin levels in untreated patients with carcinoma of the stomach, colon, and lung. Clinical material consisted of 20 patients with carcinoma of the lung, 33 cases with stomach carcinoma, and 22 with colonic carcinoma. The mean serum copper level in lung carcinoma cases was 188.2 ± 14.78 μg/100 mℓ,

171.94 ± 7.27 in stomach carcinoma, and 164.77 ± 13.40 in colonic carcinoma, compared to normal subjects of 143.03 ± 3.25 μg/100 mℓ. A significant difference ($p > 0.01$) was noted in the mean serum copper level of cancer patients as compared to the control group. An elevated serum copper level was found in 70, 67, and 59% of the patients with lung, stomach, and colon carcinoma, respectively. Evaluation of the serum ceruloplasmin levels showed a highly significant mean increase in patients with carcinoma of the stomach and lung, i.e., 78 and 73%, respectively. On the other hand, this increase was not significant in the carcinomas affecting the large intestine (57% of cases).

The correlation between serum copper and serum ceruloplasmin levels was highly significant in patients with gastric carcinoma, but it was not significant in patients with lung or colon malignancies. In 58 of 75 cases studied there was a correlation between the copper and ceruloplasmin levels in the same subjects. No statistical significant difference was found between the serum copper level and ceruloplasmin concentrations in localized disease (36 cases) and metastatic malignancies (22 patients). In conclusion, the authors stated that their investigations confirmed the importance of the serum copper and ceruloplasmin level determination as a diagnostic aid in certain types of malignancies.

Gray et al.[224] in 1977 published the monitoring of serum factors in patients with cancer. They investigated the total protein, total protein hexose, total protein hexosamine, ceruloplasmin, alpha$_1$-antitrypsin, and haptoglobin in 64 patients with various forms of malignant diseases in various stages of the disease. The authors concluded that the preliminary results indicated that the nonspecific serum factors that were studied may be of considerable value for the monitoring of cancer patients and may also possibly serve as a diagnostic screen in high-risk groups.

Pocklington and Foster[225] in 1977 studied the ceruloplasmin and iron transferrin in the blood of 213 healthy volunteers and patients with different types of solid tumors receiving various forms of treatment. The ceruloplasmin concentration showed no variation with age in either sex between 18 and 55 years, and then increased slightly in females. In addition to malignant disease, various factors such as surgery, terminal phase of malignant disease, and severe infections were shown to increase serum ceruloplasmin concentrations; 19 patients were treated surgically and showed an increase in ceruloplasmin levels after surgery, returning to pretreatment levels in about 7 days. When allowance for these factors are made, the remaining small difference in copper and iron between patients with either squamous cell carcinoma or breast carcinoma and controls appears to have no significance.

Inutsuka and Araki[226] in 1978 reported plasma copper and zinc levels in patients with malignant tumors of the digestive organs. Their clinical material consisted of 29 patients with carcinoma of the stomach, 4 cases with carcinoma of the colon, 7 cases of carcinoma of the other digestive organs, 7 patients with carcinoma of the lung, and 7 cases with carcinoma of other origins. In addition, plasma copper and zinc levels were determined in 37 cases with benign lesions and 35 subjects without disease. In the 35 cases of primary malignant tumors of the digestive organs, plasma copper level was increased (124 + 34 μg/100 mℓ) and the plasma zinc level was significantly decreased (83 ± 18 μg/100 mℓ) in comparison with the control group ($p < 0.001$). The copper:zinc ratio of 1.49 ± 0.50 was also significantly higher than in the normal group ($p > 0.001$). The changes were quite similar in all of the cancer cases (49 primary carcinomas) with the plasma copper level being 128 ± 37 μg/100 mℓ, the plasma zinc level 82 ± 17, and the copper:zinc ratio 1.57 ± 0.56. These data were all significantly different ($p < 0.001$) compared to those of the normal group. The copper:zinc ratio was even higher in cases with advanced carcinoma, especially when liver metastasis was present. Out of 49 cancer cases, 45 had a copper:zinc ratio higher than 0.98 (91.8%), and of the 35 normal subjects 31 cases (88.5%) had a lower than 0.98 ratio. On the other hand, plasma copper levels higher than some of the mean and standard deviations in the group of normal subjects were encountered in 42 of the 49 cases with malignancy (85.7%), and the plasma zinc level lower than the difference between the mean and standard deviation in the normal group was found in 30 of the cancer patients (61.2%). Therefore, the ratio discriminated the cancer patients from the normal better than the plasma copper level or plasma zinc level alone. The copper:zinc ratio appeared to reflect the stage of gastric carcinoma disease, being significantly higher in most advanced disease (1.75 ± 0.15) than in localized disease (1.18 ± 0.27). In a limited number of patients in whom plasma copper levels and plasma zinc levels were followed in the postoperative course to 5 months, increased plasma copper levels to plasma zinc levels were noted preceding clinically detectable recurrent disease. Finally, the authors stated that the prognosis of the cancer patients was analyzed by the copper:zinc ratio. The death rate within 6 months after examination was 14.8% (4 out of 27 cases) in the group in which the ratio was lower than 1.50, while it was 34.6% (9 cases of 26) in the group in which the ratio was higher than 1.50. The authors concluded that the copper:zinc ratio is a useful index in estimating the extent and prognosis of malignant tumors in the digestive organs.

In 1978, Linder and Moor[49] reported ceruloplasmin oxidase activity, antigen, and total copper concentrations in sera from 20 patients with lung cancer and a comparable number of normal individuals, heavy smokers, and patients with nonmalignant lung diseases — all males. They found that ceruloplasmin and serum copper were significantly higher in the patients with carcinoma of the lung. The group of patients with a variety of nonmalignant lung diseases also showed a significant elevation in all three parameters, but the elevation was not as great as for the cancer group, mean values being significantly lower for oxidase activity and copper. Among the cancer patients there was a relation between ceruloplasmin and disease prognosis, lower oxidase activity being associated with a good prognosis and vice versa. The authors believe that their findings imply that ceruloplasmin can distinguish most patients with lung disease (malignant and nonmalignant) from normal, healthy individuals and that very high values for oxidase activity (or oxidase-to-antigen ratios) are indicative of malignant disease, at least for the lung. Their clinical observations were supported with excellent experimental animal data which will be discussed later in connection with our animal-related studies.

Breiter et al.[227] in 1978 published their studies of serum copper and zinc levels in 18 cases with osteogenic sarcoma. There were 13 males and 5 female patients in the age range from 13 to 46 years. The serum copper level was significantly elevated (173 ± 30 μg/100 mℓ) as compared with sex- and age-matched normal controls (115 ± 16 μg/100 mℓ) at $p < 0.0001$. The serum zinc level was not significantly different in patients and controls. The elevated serum copper level:serum zinc level ratio observed in the patient group primarily reflected the elevated serum copper level found in these patients. Serum copper level and the serum copper level:serum zinc level ratio did not change significantly following curative surgery or become more abnormal in recurrent disease. The serum copper level and serum copper level:serum zinc level ratio were noted to be markedly elevated in those patients receiving BCG therapy. The authors concluded that the serum copper level, serum zinc level, and the serum copper level:serum zinc level ratio did not appear useful as markers of tumor activity in either the evaluation of therapeutic response or prediction of tumor progression in patients with osteogenic sarcoma.

Israel and Edelstein[228] reported in 1978 their studies on nonspecific blocking factors of host origin in 232 patients with various solid tumors. They observed high levels of some glycoproteins rich in sialic acids (orosomucoid, haptoglobin, alpha$_1$-antitrypsin and ceruloplasmin) in patients with solid tumors at all sites studied. This phenomenon appears to be a general characteristic of cancer, rather than a specific feature of any particular solid tumor site or type of malignancy. Significantly elevated levels were found in postoperative, supposedly "disease free" patients (who nonetheless probably had "minimal residual disease") and in patients with metastatic disease; the levels were even higher in the latter group. Their findings suggested that protein alterations occur early and increase as the disease progresses, as a response to the presence of tumor which triggers hepatic synthesis of sialoglycoproteins. These glycoproteins in turn "coat" the binding sites of both immunocompetent and tumor cells and, thereby, abrogate recognition and killing of the latter by immune system. This concept of nonspecific blocking factors of host origin has been substantiated to some extent by observations on the consequences of plasma exchange in 24 patients with metastatic tumors; 8 of these patients exhibited an objective tumor regression. The authors suggested that these studies should be extended to postoperative patients and that circulating sialoglycoprotein assays could be one of the ways of monitoring tumor growth, including growth during the nonvisible phase.

Scanni et al.[229] in 1979 published variations in the serum copper and ceruloplasmin levels in advanced gastrointestinal (GI) cancer treated with polychemotherapy. Serum copper and ceruloplasmin levels in 57 patients with advanced cancer of the stomach (35 cases) or large intestine (22 cases) treated with polychemotherapy were studied. In gastrointestinal cancer, serum copper levels, which were already high in untreated patients, had a tendency to increase further in cases of progression of the disease, while they seem to significantly decrease in cases of remission. Serum copper level during the trial with chemotherapy appeared to be correlated to the clinical evolution of the disease, particularly in the cases of stomach cancer. The patient with gastric carcinoma showed a highly significant ($p < 0.01$) decrease (from 195.30 ± 9.20 to 146.20 ± 10.66 μg/100 mℓ) in the mean values of the serum copper level in cases of regression of disease and the highly significant ($p < 0.01$) increase (from 167 ± 7.85 to 205.79 ± 6.70 μg/100 mℓ) in the mean values in cases of treatment failure and disease progression. No significant difference in the serum copper levels was found in patients with stable disease. With regard to colon carcinoma similar results were obtained. The average serum copper level also decreased from 197.57 ± 21.45 to 138.29 ± 7.22 μg/100 mℓ in case of regression and increased from 170.50 ± 10.31 to 209.70 ± 13.65 μg/100 mℓ in case of disease progression ($p < 0.05$) while they did not vary significantly in cases of stabilization (170.00 ± 15.35 vs. 179.78 ± 9.22 μg/100 mℓ). On the other hand, colonic carcinoma patients with normal serum copper levels in the pretreatment period did not show significant variations after therapy regardless of the clinical outcome of the case. The authors also investigated the serum copper changes in ten patients with GI malignancies before the beginning of therapy and after 30, 60, and 120 min, and after 24 and 72 hr from the first administration of the drugs and no significant serum copper alteration in the short-term was found. The authors are inclined to attach a diagnostic and prognostic value to the variation of serum copper levels in GI cancer.

Martin Mateo et al.[230] in 1979 reported serum copper, ceruloplasmin, lactic dehydrogenase, and alpha$_2$-globulin in 26 patients with primary lung cancer. A significant increase in the level of copper, ceruloplasmin, lactic dehydrogenase and alpha$_2$-globulins were found. The role of copper in pulmonary cancerogenesis was discussed.

Abdulla et al.[231] in 1979 investigated zinc and copper levels in whole blood and plasma in 13 patients with squamous cell carcinoma of the head and neck. Zinc in the plasma and whole blood of these patients was significantly decreased and the copper:zinc ratio in the plasma was significantly higher than in healthy controls. The plasma zinc was significantly lower in patients who did not respond to therapy and who died within 12 months than in those who responded to therapy and had a remission within 12 to 15 months. In patients who responded to therapy and alive after 12 to 15 months, zinc in the plasma and whole blood and the copper:zinc ratio became normal. The zinc in whole blood and in plasma was significantly decreased ($p < 0.001$) whereas the copper:zinc ratio in plasma was high ($p < 0.001$) prior to treatment. Most of the patients were treated with irradiation. According to these authors, their data suggested a potential screening and predicting value of zinc in the plasma and whole blood and the copper:zinc ratio in prognosis and treatment response in patients with squamous cell carcinoma.

In 1979 Andrews[232] published studies of plasma zinc, copper, ceruloplasmin, and growth hormone with special reference to carcinoma of the bronchus. The plasma zinc was higher and the plasma copper was lower in people without malignancy below the age of 30 years than they were in other age groups. It was found that about 66% of patients with carcinoma of the bronchus had plasma zinc levels of less than 115 μg/100 mℓ, but low levels were also found in 23% of other cases of malignancy and in 9% of the other patients. In carcinoma of the bronchus the low plasma zinc was found to be associated with epidermoid and anaplastic tumors and was, to some extent, related to the duration of the disease. In carcinoma of the bronchus, plasma copper was found to be higher than in all other groups and values higher than 265 μg/100 mℓ were considered to support the diagnosis of carcinoma of the bronchus. There was, however, no relationship between the increase in the plasma copper and the decrease in the plasma zinc. Raised ceruloplasmin levels above 420 mg/ℓ were found in 65% of cases of carcinoma of the bronchus and these high levels were usually associated with raised plasma copper. Surgical operations lowered plasma zinc and raised growth hormone but did not affect plasma copper. In conclusion, these authors stated that based on their material studied a raised plasma copper and raised plasma ceruloplasmin levels were useful supportive findings.

Flynn[233] in 1979 reported his observations of copper metabolism in patients with solid tumors. In 279 adults with untreated solid tumors, the mean serum copper levels were significantly higher than in 30 adults without cancer (171.5 vs. 109.3 μg/100 mℓ). Serial determinations of serum copper were obtained in 12 patients, each with breast, colorectal, and lung (nonoat cell) cancer before treatment (mean levels — 135.7, 154.2, and 167.4 μg/100 mℓ, respectively) and 1 week to 6 months after the beginning of cytotoxic chemotherapy. Each patient's response to treatment was assessed after 6 months without knowledge of the serum copper levels — 6 patients in each subgroup showed objective responses to chemotherapy and the others showed no response. The responders in each subgroup showed significantly lower serum copper levels before treatment and at 6 months after the beginning of chemotherapy. The difference between the responders and nonresponders was also significant at 3 months in patients with colorectal cancer. The greatest difference between the responders and nonresponding patients (significant after 1, 3, and 6 months) was observed in the patients with lung cancer. The normalization of serum copper levels in the responding group was a good group indicator of the efficacy of chemotherapy, but serum copper levels were not a good individual predictor of responders or nonresponders. All groups demonstrated marked differences between the responders and nonresponders prior to treatment. Normalization of the serum copper level with therapy related directly to the patient's response.

In 1980, Bhardwaj[234] et al. reported the clinical evaluation of malignant process by serum copper determinations. These authors studied 18 patients with various types of carcinomas, 2 cases with sarcomas, and 5 malignant lymphoma patients. The mean serum copper value in patients with malignant diseases was 209.6 μg/100 mℓ, as compared to a normal control of 116.4 ± 22 μg/100 mℓ (25 control subjects). Higher serum copper levels were found in generalized disease than in early stages. Subsequent to treatment there was a distinct fall in the serum copper level to 167.8 from 209.6 μg/100 mℓ before treatment. The authors concluded that based on their investigations the serum copper level is a useful test in evaluating the activity and extent of malignant disease. It could also judge the prognosis of the case and the response to the efficacy of the treatment applied.

Huhti et al.[235] in 1980 investigated the serum copper levels in patients with lung carcinoma. An increased mean serum copper level was found in 149 patients with carcinoma of the lung when compared with 19 healthy people and 24 patients with nonmalignant lung diseases. The levels seemed to reflect the stage of disease, with asymptomatic patients showing the lowest values and patients with metastatic disease the highest. In spite of the significant differences between the groups of subjects the scatter in the values was large. Hence, serum copper determinations can be of only limited importance for the differential diagnosis or in assessing the clinical stage of cancer. No differences in the copper levels were found between the groups of patients with different histological types of lung carcinoma.

Garofalo et al.[236] in 1980 reported serum zinc, copper, and the copper:zinc ratio in 80 patients with benign and malignant breast lesions — 43 of whom had infiltrating intraductal carcinoma. Of these, 17 had axillary node involvement. There were 37 patients with benign breast lesions. The range of distribution for the serum zinc level and serum copper levels were similar for all groups studied. Neither the serum zinc level, serum copper level, nor the copper:zinc ratio were of value in discriminating between control and patients with benign or malignant breast lesions. In the discussion, the authors stated that their preliminary data in patients with Hodgkin's disease confirmed the usefulness of the serum copper level in evaluating the extent of disease in addition to the disease activity and prognosis.

Jafa et al.[237] in 1980 studied trace elements in prostatic tissue and plasma in prostatic diseases of men. Serum copper was significantly elevated in malignant prostatic tissue, marginally elevated in chronic prostatitis, and reduced in benign prostatic tissue. The plasma copper concentration was significantly elevated in malignant prostatic tissue and was reduced in benign hyperplasia, fibrous prostate, and marginally elevated in chronic prostatitis. The plasma iron level was marginally increased in benign hyperplasia and chronic prostatitis and somewhat increased in malignancy and fibrous prostate.

In 1980, Habib et al.[238] published the zinc and copper content of blood leukocytes in plasma from patients with benign and malignant prostates. The zinc and copper levels in plasma had been measured in 41 patients with benign prostatic hypertrophy and 44 patients with carcinoma of the prostate, 24 of whom were receiving some form of hormonal therapy. The plasma zinc levels were not affected by age or disease whereas the plasma copper levels were significantly higher ($p < 0.01$) in the benign and malignant categories (mean = 124 μg/100 mℓ) when compared to the younger normal population (mean = 84 μg/100 mℓ). Hormonal therapy induced a rapid raise in the plasma copper concentration and a concomitant marginal fall in the zinc levels of the carcinoma patients.

Garofalo et al.[239] reported in 1980 serum zinc, copper, and the copper:zinc ratio in 50 patients with epidermoid cancers of the head and neck, with respect to site, stage of disease, and treatment response. They found no significant difference in these parameters compared to healthy normal controls. These authors could not appreciate any diagnostic or prognostic values in these parameters in patients with head and neck cancer. There was, however, a sex-related difference for the serum zinc level in men with oral cavity and women with laryngeal lesions, and a trend toward decreasing serum zinc levels and an increase in serum copper levels and the serum copper:zinc ratio with advanced stage of disease.

In 1981, Linder et al.[240] reported ceruloplasmin assays in the diagnosis and treatment of human lung, breast, and GI cancer. Ceruloplasmin was determined as an enzyme activity, as an antigen, and as total copper in serum samples from 150 male patients with carcinoma of the lung and a comparable numbers of male controls. By all three assays, the ceruloplasmin was significantly increased above normal before treatment and the degree of evaluation was related to the TNM stage [i.e., the International Union Against Cancer classification system based on the extent of primary tumor (T), condition of lymph nodes (N), and absence of presence of metastatis (M)]. Surgery had no immediate effects, but in patients with no evidence of disease for longer periods the ceruloplasmin returned to nearly normal values. High levels of ceruloplasmin was elevated in six of nine patients before tumor recurrence; two of three smokers in the first panel of sera with elevated ceruloplasmin levels subsequently developed lung cancer. The authors suggested that with regard to diagnosis, it may be fruitful to include periodic checks on serum ceruloplasmin oxidase activity determinations among the tests used to monitor the health of smokers and perhaps other groups at risk of developing primary or recurrent lung malignancy in male subjects. With regard to therapy, ceruloplasmin assay may add in the determination of the state of remission of the disease and/or of the prognosis. In GI cancer increased ceruloplasmin assays were also noted, particularly in colon cancer even though almost all the patients had grade I and II adenocarcinoma. In female patients, nonmalignant breast disease had substantial effects on the ceruloplasmin levels, therefore, no significant difference was found between malignant and nonmalignant breast conditions. The authors concluded that changes in the ceruloplasmin would appear to be merely the top of the iceberg of large events in copper metabolism that underlie the neoplastic disease process and tumor growth, and only in limited situations will the assays of ceruloplasmin add in the diagnosis, prognosis, and long-term monitoring of cancer patients.

In 1981, Fisher et al.[241] reported their observations of serum copper and zinc levels in malignant melanoma patients in various stages of disease. Patients with histories of malignant melanoma clinically free of disease at the time of the serum copper level determination had normal serum copper, serum zinc, and copper:zinc ratios. On the contrary, metastatic melanoma disease was generally associated with elevated serum copper levels. In addition to disease activity, serum copper level appeared to reflect the extent of the disease, being higher in more advanced clinical stages. Patients receiving BCG immunotherapy had higher serum copper levels than untreated or those involved in Levamisol® or placebo treatment groups. Zinc levels and serum copper:serum zinc ratios did not reflect tumor activity. A persistent elevation of the serum copper level in melanoma patients seems to be consistent with a poor prognosis. All patients whose serum copper level never decreased below 150 μg/100 mℓ died during the course of the study. Since Breiter et al.[227] found no serum zinc level alterations in the relationship between the extent or activity of osteosarcoma disease and the serum copper level and the serum zinc level or serum copper level:zinc level ratio, Fisher et al.[241] believe that this discrepancy is probably related to patient management. Most of the patients studied by Fischer et al.[241] were either untreated or treated with transfer factor, while those reported by Breiter et al.[227] received Methotrexate® with Leukovorin® rescue with or without BCG immunotherapy.

Dionigi et al.[242] reported in 1981 serum ceruloplasmin levels in surgical cancer patients with different types of malignant disease and their relationship to tumor stage and postoperative complications. Serum ceruloplasmin has been evaluated at admission in 109 surgical cancer patients (48 males and 61 females with a mean age of 57 years) divided into five groups: breast, melanoma, lymphoma, esophagus-stomach, and sigma-rectum; 33 controls (27 males, 6 females, mean age 41 years) were normal subjects or patients with minor surgical benign lesions. Tumor stage (I to IV) was evaluated following the International Union Against Cancer classification. Patients were defined septic if they presented after surgery a diagnosed focus of infection or if they had an undiagnosed fever for at least 3 days. Ceruloplasmin level was evaluated by single radial immunodiffusion technique. Cancer patients presented significant higher values of ceruloplasmin than controls. Minor differences were observed in melanoma, whereas the ceruloplasmin level was more significantly increased in breast, GI cancer, and lymphoma. Serum ceruloplasmin did not correlate with sex and age within the five groups considered. On the contrary, positive correlation was found with tumor stage. In fact the lower levels of ceruloplasmin were observed in patients at stage I, intermediate values with stages II and III, and higher levels with stage IV. The authors concluded that their investigations showed a positive relationship between ceruloplasmin levels and tumor progression suggesting that variation of ceruloplasmin may have an important clinical value for the evaluation of the extent of different malignant diseases.

In 1981 Perlin[243] reported on the recent advances toward the control of lung cancer. Improved clinical and pathological classification in staging of the disease have been made with the TNM system. The authors stated that biomarkers including carcinoembryonic antigen, serum copper levels, and thromboplastin time are useful in monitoring the progress of the disease.

Finally, Schapira and Schapira[244] reported in 1982 the ceruloplasmin levels in monitoring response and relapse in advanced malignant disease and relapse on adjuvant therapy. Ceruloplasmin elevation was found in 52 of 60 patients with carcinoma of the breast (87%), in 23 of 29 (79%) patients with carcinoma of the lung, in 24 of 26 (92%) of lymphomas, in 15 of 20 (75%) with carcinoma of the colon, 9 of 13 (69%) of patients with gastrointestinal cancer, and 12 of 19 (63%) with ovarian cancer, 4 of 4 with cervical cancer, and 6 of 6 prostatic cancer cases not on estrogens. In patients with measurable advanced disease, 19 responses to treatment were observed and the ceruloplasmin level fell by at least 30% in all patients, and over 40% in 10 of 19 patients within 2 months. There were 16 relapses noted and the ceruloplasmin level rose by at least 25% within 2 months in all patients and rose by over 40% in 10 of 16 patients in that time. Of 33 patients with carcinoma of the breast on adjuvant chemotherapy 14 patients had initially elevated ceruloplasmin levels, which normalized in 6 patients. Five patients had developed elevated ceruloplasmin levels while on or after adjuvant chemotherapy. One has developed a bone metastasis 6 months after the ceruloplasmin level became elevated and another patient developed local recurrence. Three patients followed for stage I carcinoma of the breast developed elevated levels. One was found to have metastasis in a supraclavicular node, another a new primary in the remaining breast 6 months after elevation of the ceruloplasmin level, and the third bone metastasis 5 months after the ceruloplasmin level was elevated. According to these authors, the ceruloplasmin levels appear to be useful in monitoring patients with advanced disease on treatment whose disease is not readily measurable and for following patients or adjuvant therapy, especially if the ceruloplasmin level was elevated at the start of treatment.

This review of the literature concerning serum copper and/or ceruloplasmin alterations in patients with malignant diseases, although exhaustive, is still incomplete. Our apologies are made to those authors[245-273] whose work, while equally informative and instructive, has not been included because of space limitation.

REFERENCES

1. **Underwood, E. J.,** *Trace Elements in Human and Animal Nutrition,* 4th ed., Academic Press, New York, 1977, 56—108.
2. **Scheinberg, I. H., Cook, C. D., and Murphy, J. A.,** The concentration of copper and ceruloplasmin in maternal and infant plasma at delivery, *J. Clin. Invest.,* 33, 963, 1954.
3. **Adelstein, S. J. and Vallee, B. L.,** Copper metabolism in man, *N. Engl. J. Med.,* 265, 892, 1961; 265, 941, 1961.
4. **Schenker, J. G., Jungreis, E., and Polishuk, W. Z.,** Serum copper levels in normal and pathologic pregnancies, *Am. J. Obstet. Gynecol.,* 105, 933, 1969.
5. **Schenker, J. G., Jungreis, E., and Polishuk, W. Z.,** Oral contraceptives and serum copper concentration, *Obstet. Gynecol.,* 37, 233, 1971.
6. **Carruthers, M. E., Hobbs, C. B., and Warren, R. L.,** Raised serum copper and ceruloplasmin levels in subjects taking oral contraceptives, *J. Clin. Pathol.,* 19, 498, 1966.
7. **Chuttani, H. K., Gupta, P. S., Gulati, S., and Gupta, D. W.,** Acute copper sulfate poisoning, *Am. J. Med.,* 39, 849, 1965.
8. **Davidoff, G. N., Votaw, M. L., Coon, W. W., Hultquist, D. E., Filter, B. J., and Wexler, S. A.,** Elevations in serum copper, erythrocytic copper and ceruloplasmin concentrations in smokers, *Am. J. Clin. Pathol.,* 70, 790, 1978.
9. **Bogden, J. D. and Troiano, R. A.,** Plasma calcium, copper, magnesium and zinc concentrations in patients with the alcohol withdrawal syndrome, *Clin. Chem.,* 24, 1553, 1978.
10. **Markowitz, H., Gubler, C. J., Mahoney, J. P., Cartwright, G. E., and Wintrobe, M. M.,** Studies on copper metabolism. XIV. Copper, ceruloplasmin, and oxidase activity in serum of normal human subjects, pregnant women and patients with infection, hepatolenticular degeneration and nephrotic syndrome, *J. Clin. Invest.,* 34, 1498, 1955.

11. **Adelstein, S. J., Coombs, T. L., and Vallee, B. L.,** Metallo-enzymes and myocardial infarction. I. The relation between serum copper and ceruloplasmin and its catalytic activity, *N. Engl. J. Med.,* 225, 105, 1956.
12. **Varsieck, J., Barbier, F., and Speecke, A.,** Influence of myocardial infarction on serum manganese, copper and zinc concentrations, *Clin. Chem.,* 21, 578, 1975.
13. **Rangam, C. M. and Bhagvat, A. G.,** Serum iron and copper levels in jaundice, *Indian J. Med. Sci.,* 15, 499, 1962.
14. **Sinha, S. and Gabrieli, E.,** Serum copper and zinc levels in various pathologic conditions, *Am. J. Clin. Pathol.,* 54, 570, 1970.
15. **Sass-Kortsak, A.,** Copper metabolism, *Adv. Clin. Chem.,* 8, 1—67, 1965.
16. **Evans, G. W.,** Copper homeostasis in mammalian system, *Phys. Rev.,* 53, 535, 1973.
17. **Tessmer, C. F., Krohn, W., Johnston, D., Thomas, F. B., Hrgovcic, M., and Brown, B.,** Serum copper in children (6—12 years old): an age-correction factor, *Am. J. Clin. Pathol.,* 60, 870, 1973.
18. **Williams, D. M.,** Racial differences of hemoglobin concentrations: measurement of iron, copper and zinc, *Am. J. Clin. Nutr.,* 34, 1694, 1981.
19. **Bogden, J. D., Lintz, D. I., Joselow, M. M., Charles, J., and Salaki, J. S.,** Effect of pulmonary tuberculosis on blood concentrations of copper and zinc, *Am. J. Clin. Pathol.,* 67, 251, 1977.
20. **Atkinson, R. L., Dahms, W. T., Bray, G. A., Jacob, R., and Sanstead, H. H.,** Plasma zinc and copper in obesity and after intestinal bypass, *Ann. Int. Med.,* 89, 491, 1978.
21. **Beisel, W. R.,** Trace elements in infectious processes, *Med. Clin. North Am.,* 60, 831, 1976.
22. **Cartwright, G. E., Markowitz, H., Shields, G. S., and Wintrobe, M. M.,** Studies on copper metabolism. XXIX. A critical analysis of serum copper in normal subjects, patients with Wilson's disease and relatives of patients with Wilson's disease, *Am. J. Med.,* 28, 555, 1960.
23. **Menkes, J. H., Alter, M., Steigleder, G. K., Weakley, D. R., and Sung, J. H.,** A sex-linked recessive disorder with retardation of growth, peculiar hair and focal cerebral and cerebellar degeneration, *Pediatrics,* 29, 764, 1962.
24. **Danks, D. M., Cartwright, E., and Stevens, B.,** Menkes' steely hair (kinky hair) disease, *Lancet,* 1, 891, 1973.
25. **Porter, K. L., McMaster, D., Elmes, M. E., and Love, A. H. G.,** Anaemia and low serum-copper during zinc therapy, *Lancet,* 2, 774, 1977.
26. **Cordano, A. and Graham, G. G.,** Copper deficiency complicating severe chronic intestinal malabsorption, *Pediatrics,* 38, 596, 1966.
27. **Pagliardi, E. and Giangrandi, E.,** Clinical significance of the blood copper in Hodgkin's disease, *Acta Haematol.,* 24, 201, 1960.
28. **Hrgovcic, M., Tessmer, C. F., Brown, B. W., Wilbur, J. D., Mumford, D. M., Thomas, F. B., Shullenberger, C. C., and Taylor, G.,** Serum copper studies in the lymphomas and acute leukemias, in *Progress in Clinical Cancer,* Ariel, I. M., Ed., Grune & Stratton, New York, 1973, 121—153.
29. **Hrgovcic, M., Tessmer, C. F., Thomas, F. B., Fuller, L. M., Gamble, J. E., and Shullenberger, C. C.,** Significance of serum copper levels in adult patients with Hodgkin's disease, *Cancer,* 31, 1337, 1973.
30. **Wintrobe, M. M.,** *Clinical Hematology,* 7th ed., Lea & Febiger, Philadelphia, 1976, 150—152.
31. **Shah-Reddy, I., Khilanani, P., and Bishop, R. C.,** Serum copper levels in non-Hodgkin's lymphoma, *Cancer,* 45, 215, 1980.
32. **Bariety, M. and Gajdos, A.,** Étude de la siderémie et de la cuprémie au cours des concers, particulièrement des cancers bronchiques, *Presse Méd.,* 12, 3259, 1964.
33. **Kolaric, K., Roguljic, A., and Fuss, V.,** Serum copper levels in patients with solid tumors, *Tumori,* 61, 173, 1975.
34. **Graham, G. G. and Cordano, E.,** Copper deficiency in human subjects, in *Trace Element in Health and Disease,* Vol. 1, Zinc and Copper, Prasad, A. S., Ed., Academic Press, New York, 1976, chap. 22.
35. **Naveh, Y., Hazani, A., and Berant, M.,** Copper deficiency with cow's milk diet, *Pediatrics,* 68, 397, 1981.
36. **Henkin, R. I.,** Metal-albumin-amino acid interactions: Chemical and physiological interrelationships, in *Protein-Metal Interactions,* Friedman, M., Ed., Plenum Press, New York, 1974, 299—328.
37. **Neifakh, S. A., Monakhov, N. K., Shaposhnikov, A. M., and Zubzhitski, Yu N.,** Localization of ceruloplasmin synthesis in human and monkey liver cells and its copper regulation, *Experientia,* 25, 337, 1969.
38. **Evans, G. W., Majors, P. F., and Cornatzer, W. E.,** Induction of ceruloplasmin synthesis by copper, *Biochem. Biophys. Res. Commun.,* 41, 1120, 1970.
39. **Pineda, E. P., Ravin, A. H., and Rutenburg, A. M.,** Serum ceruloplasmin: observations in patients with cancer, obstructive jaundice and other diseases, *Gastroenterology,* 43, 206, 1962.
40. **Henkin, R. I.,** Copper-zinc hormone relationship, in *Zinc and Copper in Medicine,* Karcioglu, Z. A. and Sarper, R. F., Eds., Charles C Thomas, Springfield, Ill., 1980, chap. 5.

41. **Eddington, C. L., Upchurch, H. F., and Kampschmidt, R. F.,** Effect of extracts from rabbit leukocytes on levels of acute phase globulins in rat serum, *Proc. Soc. Exp. Biol. Med.,* 136, 159, 1971.
42. **Henson, P. M.,** Mechanisms of mediator release from inflammatory cells, in *Mediators of Inflammation,* Weissman, G., Ed., Plenum Press, New York, 1974, 9—50.
43. **Kampschmidt, R. F.,** Effects of leukocytic endogenous mediator on metabolism and infection, *Ann. Okla. Acad. Sci.,* 4, 62, 1974.
44. **Pekarek, R. S., Powanda, M. C., and Wannemacher, R. W., Jr.,** The effect of leukocytic endogenous mediator (LEM) on serum copper and ceruloplasmin concentrations in the rat, *Proc. Soc. Exp. Biol. Med.,* 141, 1029, 1972.
45. **Beisel, W. R., Pekarek, R. S., and Wannemacher, R. W.,** The impact of infectious disease on trace element metabolism of the host, in *Trace Element Metabolism in Animals,* Vol. 2, Hockstray, W., Ed., University Park Press, Baltimore, 1974, 217.
46. **Eddington, C. L., Upchurch, H. F., and Kampschmidt, R. F.,** Quantitation of plasma a_2-AP globulin before and after stimulation with leukocytic extracts, *Proc. Soc. Exp. Biol. Med.,* 139, 565, 1972.
47. **Wannemacher, R. V., Pekarek, R. S., and Beisel, W. R.,** A hormone-like leukocytic endogenous mediator (LEM) which can regulate amino-acid transport and protein synthesis, *Am. J. Clin. Nutr.,* 26, 460, 1973.
48. **Sokal, J. E. and Shimaoka, A.,** Pyrogen in the urine of febrile patients with Hodgkin's disease, *Nature (London),* 215, 1183, 1967.
49. **Linder, M. C. and Moor, J. R.,** Plasma ceruloplasmin and copper in pulmonary cancer: Studies on heavy smokers and patients with malignant and non-malignant lung disease, in *Proc. 3rd Int. Symp. Detection and Prevention of Cancer, Part II,* Vol. 1, Nieburgs, H. E., Ed., Marcel Dekker, New York, 1978, 191—207.
50. **Östenberg, R.,** Therapeutic uses of copper chelating agents, in *Biological Roles of Copper,* Excerpta Medica, Amsterdam, 1980, 283—299.
51. **Sternlieb, I. and Scheinberg, I. H.,** Wilson's disease, in *Liver and Biliary disease,* Wright, R., Alberti, K. G., Karen, S., and Milward-Sadler, G. H., Eds., W. B. Saunders, Philadelphia, 1979, chap. 33.
52. **Henkin, R. I.,** On the role of adrenocorticosteroids in the control of zinc and copper metabolism, in *Trace Element Metabolism in Animals,* Hoekstra, W. G., Suttie, J. W., Ganther, H., and Mertz, W., Eds., University Park Press, Baltimore, 1974, chap. 91.
53. **Prasad, A. S., Brewer, G. J., Schoomaker, E. B., and Rabboni, P.,** Hypocupremia induced by zinc therapy in adults, *J. Am. Med. Assoc.,* 240, 2166, 1978.
54. **Rice, E. W.,** Correlation between serum copper, ceruloplasmin activity and C-reactive protein, *Clin. Chim. Acta,* 5, 632, 1960.
55. **Meyer, B. J., Meyer, A. C., and Horwitt, M. K.,** Effect of triiodothyronine on serum copper and basal metabolism in schizophrenic patients, *Arch. Gen. Psychiatry,* 1, 372, 1959.
56. **Taylor, J. D., Krahn, P. M., and Higgins, T. N.,** Serum copper levels and diphenylhydantoin, *Am. J. Clin. Pathol.,* 61, 577, 1974.
57. **Cartwright, G. E., Gubler, C. J., and Wintrobe, M. M.,** Studies on copper metabolism. XI. Copper and iron metabolism in the nephrotic syndrome, *J. Clin. Invest.,* 33, 685, 1954.
58. **Lahey, M. E., Gubler, C. J., Cartwright, G. E., and Wintrobe, M. M.,** Studies on copper metabolism. VII. Blood copper in pregnancy and various pathological states, *J. Clin. Invest.,* 32, 329, 1953.
59. **Ravin, H. A.,** An improved colormetric enzymatic assay of ceruloplasmin, *J. Lab. Clin. Med.,* 58, 161, 1961.
60. **Lyko, J.,** Serum ceruloplasmin levels in healthy persons and in certain pathological conditions, *Acta Med. Polona,* 8, 269, 1967.
61. **Delves, H. T.,** Dietary sources of copper, in *Biological Roles of Copper (Ciba Found. Symp. 79),* Excerpta Medica, Amsterdam, 1980, 5—22.
62. **Danks, D. M.,** Copper deficiency in humans, in *Biological Roles of Copper (Ciba Found. Symp. 79),* Excerpta Medica, Amsterdam, 1980, 209—225.
63. **Sarkar, B.,** Transport of copper, in *Metal Ions in Biological Systems,* Vol. 12, Properties of Copper, Sigel, H., Ed., Marcel Dekker, New York, 1981, chap. 6.
64. **Guthrie, B. E., McKenzie, J. M., and Casey, C. C.,** Copper status of New Zealanders, in *Trace Element Metabolism in Man and Animals,* Kirchgessner, M., Ed., Freising-Weihenstephan, 3, 304, 1978.
65. **Klevay, L. M., Reck, S. J., and Barcome, D. F.,** Evidence of dietary copper and zinc deficiencies, *J. Am. Med. Assoc.,* 241, 1916, 1979.
66. **Dunlap, W. M., James, G. W., and Hume, D. M.,** Anemia and neutropenia caused by copper deficiency, *Ann. Intern. Med.,* 80, 470, 1974.
67. **Prasad, A. S.,** *Trace Elements and Iron in Human Metabolism,* Plenum Press, New York, 1978, chap. 2.
68. **Linder, M. C. and Munro, H. N.,** Iron and copper metabolism during development, *Enzyme,* 15, 111, 1973.

69. **Widdowson, E. M.,** Trace elements in human development, in *Mineral Metabolism in Pediatrics,* Barltrop, D. and Burland, W. L., Eds., Blackwell Scientific, Oxford, 1969, 85—97.
70. **Zackheim, H. F. and Wolf, P.,** Serum copper in psoriasis and other dermatoses, *J. Invest. Dermatol.,* 58, 28, 1972.
71. **Lifschitz, M. D. and Henkin, R. I.,** Circadian variation in copper and zinc in man, *J. Appl. Physiol.,* 31, 88, 1971.
72. **Munch-Petersen, S.,** The variations in serum copper in the course of 24 hours, *Scand. J. Clin. Lab. Invest.,* 2, 48, 1950.
73. **von Studnitz, W. and Berezin, D.,** Studies on serum copper during pregnancy, the menstrual cycle, and after the administration of oestrogens, *Acta Endocrinol. Copenhagen,* 27, 245, 1958.
74. **Haralambie, G. and Kreul, J.,** Das Verhalten von Serum-Coeruloplasmin und Kupfer bei langauernder Körperbelastung, *Arzneim. Forsch.,* 24, 112, 1970.
75. **Krebs, H. A.,** Über das kupfer in menschlichen Blutserum, *Klin. Wochenschr.,* 7, 584, 1928.
76. **Nielsen, A. L.,** On serum copper. IV. Pregnancy and parturition, *Acta Med. Scand.,* 118, 92, 1944b.
77. **Russ, E. M. and Raymunt, J.,** Influence of estrogens on total serum copper and ceruloplasmin, *Proc. Soc. Exp. Biol. Med.,* 92, 465, 1956.
78. **O'Leary, J. A. and Spotlacy, W. N.,** Serum copper alteration after ingestion of an oral contraceptive, *Science,* 168, 682, 1968.
79. **Briggs, M., Austin, J., and Staniford, M.,** Oral contraceptives and copper metabolism, *Nature (London),* 225, 81, 1970.
80. **Hefnawi, F., Kandil, O., Askalani, H., Zaki, K., Nasr, F., and Mousa, M.,** Copper levels in women using intrauterine devices or oral contraceptives, *Fertil. Steril.,* 25, 556, 1974.
81. **Halsted, J. A., Hackley, B. M., and Smith, J. C.,** Plasma zinc and copper in pregnancy and after oral contraceptives, *Lancet,* 2, 278, 1968.
82. **Rubinfeld, Y., Maor, Y., and Modai, D.,** A progressive rise in serum copper levels in women taking oral contraceptives: a potential hazard? *Fertil. Steril.,* 32, 599, 1979.
83. **Johnson, N. C., Kheim, T., and Kountz, W. B.,** Influence of sex hormones on total serum copper, *Proc. Soc. Exp. Biol. Med.,* 102, 98, 1959.
84. **Bailar, J. C.,** Thromboembolism and oestrogen therapy, *Lancet,* 2, 560, 1967.
85. **Ziel, H. K. and Finkle, W. D.,** Increased risk of endometrial carcinoma among users of conjugated estrogens, *N. Engl. J. Med.,* 293, 1167, 1975.
86. **Smith, D. C., Prentic, R., Thompson, D. J., and Hermann, W. L.,** Association of exogenous estrogen and endometrial carcinoma, *N. Engl. J. Med.,* 293, 1164, 1975.
87. **Baum, J., Holtz, J., Bookstein, J., and Klein, E. W.,** Possible association of benign hepatomas and oral contraceptives, *Lancet,* 2, 926, 1973.
88. **Mays, E. T., Christopherson, W. M., Mahr, M. M., and Williams, H. C.,** Hepatic changes in young women ingesting contraceptive steroids: hepatic hemorrhage and primary hepatic tumors, *J. Am. Med. Assoc.,* 235, 730, 1976.
89. **Edmondson, H. A., Henderson, B., and Benton, B.,** Liver cell adenomas associated with the use of oral contraceptives, *N. Engl. J. Med.,* 294, 470, 1976.
90. **Herbst, A. L., Ulfelder, H., and Poskanzer, D. C.,** Adenocarcinoma of vagina, *N. Engl. J. Med.,* 284, 878, 1971.
91. **Greenwald, P. and Nasca, R. C.,** Transplacental induction of vaginal cancer by synthetic estrogens, in *Progress in Clinical Cancer,* Ariel, M. E., Ed., Grune & Straton, New York, 1973, 13—19.
92. **Boston Collaborative Drug Surveillance Program,** Oral contraceptives and venous thromboembolic disease, surgically confirmed gallbladder disease, and breast tumors, *Lancet,* 1, 1399, 1973.
93. **Boston Collaborative Drug Surveillance Program,** Surgically confirmed gallbladder disease, venous thromboembolism, and breast tumors in relation to post-menopausal estrogen therapy. *N. Engl. J. Med.,* 290, 15, 1974.
94. **Hoover, R., Gray, L. A., Sr., Cole, P., and MacMahon, B.,** Menopausal estrogens and breast cancer, *N. Engl. J. Med.,* 295, 401, 1976.
95. **Sternlieb, I.,** Copper and the liver, *Gastroenterology,* 78, 1615, 1980.
96. **Walshe, J. M.,** Wilson's disease, in *The Biochemistry of Copper,* Peisach, J., Aisen, P., and Blumberg, W. E., Eds., Academic Press, New York, 1966, 475—493.
97. **Scheinberg, I. H. and Sternlieb, I.,** Copper toxicity and Wilson's disease, in *Trace Elements in Human Health and Disease,* Vol. 1, Prasad, A. S., Ed., Academic Press, New York, 1976, 415.
98. **Owen, C. A.,** Copper and hepatic function, in *Biological Roles of Copper (Ciba Found. Symp. 79),* Excerpta Medica, Amsterdam, 1980, 267—282.
99. **Sass-Kortsak, A.,** Wilson's disease: a treatable liver disease in children, *Pediatr. Clin. North Am.,* 22, 963, 1975.
100. **Danks, D. M., Cambell, P. E., Stevens, B. J., Mayne, V., and Cartwright, E.,** Menkes' kinky hair syndrome: an inherited defect in copper absorption with sidespread effects, *Pediatrics,* 50, 188, 1972a.

101. **Danks, D. M., Campbell, P. E., Walker-Smith, J., Stevens, B. J., Gillespie, J. M., Blomfield, J., and Turner, B.**, Menkes' kinky hair syndrome, *Lancet*, 1, 1100, 1972b.
102. **Evans, G. W., Dubois, R., and Hambidge, K. M.**, Wilson's disease: identification of an abnormal copper-binding protein, *Science*, 181, 1175, 1973.
103. **Scheinberg, I. H.**, Wilson's disease and copper binding proteins, *Science*, 185, 1184, 1974.
104. **Sternlieb, I., Van Den Hamer, C. J. A., Morell, A. G., Alpert, S., Gregoriadis, G., and Scheinberg, I. H.**, Lysosomal defect of hepatic copper excretion in Wilson's disease (hepatolenticular degeneration), *Gastroenterology*, 64, 93, 1973.
105. **Lough, J. and Wigglesworth, F. W.**, Wilson's disease: comparative ultrastructure in sibship of 9, *Arch. Pathol.*, 100, 659, 1976.
106. **Gibbs, K. and Walshe, J. M.**, A study of caeruloplasmin concentrations found in 75 patients with Wilson's disease, their kinships and various control groups, *Q. J. Med.*, 48, 447, 1979.
107. **Sternlieb, I. and Scheinberg, I. H.**, The role of radiocopper in the diagnosis of Wilson's disease, *Gastroenterology*, 77, 138, 1979.
108. **Sternlieb, I.**, Differential diagnosis of chronic hepatitis and Wilson's disease, *IM (Intern. Med. Specialist)*, 30, 33, 1982.
109. **Walshe, J. M.**, Copper chelation in patients with Wilson's disease. A comparison of penicillamine and triethylene tetramine, *Q. J. Med.*, 62, 441, 1973.
110. **Walshe, J. M.**, Brief observations on the management of Wilson's disease, *Proc. R. Soc. Med.*, 70 (Suppl. 3), 1, 1977.
111. **Sternlieb, I.**, Present status of diagnosis and prophylaxis of asymptomatic patients with Wilson's disease, in *Progress in Liver and Biliary Tract Diseases*, Leevy, C. M., Ed., S. Karger, New York, 1976, 137—142.
112. **Grand, R. J. and Vawter, G. F.**, Juvenile Wilson's disease: histologic and functional studies during penicillamine therapy, *J. Pediatr.*, 87, 1161, 1975.
113. **Dekaban, A. S., Aamodt, R., Rumble, W. F., Johnstone, G. S., and O'Reilly, S.**, Kinky hair disease: study of copper metabolism with use of ^{67}Cu, *Arch. Neurol.*, 32, 672, 1975.
114. **Solomons, N. W.**, Zinc and copper in the gastrointestinal system, in *Zinc and Copper in Medicine*, Karcioglu, Z. A. and Sarper, R. F., Eds., Charles C Thomas, Springfield, Ill., 1980, chap. 9.
115. **Danks, D. M.**, Steely hair, mottled mice, and copper metabolism, *N. Engl. J. Med.*, 293, 1147, 1975.
116. **Danks, D. M., Stevens, J. J., Campbell, P. E., Cartwright, E. C., Gillespie, J. M., Townley, R. R. W., Walker-Smith, J. A., Blomfield, J., Turner, B. B., and Mayne, V.**, Menkes' kinky-hair syndrome: an inherited defect in the intestinal absorption of copper with widespread effects, *Birth Defects*, 10, 132, 1974.
117. **Holtzman, N. A.**, Menkes' kinky hair syndrome: a genetic disease involving copper, *Fed. Proc.*, 35, 2276, 1976.
118. **Garnica, A. D., Frias, J. L., and Rennert, O. M.**, Menkes' kinky hair syndrome: is it a treatable disorder? *Clin. Genet.*, 11, 154, 1977.
119. **Horn, N.**, Copper incorporation studies on cultured cells for prenatal diagnosis of Menkes' disease, *Lancet*, 1, 1156, 1976.
120. **Zimmerman, A. W., Mattieu, J.-M., Quareles, R. H., Brady, R. O., and Hsu, J. M.**, Hypomyelination in copper-deficient rats, pre-natal and postnatal copper replacement, *Arch. Neurol.*, 33, 111, 1976.
121. **Oakes, B. W., Danks, D. M., and Campbell, P. E.**, Human copper deficiency: ultrastructural studies of the aorta and skin in a child with Menkes' syndrome, *Exp. Mol. Pathol.*, 25, 82, 1976.
122. **Haas, R. H., Robinson, A., Lascelles, P. T., and Dubowitz, V.**, An x-linked disease of the nervous system with disordered copper metabolism and features differing from Menkes' disease, *Neurology*, 31, 852, 1981.
123. **Brendstrup, P.**, Serum copper, serum iron and total iron binding capacity of serum in acute and chronic infections, *Acta Med. Scand.*, 145, 315, 1953.
124. **Oleske, J. M., Valentine, J. L., and Minnefore, A. B.**, The effects of acute infection on blood lead, copper and zinc levels in children, *Health Lab. Sci.*, 12, 230, 1975.
125. **Holmberg, G. G. and Laurell, C. B.**, Oxidase reactions in human plasma caused by coeruloplasmin, *Scand. J. Clin. Lab. Invest.*, 3, 103, 1951.
126. **Zwicker, M.**, Post-Operative Serum Kupferspiegelveranderungen, *Klin. Wochenschr.*, 37, 933, 1959.
127. **Foster, M., Picklington, T., and Dawson, A. A.**, Ceruloplasmin and iron transferrin in human malignant disease, in *Metal Ions in Biological Systems*, Vol. 10, Siegel, H., Ed., Marcel Dekker, New York, 1981, chap. 5.
128. **Brendstrup, P.**, Serum iron, total iron-binding capacity of serum and serum copper in acute hepatitis, *Acta Med. Scand.*, 146, 107, 1953.
129. **Ritland, S., Steinnes, E., Skrede, S.**, Hepatic copper content, urinary copper excretion and serum ceruloplasmin in liver disease, *Scand. J. Gastroenterol.*, 12, 81, 1977.

130. **Gubler, C. J., Brown, H., Markowitz, H., Cartwright, G. E., and Wintrobe, M. M.,** Studies on copper metabolism. XXIII. Portal (Laennec's) cirrhosis of liver, *J. Clin. Invest.*, 36, 1208, 1957.
131. **Gewurz, H., Mold, C., Siegel, J., and Fiedel, B.,** C-reactive protein and the acute phase response, *Adv. Intern. Med.*, 27, 345, 1982.
132. **Kushner, I.,** The acute phase reactants and the erythrocyte sedimentation rate, in *Textbook of Rheumatology*, Kelley, W. N. et al., Eds., W. B. Saunders, Philadelphia, 1981, 669.
133. **Weissmann, G., Smolen, J. E., and Hoffstein, S.,** Polymorphonuclear leukocytes as secretory organs of inflammation, *J. Invest. Derm.*, 71, 95, 1978.
134. **Weissmann, G.,** *Prostaglandin in Acute Inflammation,* The Upjohn Company, Kalamazoo, Mich., 1980.
135. **Versieck, Barbier, F., Speeck, A., and Hoste, J.,** Manganese, copper and zinc concentrations in serum and packed blood cells during acute hepatitis, chronic hepatitis and post-necrotic cirrhosis, *Clin. Chem.*, 20, 1141, 1974.
136. **Prasad, A. A., Ortega, J., Brewer, G. J., Oberleas, D., and Schoomaker, E. B.,** Trace elements in sickle cell disease, *J. Am. Med. Assoc.*, 235, 2396, 1976.
137. **Williams, D. M., Atkins, C. L., Frens, D. B., and Bray, P. F.,** Menkes' kinky hair syndrome: studies of copper metabolism and long term copper therapy, *Pediatr. Rev.*, 11, 823, 1977.
138. **Shields, G. S., Markowitz, S. M., Klassen, W. H., Cartwright, G. E., and Wintrobe, M. M.,** Studies on copper metabolism. XXXI. Erythrocytic copper, *J. Clin. Invest.*, 40, 200, 1960.
139. **Henkin, R. I.,** Trace metals in endocrinology, *Med. Clin. North Am.*, 60, 779, 1976.
140. **Sachs, A., Levine. V. E., and Griffith, W. O.,** Blood copper and iron in Addison's disease, *Proc. Soc. Exp. Biol. Med.*, 37, 186, 1937.
141. **Locke, A., Main, E. R., and Rosbash, D. O.,** The copper and nonhemoglobinous iron contents of the blood serum in disease, *J. Clin. Invest.*, 11, 527, 1932.
142. **Nielsen, A. L.,** Serum copper. V. Thyrotoxicosis and myxoedema, *Acta Med. Scand.*, 118, 431, 1944.
143. **McCall, J. T., Goldstein, N. P., and Smith, L. H.,** Implications of trace elements in human diseases, *Fed. Proc.*, 30, 1011, 1971.
144. **Canelas, H. M., Assis, L. M., DeJorge, F. B., Tolosa, A. P., and Cintra, A. B.,** Disorders of copper metabolism in epilepsy, *Acta Neurol. Scand.*, 40, 97, 1964.
145. **Niedermeyer, W. and Griggs, J.,** Trace metal composition of synovial fluid and blood serum of patients with rheumatoid arthritis, *J. Chronic Dis.*, 23, 527, 1971.
146. **Ellis, P. P.,** Ocular deposition of copper in hypercupremia, *Am. J. Ophthalmol.*, 68, 423, 1969.
147. **Hrgovcic, M., Tessmer, C. F., Minckler, T. M., Mosier, B., and Taylor, G. H.,** Serum copper levels in lymphoma and leukemia, *Cancer,* 21, 743, 1968.
148. **Goodman, S. I., Rodgerson, D. O., and Kauffman, J.,** Hypercupremia in a patient with multiple myeloma, *J. Lab. Clin. Med.*, 70, 57, 1967.
149. **Lewis, R. A., Hultgwist, D. E., Baker, B. L., Falls, H. F., Gershowitz, H., and Penner, J. A.,** Hypercupremia associated with a monoclonal immunoglobulin, *J. Lab. Clin. Med.*, 88, 375, 1976.
150. **Sandberg, M., Gross, H., and Holly, O. M.,** Changes in retention of copper and iron in liver and spleen in chronic diseases accompanied by secondary anemia, *Arch. Pathol.*, 33, 834, 1942.
151. **Keiderling, W. and Scharpf, H.,** Über die Klinische Bedeutung der Serum Kupfer-und Serumeisenbestimmung bei Neoplastischen Krankheitszustanden, *Muench. Med. Wochenschr.*, 95, 437, 1953.
152. **Rechenberger, J.,** Serumeisen and Serumkupfer bei aukten and chronischen Leukämien sowie bei morbus Hodgkin, *Dtsch. Z. für Verdau. Stoffwechselkr.*, 17, 79, 1957.
153. **Koch, H. J., Smith, E. R., and McNeeley, J.,** Analysis of trace elements in human tissues. II. The lymphomatous diseases, *Cancer,* 10, 151, 1957.
154. **Pagliardi, E. and Giangrandi, E.,** Clinical significance of the blood copper in Hodgkin's disease, *Acta Haematol.*, 24, 201, 1960.
155. **Herring, W. B., Leavell, B. S., Paixao, L. M., and Yoe, J. H.,** A study of magnesium, chromium, nickel, copper and zinc. II. Observations of patients with some hematologic diseases, *Am. J. Clin. Nutr.*, 8, 855, 1960.
156. **Cherry, N. H., Kalas, J. P., and Zarafonetis, C. J. D.,** Study of plasma copper levels in patients with acute leukemia, *J. Einstein Med. Cent.*, 9, 24, 1961.
157. **Pagliardi, E., Cravario, A., Brusa, L., Cantino, D., and Giangrandi, E.,** Indagine sul metabolismo del rame nel morbo di Hodgkin, *Hematologica,* 481, 209, 1963.
158. **Preece, A. W., Light, P. A., Evans, P. A., and Nunn, A. D.,** Serum-copper, penicillamine and cytotoxic therapy, *Lancet,* 1, 953, 1977.
159. **Jensen, K. B., Thorling, E. G., and Anderson, C. F.,** Serum copper in Hodgkin's disease, *Scand. J. Haematol.*, 1, 62, 1964.
160. **Valberg, L. S., Holt, J. M., and Card, R. T.,** Erythrocyte magnesium, copper and zinc in malignant diseases affecting the hemopoietic system, *Cancer,* 19, 1833, 1966.

161. **Hobbs, C. B., Warren, R. L., and Jeliffe, A. M.,** The flame spectrometric measurement of serum copper and its use in the assessment of patients with malignant lymphoma, *Proc. Assoc. Clin. Biochem.*, 4, 197, 1967.
162. **Pisi, E., Di Feliciantonio, R., Figus, E., and Ferri, S.,** Comportamento e significato prognostico della cerulopplasmina sierica in relazione al quadro isto-patologico nella malattia di Hodgkin, *Minerva Med.*, 59, 944, 1968.
163. **Warren, L. R., Jeliffe, A. M., Watson, J. V., and Hobbs, C. B.,** Prolonged observations on variations in the serum copper in Hodgkin's disease, *Clin. Radiol.*, 20, 247, 1969.
164. **Tura, S., Bernardi, L., Baccarani, M., Sanguinetti, F., and Branzi, A.,** La cupremia e la ceruloplasminemia nelle emolinfopatie. Nota I. I linfomi Maligni, *G. Clin. Med. (Bologna)*, 49, 1090, 1968.
165. **Ilicin, G.,** Serum copper and magnesium levels in leukemia and malignant lymphoma, *Lancet*, 1, 1036, 1971.
166. **Alexander, F. W., Delves, H. T., and Lay, H.,** Plasma copper and zinc in acute leukemia (abstr.), *Arch. Dis. Child.*, 47, 671, 1972.
167. **Mortazavi, S. H., Bani-Hashemi, A., Mozafari, M., and Raffi, E.,** Values of serum copper measurement in lymphomas and several other malignancies, *Cancer*, 29, 1193, 1972.
168. **Jelliffe, A. M.,** Invited discussion: value of fluctuations in the serum copper level in the control of patients with Hodgkin's disease, *Natl. Cancer Inst. Monogr.*, 36, 325, 1973.
169. **Delves, H. T., Alexander, F. W., and Lay, H.,** Copper and zinc concentrations in the plasma of leukemic children, *Br. J. Haematol.*, 24, 525, 1973.
170. **Ray, G. R., Wolf, P. H., and Kaplan, H. S.,** Value of laboratory indicators in Hodgkin's disease: preliminary results, *Natl. Cancer Inst. Mongr.*, 36, 315, 1973.
171. **Masi, M., Vecchi, V., Vivarelli, F., and Paolucci, P.,** La ceruloplasminemia nella leucosi acuta linfoblastica e nel morbo di Hodgkin della infanzia, *Minerva Pediatr.*, 27, 1223, 1975.
172. **Capellaere, P., Sulman, Ch., Chechan, Ch., and Gosselin-Delaquais, P.,** Les concentrations, plasmatiques du cuivre et du zinc au cours de la maladie de Hodgkin, *Lille Med.*, 20, 904, 1975.
173. **Thorling, E. B. and Thorling, K.,** The clinical usefulness of serum copper determinations in Hodgkin's disease, *Cancer*, 38, 255, 1976.
174. **Kesava Rao, K. V., Shetty, P. A., Bapat, C. V., and Jussawalla, D. J.,** Serum copper assay as a biochemical marker to assess the response to therapy in Hodgkin's disease, *Indian J. Cancer*, 14(4), 320, 1977.
175. **El-Haddad, S., Mahfouz, M., Magahed, Y., Mahmoud, F., Kamel, M., Ali, M. A.,** Value of serum copper measurement in acute leukaemia of childhood, *Gaz. Egypt Paediatr. Assoc.*, 26, 67, 1977.
176. **Bucher, W. C. and Jones, S. E.,** Serum copper-zinc ratio in patients with malignant lymphoma (abstr.), *Am. J. Clin. Pathol.*, 68(1), 104, 1977.
177. **Rawat, M. and Vijayvargiya, R.,** Serum copper estimation in lymphomas, *Indian J. Med. Res.*, 66, 815, 1977.
178. **Davidoff, G. N., Votaw, M. L., Coon, W. W., Hecker, L. H., Richardson, R. J., Finkel, J. D., and Weitz, J.,** Serum copper and ceruloplasmin levels as an adjunct to diagnosis of Hodgkin's and non-Hodgkin's lymphoma (abstr.), *Clin. Res.*, 25(4), 636A, 1977.
179. **Foster, M., Fell, L., Pocklington, T., Akinsete, A. D., Dawson, A., Hutchinson, M. S., and Mallard, J. R.,** Electron spin resonance as a useful technique in the management of Hodgkin's disease, *Clin. Radiol.*, 28, 15, 1977.
180. **Foster, M., Dawson, A., Pocklington, T., and Fell, L.,** Electron spin resonance measurements of blood ceruloplasmin and iron transferrin levels in patients with non-Hodgkin's lymphoma, *Clin. Radiol.*, 28, 23, 1977.
181. **Pizzolo, G., Savarin, T., Molino, A. M., Ambroseti, A., Todeschini, G., and Vettore, L.,** The diagnostic value of serum copper levels and other hematochemical parameters in malignancies, *Tumori*, 64, 55, 1978.
182. **Legutko, L.,** Serum copper investigations in children with acute lymphoblastic leukemia, *Folia Hematol. (Leipzig)*, 105, 248, 1978.
183. **Williams, J., Thompson, E., and Smith, K. L.,** Value of serum copper levels and erythrocyte sedimentation rates as indicators of disease activity in children with Hodgkin's disease, *Cancer*, 42, 1929, 1978.
184. **Subrahmaniyam, K., Rao, B. N., Shukla, P. K., Rastogi, B. L., Roy, S. K., Sanyal, B., Pant, G. C., and Khanna, N. N.,** Serum copper in patients in non-Hodgkin's lymphomas — a preliminary appraisal of its significance in management, *Indian J. Radiol.*, 33(2), 101, 1979.
185. **Sirsat, A. V.,** Serum ceruloplasmin levels in Hodgkin's disease and malignant lymphomas in correlation with serum copper levels, *Indian J. Cancer*, 16, 32, 1979.
186. **Mitta, S. K. and Tan, C.,** Serum copper levels in children with Hodgkin's disease, in *Proc. AACR and ASCO*, Waverly Press, Baltimore, 1979, C-377 (abstract).
187. **Asbjörnsen, G.,** Serum copper compared to erythrocyte sedimentation rate as indicator of disease activity in Hodgkin's disease, *Scand. J. Haematol.*, 22, 193, 1979.

188. **Horn, R. A., Friesen, E. J., Stephens, R. L., Hedrick, W. R., and Zimbrick, J. D.,** Electron spin resonance studies on properties of ceruloplasmin and transferrin in blood from normal human subjects and cancer patients, *Cancer,* 43(6), 2392, 1979.
189. **Hamberg, S. J.,** Electron paramagnetic resonance and X-ray fluorescence determinations of serum copper in patients with Hodgkin's disease, leukemia, and lung cancer, *Diss. Abstr. Int. B,* 39 (11), 5203, 1979.
190. **Gobbi, P. G., Scarpellini, M., Minoia, C., Pozzoli, L., and Parugini, S.,** Plasma iron and copper in Hodgkin's disease a comparison with other laboratory indicators, *Haematologica (Pavia),* 64(4), 416, 1979.
191. **De Bellis, R., Boulard, M. R., Kasdorf, H., Rodriguez, I., Ferrando, R., Di Landro, J., Ferrari, M., Sanguinett, C. M., and Tanzer, J.,** Metabolic changes in red blood cells in malignant lymphomas, *Br. J. Haematol.,* 42, 35, 1979.
192. **Wolf, P. L., Ray, G., and Kaplan, H. S.,** Evaluation of copper oxidase (ceruloplasmin) and related tests in Hodgkin's disease, *Clin. Biochem.,* 12, 202, 1979.
193. **Green, J. A., Pocklington, T., Dawson, A. A., and Foster, M.,** Electron spin resonance studies on caeruloplasmin and iron transferrin in patients with chronic lymphocytic leukaemia, *Br. J. Cancer,* 41(3), 356, 1980.
194. **Larsen, B., Heron, I., and Thorling, B.,** Elevated serum Cu in Hodgkin's disease and inhibitory effects of ceruloplasmin on lymphocyte response in vitro, *Eur. J. Cancer,* 16, 415, 1980.
195. **Kaplan, H. S.,** *Hodgkin's Disease,* Harvard University Press, Cambridge, Mass., 1980, 131.
196. **Cohen, Y., Haim, N., and Zinder, O.,** Serum copper levels in non-Hodgkin's lymphoma (abstr.), *Proc. Am. Assoc. Cancer Res.,* 22, 184, 1981.
197. **Olson, K. B., Heggen, G. E., and Edwards, C. F.,** Analysis of five trace elements in liver of patients dying of cancer and non-cancerous disease, *Cancer,* 11, 554, 1958.
198. **Mulay, I. L., Roy, R., Knox, B. E., Suhr, N. H., and Delaney, W. E.,** Trace metal analysis of cancerous and noncancerous human tissues, *J. Natl. Cancer Inst.,* 47, 1, 1971.
199. **Morgan, J. M.,** Hepatic copper, manganeze and chromium content in bronchiogenic carcinoma, *Cancer,* 29, 710, 1971.
200. **Schwartz, A. E., Leddicott, G. W., Fink, R. W., and Friedman, E. W.,** Trace elements in normal and malignant human breast tissue, *Surgery,* 76, 325, 1974.
201. **Janes, J. M., McCall, J. T., and Elveback, Z. R.,** Trace metals in human osteogenic sarcoma, *Mayo Clin. Proc.,* 47, 476, 1972.
202. **Santoliquido, P. M., Southwick, H. W., and Olwin, J. H.,** Trace metal levels in cancer of the breast, *Surg. Gynecol. Obstet.,* 142, 65, 1976.
203. **Karcioglu, Z. A., Sarper, M. R., Van Rinsvelt, H. A., Guffey, J. A., and Fink, R. W.,** Trace element concentrations in renal cell carcinoma, *Cancer,* 42, 1330, 1978.
204. **Roguljic, A., Roth, A., Kolaric, K., and Maricic, Z.,** Iron, copper, and zinc liver tissue in patients with malignant lymphoma, *Cancer,* 461, 565, 1980.
205. **Pirrie, R.,** Serum copper and its relationship to serum iron in patients with neoplastic diseases, *J. Clin. Pathol.,* 5, 190, 1952.
206. **Graf, H.,** Ceruloplasmin bei Kollumkarzinomen verschiedener Stadien, *Med. Welt.,* 19, 1059, 1965.
207. **DeJorge, F. B., Paiva, L., Mion, D., and Da Nova, R.,** Biochemical studies on copper, copper oxidase, magnesium, sulfur, calcium, and phosphorus in cancer of the larynx, *Acta Otolaryngol.,* 61, 454, 1966.
208. **DeJorge, F. B., Sampaio Goes, Jr., J., Guedes, J. L., and De Ulhoa Cintra, A. B.,** Biochemical studies on copper, copper oxidase, and magnesium, phosphorus in cancer of the breast, *Clin. Chim. Acta,* 12, 403, 1965.
209. **Wu, C. C.,** Serum iron and copper in hepatobiliary diseases, *J. Formosan Med. Assoc.,* 64, 257, 1965.
210. **Zamello, J.,** Copper and iron levels in patients with cancer of the reproductive organs, *Chem. Abstr.,* 65, 1165b, 1966.
211. **Maracek, Z. and Hoenigova, J.,** Changes in copper and ceruloplasmin in malignant disease, *Chem. Abstr.,* 71, 47610c, 1969.
212. **Wiederanders, R. E. and Evans, G. W.,** The copper concentration of hyperplastic and cancerous prostates, *Invest. Urol.,* 6, 531, 1969.
213. **O'Leary, J. A. and Feldman, M.,** Serum copper alterations in genital cancer, *Surg. Forum,* 21, 411, 1970.
214. **Bélanger, L.,** Evaluation du taux de céruloplasmine serique au cours des maladies de la prostata, *L'Union Med. Can.,* 100, 1554, 1971.
215. **Tani, P. and Kokkola, K.,** Serum iron, copper and iron-binding capacity in bronchogenic pulmonary carcinoma, *Scand. J. Respir. Dis. Suppl.,* 80, 121, 1972.
216. **Schapira, M.,** The presence of ceruloplasma in cancer of the breast and female genital organs, *J. R. Coll. Gen. Pract.,* 22, 383, 1972.
217. **Hughes, N. R.,** Serum transferrin and ceruloplasmin concentrations in patients with carcinoma, melanoma, sarcoma and cancers of hematopoietic tissues, *Aust. J. Exp. Biol. Med. Sci.,* 50, 97, 1972.

218. **Dines, D. E., Elveback, L. R., and McCall, J. T.,** Zinc, copper and iron contents of pleural fluid in benign and neoplastic disease, *Mayo Clin. Proc.*, 49, 102, 1974.
219. **Kolaric, K., Roguljic, A., and Fuss, V.,** Serum copper levels in patients with solid tumors, *Tumori,* 61, 173, 1975.
220. **Monari, M., Retamal, C., Galvez, S., and Cordero, I.,** Normal values of ceruloplasmin concentration in the serum and in cerebrospinal fluid. Variations in some neurological diseases and in neoplastic affection, *Neurocirugia,* 34 (1/2), 5, 1976.
221. **Fisher, G. L., Byers, V. S., Shifrine, M., and Levin, A. S.,** Copper and zinc levels in serum from human patients with sarcomas, *Cancer,* 37, 356, 1976.
222. **Shifrine, M. and Fisher, G. L.,** Ceruloplasmin levels in sera from human patients with osteosarcoma, *Cancer,* 38, 244, 1976.
223. **Scanni, A., Licciardello, L., Trovato, M., Tomirotti, M., and Biraghi, M.,** Serum copper and ceruloplasmin levels in patients with neoplasias localized in the stomach, large intestine or lung, *Tumori,* 63(2), 175, 1977.
224. **Gray, B. N., Walker, C., and Bennett, R. C.,** The monitoring of serum factors in patients with cancer, *Aust. N.Z. J. Surg.*, 47, 648, 1977.
225. **Pocklington, T. and Foster, M. A.,** Electron spin resonance of caeruloplasmin and iron transferrin in blood of patients with various malignant diseases, *Br. J. Cancer,* 36, 369, 1977.
226. **Inutsuka, S. and Araki, S.,** Plasma copper and zinc levels in patients with malignant tumors of digestive organs, *Cancer,* 42, 626, 1978.
227. **Breiter, D. N., Diasio, R. B., Neifeld, J. P., Roush, M. L., and Rosenberg, S. A.,** Serum copper and zinc measurements in patients with osteogenic sarcoma, *Cancer,* 42, 598, 1978.
228. **Israel, L. and Edelstein, R.,** In vivo and in vitro studies on nonspecific blocking factors of host origin in cancer patients, *Isr. J. Med. Sci.*, 14, 105, 1978.
229. **Scanni, A., Tomirotti, M., Licciardello, L., Annibali, E., Biraghi, M., Trovato, M., Fittipaldi, M., Adamoli, P., and Curtarelli, G.,** Variations in serum copper and ceruloplasmin levels in advanced gastrointestinal cancer treated with polychemotherapy, *Tumori,* 65, 331, 1979.
230. **Martin Mateo, M. C., Bustamante, J., and Arellano, I. F.,** Serum copper, ceruloplasmin, lactic dehydrogenase, and α_2-globulin in lung cancer, *Biomedicine,* 31, 66, 1979.
231. **Abdulla, M., Biörklund, A., Mathur, A., and Wallenius, K.,** Zinc and copper levels in whole blood and plasma from patients with squamous cell carcinomas of head and neck, *J. Surg. Oncol.*, 12, 107, 1979.
232. **Andrews, G. S.,** Studies on plasma zinc, copper, caeruloplasmin and growth hormone, *J. Clin. Pathol.*, 32, 325, 1979.
233. **Flynn, A.,** Copper metabolism in patients with solid tumors, in *Nutrition and Cancer,* Vol. 2, Van Eys, J., Nichols, B. L., and Seelig, M. S., Eds., S. P. Scientific and Medical Books, New York, 1979, 297.
234. **Bhardwaj, D. N., Chandler, J., Singh, R. P., and Kaur, K.,** Clinical evaluation of malignant process by serum copper estimation, *J. Indian Med. Assoc.*, 75, 119, 1980.
235. **Huhti, E., Pokkula, A., and Uksila, E.,** Serum copper levels in patients with lung cancer, *Respiration,* 40, 112, 1980.
236. **Garofalo, J. A., Ashikari, H., Lesser, M. L., Menendez-Bodet, C., Cunningham-Rundles, S., Schwartz, M. K., and Good, R. A.,** Serum zinc, copper and the Cu/Zn ratio in patients with benign and malignant breast lesions, *Cancer,* 46, 2682, 1980.
237. **Jafa, A., Mahendra, A. R., Chowdhury, A. R., and Kamboj, V. P.,** Trace elements in prostatic tissue and plasma in prostatic diseases of man, *Indian J. Cancer,* 17, 34, 1980.
238. **Habib, F. K., Dembinski, T. C., and Stich, S. R.,** The zinc and copper of blood leukocytes and plasma from patients with benign and malignant prostates, *Clin. Chim. Acta,* 104, 329, 1980.
239. **Garofalo, J. A., Erlandson, E., Strong, E. W., Lesser, M., Garold, F., Spiro, R., Schwartz, M., and Good, R. A.,** Serum zinc, serum copper, and the Cu/Zn ratio in patients with epidermoid cancers of the head and neck, *J. Surg. Oncol.*, 15, 381, 1980.
240. **Linder, M. C., Moor, J. R., and Wright, K.,** Ceruloplasmin assays in diagnosis and treatment of human lung, breast and gastrointestinal cancers, *J. Natl. Cancer Inst.*, 67, 263, 1981.
241. **Fisher, G. L., Spitler, L. E., McNeill, K. L., and Rosenblatt, L. S.,** Serum copper and zinc levels in melanoma patients, *Cancer,* 47, 1838, 1981.
242. **Dionigi, P., Dionigi, R., Pavesi, F., Mazari, S., and Goggi, D.,** Serum ceruloplasmin levels in surgical cancer patients: relationship with tumor stage and postoperative complications (meet. abstr.), *Eur. Surg. Res.*, 13, 31, 1981.
243. **Perlin, E.,** Carcinoma of the lung: a perspective, *Jefferson Med. Coll. Alumni Bull.*, 30, 32, 1981.
244. **Schapira, D. V. and Schapira, M.,** Ceruloplasmin levels to monitor response and relapse on adjuvant therapy, in *Proc. ASCO,* 1982, C_2 (abstract).
245. **Samsahl, K., Brune, D., and Wester, P. O.,** Simultaneous determination of 30 trace elements in cancerous and noncancerous human tissue samples by NAA, *Int. J. Appl. Radiol.*, 16, 273, 1965.

246. **Tsilyunyk, T. I., Ishchenko, M. M., and Rusenko, S. V.**, Content of copper and manganese in the blood serum of patients with pulmonary cancer (in Russian), *Klin. Med.*, 43, 18, 1965.
247. **Thorling, E. B., Jensen, B. K., Andersen, C. J., and Jensen, Y.**, Cobre serico y ceruloplasmina en la enfermedad de Hodgkin, *Folia Clin. Int. (Barcelona)*, 15, 2, 1965.
248. **Cravario, A., Brusa, L., De Filippi, P. G., and Giangrandi, E.**, La frazioni sieriche del rame nel decorso del morbo di Hodgkin, *Cancro*, 19, 399, 1966.
249. **Wiljasalo, M. A. and Haikonen, M.**, Blood plasma copper in pulmonary cancer and in chronic pulmonary infections, *Ann. Med. Intern. Fenn.*, 55, 107, 1966.
250. **Weissleder, D. H. and Marongiu, F.**, Enfermedad de Hodgkin. Linfoadenografia, cupremia y sideremia, *Prensa Med. Argent.*, 53, 1001, 1966.
251. **Kolomiitseva, M. G. and Neimark, I. I.**, Copper and cobalt content in the blood and in tumor tissue of patients with pulmonary and gastric cancer, *Chem. Abstr.*, 64, 1133f, 1966.
252. **Laparevich, Z. V.**, Effect of radiation on the level of zinc and copper in the blood of patients with cervical carcinoma, *Chem. Abstr.*, 68, 27382S, 1968.
253. **LeBorgne de Kaouel, C., Aubert, C., and Juret, P.**, Etude de la cuprémie et de la sidérémie chez les femmes atteintes d'affections mammaires benignes ou malignes, *Pathol. Biol.*, 16, 85, 1968.
254. **Sischy, B., Capacho-Delgado, L., Carella, R. J., and Newall, J.**, The level of serum copper as an indication of activity, prognosis and response to therapy in Hodgkin's disease (abstr.), in *10th Int. Cancer Cong.*, Houston, Tex., May 1970, 739.
255. **DeJorge, F. G., De Zevedo, M. L., Jose, W. K., and Sanchez, N.**, Biochemical aspects of eye tumours: mineral content in neoplasia and in the ocular tissues, *Rev. Hosp. Clin. Fac. Med. (Sao Paulo)*, 24, 35, 1969.
256. **Danielsen, A. and Steinnes, E.**, A study of some selected trace elements in normal and cancerous tissue by neutron activation analysis, *J. Nucl. Med.*, 11, 260, 1970.
257. **Roguljic, A. and Ivankovic, G.**, Serum copper and zinc changes in patients with malignant diseases (in Croatian). III, *Yugoslav Oncol. Cong.*, Zagreb, 1971.
258. **Donath, I.**, Copper and cancer, *Prot. Vitae*, 16, 11, 1971.
259. **Albert, L., Hienzsch, E., Arndt, J., und Kriester, A.**, Bedeutung und Veränderungen des Serum-Kupferspiegels während und nach der Bestrahlung von Harnblasenkarzinomen, *J. Urol.*, 8, 561, 1972.
260. **Mailer, C., Swartz, H. M., Konieczny, M., Ambeganonkar, S., and Moore, V. L.**, Identity of the paramagnetic element found in increased concentrations in plasma of cancer patients and its relationship to other pathological processes, *Cancer Res.*, 34, 637, 1974.
261. **Mansour, E. G., Pories, W. J., Strain, W. H., and Flynn, A.**, Chemotherapy response in cancer patients evaluated by serum copper zinc changes, in *Proc. 10th Ann. Meet. Am. Soc. Clin. Oncol.*, March 1974.
262. **Sirbu, P., Motoiu, V., and Butte, V.**, Etude de la cuprémie comme test de dépistage précoce dans les dysplasies aggravées et dans le cancer du col, *Rev. Fr. Gynecol.*, 67, 565, 1972.
263. **Hahib, F. K. and Stitch, S. R.**, The inter-relationship of the metal and androgen binding proteins in normal and cancerous human prostatic tissues, *Acta Endocrinol.*, Suppl. 199, 129, 1975.
264. **Androwikashvihi, E. L. and Mosullishvili, L. M.**, Human leukemia and trace elements, in *Metal Ions in Biological Systems*, Vol. 10, Sigel, H., Ed., Marcel Dekker, New York, 1980, chap. 6.
265. **Gross, H., Sandberg, M., and Holly, O. M.**, Changes in copper and iron retention in chronic diseases accompanied by secondary anaemia. II. Changes in liver, spleen and stomach, *Am. J. Med. Sci.*, 204, 201, 1942.
266. **Gisinger, E.**, Serumeisen, Transferrin und Serumkupfer bei malignen Erkrankungen, *Krebsarzt*, 13, 105, 1958.
267. **Rummel, A. C.**, Evaluation of iron and copper level in normal and sick women. A contribution to the significance of heavy metals in abdominal cancer in women, *Medizinische*, 22, 1062—1067, 1959.
268. **Kautzch. E.**, Eisen und Kupferstoffwechsel bei malignen neoplasien, *Med. Klin.*, 40, 1851, 1959.
269. **Piskazeck, K., Billek, K., and Rothe, K.**, Copper in carcinoma of female genitalia in relation to the localization and the form of therapy, *Arch. Gynaekol.*, 196, 447, 1962.
270. **Stegmann, H., Foehlisch, F., and Clotten, R.**, Der Serum-kupferspiegel bei Frauen mit Genitalkarzinom, *Klin. Wochenschr.*, 40, 1120, 1962.
271. **Wysocki, K.**, Diagnostic significance of low values of serum copper, *Polish Med. Hist. Sci.*, 2, 19, 1959.
272. **Vescovo, R. and Lorenzoni, L.**, Serum copper in patients with uterine new growths, with special reference to cervical localization, *Excerpta Med.*, 18, 361, 1965.
273. **Lee, G. R.**, The anemia of chronic disease, in *Seminars in Hematology*, Vol. 20, Metal Metabolism in Hematologic Disorders, Miescher, P. A. and Jaffe, E. R., Eds., Grune & Stratton, New York, 1983, 61-80.

Chapter 7

PATHOLOGY AND CLASSIFICATION OF MALIGNANT LYMPHOMA

I. INTRODUCTION

The lymphomas represent a malignant neoplastic proliferation of cells of the lymphoreticular tissue. They arise in lymph node tissue and may involve other lymphoreticular organs such as the spleen, tonsils, thymus, and bone marrow. Involvement of the bone marrow and peripheral blood may result in a leukemia of the corresponding cell type. On rare occasions lymphomas may arise in extranodal sites such as the lungs, GI tract, brain, gonads, and thyroid.

The first pathological description of the lymphomas can be traced to Hodgkin,[1] who, in 1832 published his observations on seven patients with tumors of the "absorbent glands" and spleen. In the 150 years since then, malignant lymphomas have been studied extensively and various clinicopathological subtypes described. Virchow,[2] in 1845, distinguished lymphosarcoma from leukemia. The term malignant lymphoma was coined by Bilroth[3] in 1871. The first histological description of Hodgkin's disease was made in 1872 (Langhans[4]). Sternberg[5] (1898) and Reed[6] (1902) gave definitive descriptions of Hodgkin's disease and clearly illustrated the characteristic giant cells that bear their names. Since then, Hodgkin's disease has been generally accepted as a clinicopathological entity distinct from the non-Hodgkin's lymphomas.

Further classification of Hodgkin's disease into three histological subtypes was devised by Jackson and Parker[7] (1947). This classification into paragranuloma, granuloma, and sarcoma had a sound histological basis and reflected increasing biological aggressiveness. The clinical usefulness of this classification was limited, however, because 90% of the cases fell into the broad category of Hodgkin's granuloma. The current classification was first suggested by Lukes and Butler[8,9] in 1966 and modified to a simpler form (Rye subclassification[10]) by a special nomenclature committee. This will be discussed in detail later.

The classification of non-Hodgkin's lymphomas has been controversial largely because of the nomenclature of the various subtypes that have been described. Histologic criteria for lymphosarcoma were described by Dreschfeld[11] in 1892. Brill et al.[12] (1925) and Symmers[13] (1927) described giant follicular lymphoma and considered it a benign disease. Roulet[14] in 1930 developed histological criteria for diagnosing reticulum cell sarcoma which was thought to arise from the supporting cells of the lymph node. The term reticulum cell sarcoma was used for all large cell malignant lymphomas and was popular for a number of years. In 1942 Gall and Mallory[15] introduced the terms clasmatocytic lymphomas and stem cell lymphomas for the large-cell lymphoma, the latter composed of primitive undifferentiated cells. They[15] also developed criteria for distinguishing benign follicular hyperplasia from malignant follicular lymphoma and introduced the concept of lymphocytic and lymphoblastic lymphoma representing tumors of well-differentiated and poorly differentiated lymphocytes, respectively. Rappaport et al.[16] clearly defined follicular lymphoma in the spectrum of malignant lymphoma in 1956 and introduced the term nodular lymphoma. The significance of the nodular pattern in a lymphoma was further stressed by Rappaport[17] in a classification proposed in 1966 (Table 1). The term histiocytic lymphoma was also introduced in this classification, replacing the reticulum cell sarcoma of the past, implying that these large cells arose not from the reticular framework but from the histiocytic cells of the nodal tissue. This classification also defined the undifferentiated cell lymphomas, dividing them into two groups, the pleomorphic group and the Burkitt's type. The latter was a unique malignant lymphoma

Table 1
RAPPAPORT CLASSIFICATION OF MALIGNANT LYMPHOMA (1966)

Nodular → Diffuse

Well-differentiated lymphocytic
Poorly differentiated lymphocytic
Mixed lymphocytic-histiocytic
Histiocytic
Undifferentiated (including Burkitt's tumor)

of the jaw seen in Africa, described by Burkitt[18] in 1958. Histologically similar tumors were recognized in the U.S. later, and in 1969 Berard et al.[19] gave a histopathological definition of Burkitt's tumor. The Rappaport classification has gained wide acceptance and is still in use in many institutions. This classification has also been modified over the years to include more recently described entities such as convoluted cell lymphoma[20] and nonconvoluted lymphoblastic lymphoma.[21] Although this classification has proved its clinical usefulness, the terminology appears to be inappropriate in the light of modern concepts of immunology and lymphocyte function. In 1974 Lukes and Collins[22] detailed the immunological characterization of the neoplastic cells in malignant lymphomas and proposed a new classification correlating immunologic and morphologic aspects of the cell of origin. Three other major classifications were also proposed around this time — Dorfman,[23] Bennett et al.[24] (British classification), and Gerrard-Marchant et al.[25] and Lennert et al.[26] (Kiel classification). In 1976, the World Health Organization (WHO) published an international classification of lymphomas.[27] With at least six well-described histopathologic systems in use for classifying lymphomas, a need for a multi-institutional study to resolve the issue objectively was perceived. Such a study was planned and sponsored in 1976 by the National Cancer Institute (NCI) with a panel of experts composed of the proponents of the six classification systems in use. Other experts in the field also contributed to the evaluation of the six systems. This study[28] resulted in the development of a working formulation (Table 4) of the non-Hodgkin's lymphomas for clinical usage. The formulation (published in 1982) was proposed as a means for translation of terminology between current classifications and for comparison of clinical therapeutic trials, and not as a new classification in itself. The new formulation lists the subtypes in groups of increasing grades of malignancy. It also recognizes the significance of a nodular (follicular) pattern and recognizes the fact, borne out by immunological studies, that most large cell lymphomas previously termed reticulum cell sarcoma or histiocytic lymphomas are, in fact, composed of lymphoid cells (either B or T cells) that have undergone transformation.

II. PATHOLOGY OF HODGKIN'S DISEASE

Hodgkin's disease is a unique form of lymphoma characterized histologically by a proliferation of malignant cells (Reed-Sternberg cells and their mononuclear variants) mixed with varying amounts of fibrous tissue, lymphocytes, histiocytes, plasma cells, and eosinophils. The Reed-Sternberg cells are large cells with two or more nuclei, each containing a single prominent nucleolus. Presence of these cells (or the mononuclear variants) and an appropriate cellular environment as described by Lukes and Butler[8] are necessary for the diagnosis of Hodgkin's disease. Lukes and Butler[8] recognized six different histological patterns which appeared to correlate well with clinical stage and aggressiveness of the disease. This classification reflected the inverse relationship existing between the number of lym-

Table 2
COMPARISON OF DIFFERENT CLASSIFICATIONS OF HODGKIN'S DISEASE

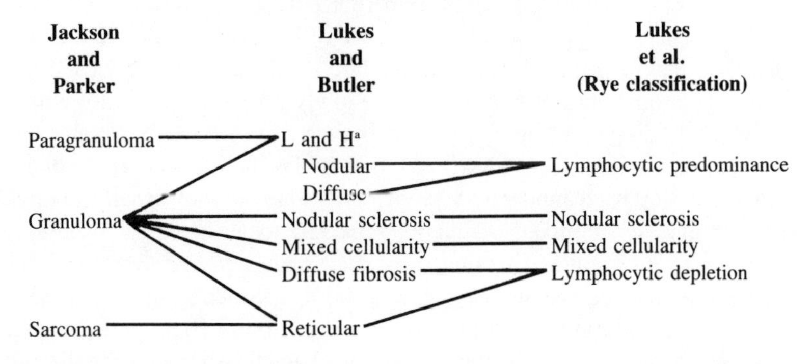

[a] Lymphocytic and/or histiocytic.

Reproduced with permission from Butler, J. J., Histopathology of Malignant Lymphomas and Hodgkin's Disease, in M. D. Anderson Hospital and Tumor Institute, *Leukemia-Lymphoma*. Copyright © 1970 by Year Book Medical Publishers, Inc., Chicago.

phocytes and the number of Reed-Sternberg cells in Hodgkin's disease and also recognized the unique pattern of nodular sclerosing Hodgkin's disease initially described by Lukes and Butler[29] in 1958. At the Rye symposium in 1965, this classification was modified to four subtypes of Hodgkin's disease. The relationship between these two classifications and the Jackson and Parker classification is depicted in Table 2.

Lymphocytic predominance type — This type of Hodgkin's disease is characterized by either a nodular or a diffuse pattern of mixtures of lymphocytes and histiocytes with only rare Reed-Sternberg cells. The diffuse type may show lymphocytes exclusively and may be mistaken for lymphocytic lymphoma if the rare Reed-Sternberg cells are not identified. Also seen in this pattern are atypical large cells with multilobulated nuclei (L and H cells), probably representing precursors of R-S cells. Necrosis and fibrosis are not seen.

Mixed cellularity — This type is characterized by a decrease in the number of lymphocytes, increase in the number of Reed-Sternberg cells and, usually, a significant infiltration of eosinophils. Small areas of necrosis may be seen.

Lymphocytic depletion — This has two patterns. The lymphocytes are decreased in both. In one type (diffuse fibrosis) there is a progressive decrease in cellularity, especially of lymphocytes, resulting in an acellular pattern with deposition of proteinaceous material. Rare Reed-Sternberg cells are present. In the second type (reticular), there is a marked increase in Reed-Sternberg cells and their mononuclear variants. Some of the cases show many bizarre pleomorphic cells. Distinction from pleomorphic large cell lymphoma may be difficult in this subtype.

The above three types represent histologic stages with possible progression from one to the next stage.

Nodular sclerosis type — Diagnosis of this type of Hodgkin's disease is based on the presence of broad septa of collagenous tissue subdividing the lymph node into nodules containing atypical cells lying in spaces or lacunae. These cells, called lacunar cells, are characteristic of this type of Hodgkin's disease and probably represent a variant of Reed-Sternberg cells with the appearance of cells lying in a lacunar space probably representing an artifact of formalin fixation. In some cases, fibrosis may not be evident and large numbers of lacunar cells may be seen, suggesting an early ''cellular'' phase of Hodgkin's disease of the nodular sclerosis type. Classic Reed-Sternberg cells are often difficult to identify. Eosinophils may be seen in large numbers. Large areas of necrosis may also be seen.

This separation of Hodgkin's disease into histologic subtypes is subjective and can, at times, be extremely difficult.[30,31] Attempts are made to keep the prognostically significant groups (LP, NS, and LD) as pure as possible, leaving an admittedly less clearly defined mixed cellularity group. The classification into four subtypes is important for the clinical implication of the separate types. Hodgkin's disease of lymphocytic predominance type (10 to 15%) is usually seen in young males, is frequently localized in the cervical region, is not associated with symptoms, and usually reflects a slowly progressive disease with favorable prognosis. Hodgkin's disease of nodular sclerosis type (40 to 50%) is usually localized to lower cervical, supraclavicular, or mediastinal lymph nodes, is seen usually in young females, and is considered a slowly to moderately progressive disease with usually favorable prognosis. Hodgkin's disease of mixed cellularity type (20 to 40%) tends to be widespread, occurs most often in older males, is often associated with systemic symptoms and is considered a moderately progressive disease with a good-to-guarded prognosis. Hodgkin's disease of lymphocytic depletion type (5 to 10%) is seen in older patients with symptomatic, widespread disease at the time of diagnosis. This is characteristically a rapidly progressive disease with a generally poor prognosis.

III. PATHOLOGY OF NON-HODGKIN'S LYMPHOMA

As mentioned above, the Rappaport system of classification and terminology is still widely used and serves for purposes of clinical correlation with histologic subtypes. A modification of this system[30] (Table 3) was used in the studies of copper and lymphomas carried out at the University of Texas M. D. Anderson Hospital detailed elsewhere in this book. This classification is still valid, with the exception of the term histiocytic lymphoma. As mentioned before, most of these large cell lymphomas are not of histiocytic origin but represent neoplasms of transformed lymphocytes. The term ''large cell lymphomas'' is used as a noncomittal term with further characterization such as follicle center cell type, immunoblastic type, or T-cell type suggested, only if possible on the basis of routine morphology. All current classifications recognize the fact that identification of a nodular (''follicular'') pattern in a lymphoma is of prognostic importance even when nodularity is seen only in portions of the tumor. Diagnosis of the histological type of lymphoma requires, first, a characterization of the pattern either as nodular or diffuse, and then a recognition of the cell type. Familiarity with different morphologic and functional cell types seen in normal reactive lymph nodes is essential for identification of the cell of origin of a malignant lymphoma.

Application of immunological methods to study cell membranes has led to identification of two functionally distinct types of lymphocytes, the T cells and the B cells, the former involved in cell-mediated immunity and the latter in humoral immunity. From an immunological standpoint, the lymphoid system can be considered to have three components:

1. The bone marrow (stem cell) which is the site of origin of lymphocytes.
2. The primary (or central) lymphoid organ (thymus or Bursa equivalent) which is the site of lymphocyte maturation (T and B cells).
3. The peripheral lymphoid organ (lymph nodes, spleen, and gut-associated lymphoid tissue) which is the site of antigen recognition with production of appropriate response.

Lymphocytes are derived from a pleuripotential stem cell in the bone marrow. This cell contains a DNA-replicative enzyme, terminal deoxynucleotidyl transferase (TdT), but does not show cytoplasmic or surface immunoglobulins. The bone marrow also shows smaller numbers of TdT positive cells with cytoplasmic immunoglobulins (pre-B cells). Also seen are some TdT positive cells probably representing pre-T cells. Some of the cells from the bone marrow travel to the thymus where they differentiate into thymus-dependent lympho-

Table 3
M.D. ANDERSON HOSPITAL CLASSIFICATION (1969)

Malignant Lymphoma

Nodular
- Lymphocytic
  - Well differentiated
  - Poorly differentiated
- Mixed
- Histiocytic (RCS)
- Undifferentiated (RCS)
  - Type A (Burkitt's)

Diffuse
- Lymphocytic
  - Well differentiated
  - Poorly differentiated
- Histiocytic (RCS)
- Undifferentiated (RCS)
  - Type A (Burkitt's)
  - Type B

Note: RCS = reticulum cell sarcoma.

Reproduced with permission from Butler, J. J., Histopathology of Malignant Lymphomas and Hodgkin's Disease, in M. D. Anderson Hospital and Tumor Institute, *Leukemia-Lymphoma*. Copyright © 1970 by Year Book Medical Publishers, Inc., Chicago.

cytes (T cells) and acquire T-cell surface markers. The T lymphocytes leave the thymus and reach the peripheral lymphoid organs. In the lymph nodes (Figure 1), the T cells are concentrated in the deeper paracortical areas; in the spleen they reach the periarterial lymphoid sheaths.

Some of the TdT-positive cells from the bone marrow mature not in the thymus but in the mammalian equivalent of the Bursa of Fabricius found in birds. This primary lymphoid organ is not clearly identified, but may be represented in the bone marrow itself or in the gut-associated lymphoid tissue. These ''B'' cells lose the TdT marker, but acquire unique surface markers and end up in the follicles and medullary cords of the lymph node (Figure 1) and the follicles of the splenic white pulp. Under antigenic stimulation, the B cells are transformed into larger cells (the immunoblasts) and eventually into plasma cells which secrete immunoglobulins (antibodies) of various classes. Subsets of T cells with suppressor and helper effect on antibody production by B cells have also been identified. The T and B cells can be distinguished by surface marker studies, by specific antisera, and by demonstration of immunoglobulins by fluorescent microscopy. Some lymphocytes, especially in some forms of acute lymphocytic leukemia, show no surface markers and have been called Null cells. Approximately 70 to 80% of the normal peripheral blood lymphocytes are T cells and 15 to 20% B cells.

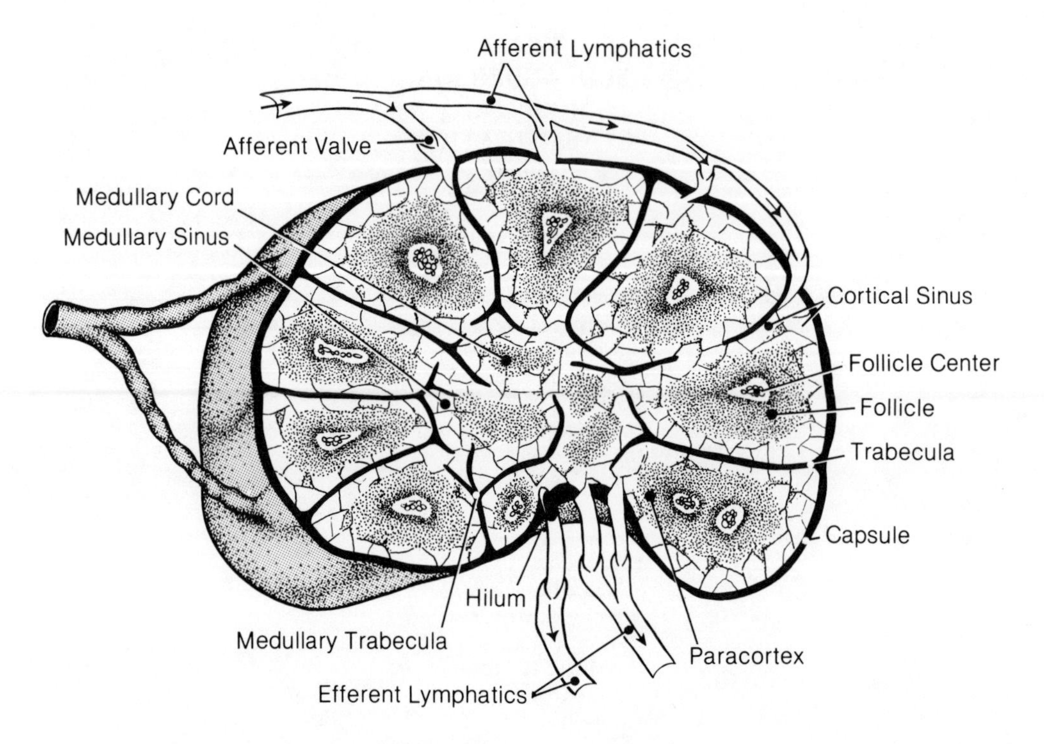

FIGURE 1. Diagrammatic representation of a lymph node.

In addition to T and B cells, the lymph node tissue contains dendritic reticulum cells in the follicles and phagocytic mononuclear cells (histiocytes, macrophages) in the follicles, in the paracortex, and in the sinuses. Histiocytes have some of the surface markers that are on B lymphocytes but can be distinguished by electron microscopy and by histochemical demonstration of lysosomal enzymes in their cytoplasm. The histiocytes play a role in antigen processing and act as effector cells for the lymphocytes. The dendritic reticulum cells probably represent cells of the reticular framework of the follicles and are associated with production of reticulin fibers. It is felt that these cells help trap the antigen within the lymphoid follicle. They can be identified by special histochemical means and by electron microscopy. Presence of these cells within lymphomatous tissue has been used by some as proof of follicular origin of the lymphoma.

Details of the various markers present in and on the lymphoid cells are beyond the scope of this book and are available in several recent reviews of the topic.[32-35] Application of T and B cell-typing techniques to cells of malignant lymphomas has resulted in several observations. Most of the non-Hodgkin's lymphomas are B cell neoplasms. Some of these such as the well-differentiated lymphocytic lymphoma (chronic lymphocytic leukemia), probably arise from the B cells in the medullary cords and have a diffuse pattern. Most other B cell lymphomas arise within the follicles and have, at least initially, a nodular ("follicular") pattern. Morphologically, these lymphoma cells show a variation in nuclear shape and size corresponding to that seen in normal reactive follicles. The terms cleaved and noncleaved cells (small and large) were given by Lukes and Collins[22] to these cells. The cleaved cells show a nuclear cleft and do not show nucleoli and the noncleaved cells have round or oval nuclei with one or two nucleoli. These cells represent various stages of transformation of the B lymphocytes within the follicles and do not indicate stages of differentiation of the tumor. Transformation of small lymphoid cells into large cells with pyroninophilic cytoplasm (immunoblasts) has been observed in normal nodes in centers of follicles, and, more frequently, in the interfollicular tissue. Some large cell lymphomas

Table 4
WORKING FORMULATION OF NCI FOR CLINICAL USAGE

Low-grade
- Malignant lymphoma, small lymphocytic
- Malignant lymphoma, follicular, predominantly small-cleaved cell
- Malignant lymphoma, follicular, mixed small-cleaved and large-cell

Intermediate-grade
- Malignant lymphoma, follicular, predominantly large-cell
- Malignant lymphoma, diffuse, small-cleaved cell
- Malignant lymphoma, diffuse, mixed small- and large-cell
- Malignant lymphoma, diffuse, large-cell

High-grade
- Malignant lymphoma, large-cell, immunoblastic
- Malignant lymphoma, lymphoblastic
- Malignant lymphoma, small noncleaved cell

Miscellaneous
- Composite malignant lymphoma
- Mycosis fungoides
- Extramedullary plasmacytoma
- Unclassified
- Other

composed largely of interfollicular immunoblasts have been called immunoblastic sarcoma. Malignant lymphomas of T-cell type are relatively uncommon. About 1% of the small well-differentiated lymphocytic lymphomas are of T-cell type and show a more aggressive clinical behavior than the B-cell type with frequent skin involvement. Sezary syndrome (a T-cell leukemia with erythroderma) and mycosis fungoides are T-cell neoplasms. A T-cell lymphoma of lymph nodes (peripheral T-cell lymphoma) has also been described. Although surface marker studies are generally needed for distinction between T- and B-cell lymphomas, morphologic differences can be appreciated on light microscopy in many cases and better appreciated on electron microscopy of the tumor cells.[36]

A. Subtypes Of Non-Hodgkin's Lymphoma

The individual subtypes of lymphoma will be discussed in the order listed in the NCI formulation (Table 4).

1. Low-Grade Lymphomas

Small lymphocytic type — This is a diffuse lymphoma composed of B cells probably arising in medullary cords and representing the tissue counterpart of chronic lymphocytic leukemia. These cells are small- to medium-sized round lymphocytes. Some cases may show transformation to plasma cells (plasmacytoid type) and be associated with monoclonal gammopathy; 1% of these may be of T-cell type.

Follicular, predominantly small-cleaved cell type — This is the commonest type of lymphoma (22.5%). This is a nodular lymphoma which may have diffuse areas and also areas of sclerosis. It is composed of B cells arising in follicles showing morphological features of small lymphoid cells with ''cleaved'' nuclei. Large noncleaved cells may be present in small numbers.

Follicular, mixed small-cleaved and large-cell type — This is a nodular lymphoma which may have diffuse areas and/or sclerosis. It arises from follicular B cells, has a mixture of cells with small-cleaved nuclei and large-cleaved or noncleaved nuclei without preponderance of one cell type.

2. Intermediate-Grade Lymphomas

Follicular, predominantly large-cell type — In this nodular lymphoma of follicular B-cell origin there is a predominance of large cleaved or, more often, noncleaved nuclei. Diffuse areas and areas of sclerosis are often seen.

Diffuse small-cleaved cell type — This is a diffuse lymphoma of follicular B-cell origin composed of small-cleaved cells probably representing an advanced stage of the follicular, predominantly small-cleaved cell type.

Diffuse mixed small- and large-cell type — This is a heterogenous group, some representing diffuse stage of the follicular, mixed small-cleaved and large-cell type, and some others composed of small cells and large cells with noncleaved nuclei with T cell markers (probably arising in the paracortical region).

Diffuse large-cell type — About 20% of the non-Hodgkin's lymphomas are of this type. There is a diffuse proliferation of large cells with cleaved or noncleaved nuclei (of follicular origin) reflecting a B-cell neoplasm. Small numbers of small-cleaved cells may be present. Sclerosis is prominent in some cases, resulting in a pattern suggestive of an epithelial neoplasm.

3. High-Grade Malignant Lymphomas

Large-cell, immunoblastic type — This lymphoma shows a diffuse pattern of proliferating immunoblasts probably not of follicular origin. Most are of B-cell type (interfollicular) and have plasmacytoid features. Most of the lymphomas arising in previous polyclonal or monoclonal lymphoproliferative disorders such as alpha-chain disease, angio-immunoblastic lymphadenopathy, and Sjogren's syndrome are of this type. Lymphomas arising in immunosuppressed transplant recipients are also of this type. In addition to the plasmacytoid type, clear cell and polymorphous variants are also seen, probably reflecting T-cell immunoblastic proliferation. These T-cell lymphomas show a mixture of small cells with twisted nuclei, larger clear cells, and pleomorphic cells with hyperlobated or multiple nuclei. Some of these large cells may resemble Reed-Sternberg cells. The T-cell lymphomas in this group are also called peripheral T-cell lymphoma to be distinguished from the "central" T-cell lymphomas originating in the thymus that are described later. When associated with a prominent epithelioid histiocytic component, the polymorphous variant may be indistinguishable from the so-called Lennert's lymphoma as defined by Burke and Butler.[37]

Lymphoblastic lymphoma[21] — This is a diffuse lymphoma composed of lymphoblasts corresponding to the immature lymphoid cells of acute lymphocytic leukemia. Some cases show cells with T-cell markers and some other have no surface markers. Morphologically the majority of the cases show convoluted nuclei reflecting T-cell origin. Prominent mediastinal involvement in these cases suggests a thymic (central) T-cell lymphoma. The nonconvoluted cell variant shows cells with round nuclei having fine linear subdivisions. These lymphomas have a high propensity for leukemic transformation and are usually seen in young adults.

Small noncleaved cell type — There are two morphologic cell types. Some with uniform nuclei the size of macrophage nuclei fulfill the morphologic requirements for Burkitt's tumor.[18,19] These appear to be composed of B cells of follicular origin although a nodular pattern is rare. The others show nuclear pleomorphism and appear to be tumors of undifferentiated cells with no immunologic markers.

4. Miscellaneous Group

Composite lymphoma — A term used for the presence of two distinct forms of non-Hodgkin's lymphoma (or, rarely, non-Hodgkin's lymphoma and Hodgkin's disease) in a single organ or tissue.

Mycosis fungoides — A T-cell lymphoma with skin involvement as a prominent clinical feature. Sezary syndrome is a variant with leukemic picture not included among lymphomas.

Extramedullary plasmacytomas — May arise in lymphoid tissue. Although multiple myeloma is a B-cell neoplasm, it is not included in the formulation of lymphomas.

True histiocytic lymphomas — Are extremely rare and require electron microscopy and identification of appropriate cell markers by immunological and histochemical means. Histiocytic medullary reticulosis, a true histiocytic neoplasm, is not included in this formulation.

Unclassified lymphomas — Are lymphomas that cannot be classified by morphologic or other means.

REFERENCES

1. **Hodgkin, T.,** On some morbid appearances of the absorbent glands and spleen, *Trans. Med. Chir. Soc. London,* 17, 68, 1832.
2. **Virchow, R.,** Weisses blut, *Neue Notizen aus den Gebiete der Natur und Heilkunde (Froriep's neue Notizen),* 36, 151-156, 1845.
3. **Bilroth, T.,** Multiple lymphome. Erfolgreiche Behandling mit arsenik, *Wein Med. Wochengchie,* 21, 1066-1067, 1871.
4. **Langhans, T.,** Das Maligne Lymphosarkom (Pseudoleukamie), *Virchow's Arch. Pathol. Anat.,* 54, 509-537, 1872.
5. **Sternberg, C.,** Ueber eine eigenartige unter dem Bilde der Pseudoleukamie verlaufende Tuberculose des lymphatischen Apparates, *Heilkunde (Berlin),* 19, 21, 1898.
6. **Reed, D. M.,** On the pathological changes in Hodgkin's disease, with special reference to its relationship to tuberculosis, *Johns Hopkins Hosp. Rep.,* 10, 133, 1902.
7. **Jackson, H., Jr. and Parker, F., Jr.,** *Hodgkin's Disease and Allied Disorders,* Oxford University Press, New York, 1947.
8. **Lukes, R. J. and Butler, J. J.,** The pathology and nomenclature of Hodgkin's disease, *Cancer Res.,* 26, 1063, 1966.
9. **Lukes, R. J., Butler, J. J., and Hicks, E. B.,** Natural history of Hodgkin's disease as related to its pathologic picture, *Cancer,* 19, 317, 1966.
10. **Lukes, R. J., Craver, L. F., Hall, T. C., Rappaport, H., and Ruben, P.,** Report of the nomenclature committee, *Cancer Res.,* 26, 1311, 1966.
11. **Dreschfeld, J.,** Clinical lecture on acute Hodgkin's (or pseudoleucocythemia), *Br. Med. J.,* 1, 893-896, 1892.
12. **Brill, N. E., Baehr, G., and Rosenthal, N.,** Generalized giant follicle hyperplasia of lymph nodes and spleen, a hitherto undescribed type, *J. Am. Med. Assoc.,* 84, 668-671, 1925.
13. **Symmers, D.,** Follicular lymphadenopathy with splenomegaly. A newly recognized disease of lymphatic system, *Arch. Pathol. Lab. Med.,* 3, 816-820, 1927.
14. **Roulet, F.,** Dasprinare Retothelsarkom der Lymphkonten, *Virchow's Arch. Pathol. Anat.,* 277, 15-47, 1930.
15. **Gall, E. A. and Mallory, T. B.,** Malignant lymphoma: a clinico-pathological survey of 618 cases, *Am. J. Pathol.,* 18, 381-429, 1942.
16. **Rappaport, H., Winter, W. J., and Hicks, E. B.,** Follicular lymphoma: a re-evaluation of its position in the scheme of malignant lymphoma, based on a survey of 253 cases, *Cancer,* 9, 792-821, 1956.
17. **Rappaport, H.,** Tumors of the hematopoietic system, in *Atlas of Tumor Pathology,* Sect. III, Fasicle 8, Armed Forces Institute of Pathology, Washington, D.C., 1966, 97-161.
18. **Burkitt, D.,** A sarcoma involving the jaws in African children, *Br. J. Surg.,* 46, 218-223, 1958.
19. **Berard, C. W., O'Conor, G. T., Thomas, L. B., and Torloni, H., Eds.,** Histopathological definition of Burkitt's tumor, *Bull. WHO,* 40, 601-607, 1969.
20. **Barcos, M. P. and Lukes, R. J.,** Malignant lymphoma of convoluted lymphocytes — A new entity of possible T cell type, in *Conflicts in Childhood Cancer, An Evaluation of Current Management,* Vol. 4, Alan R. Liss, New York, 1975, 147-178.
21. **Nathwani, B. N., Kim, H., and Rappaport, H.,** Malignant lymphoma, lymphoblastic, *Cancer,* 38, 964-983, 1975.
22. **Lukes, R. J. and Collins, R. D.,** Immunological characterization of human malignant lymphomas, *Cancer,* 34, 1488-1505, 1974.
23. **Dorfman, R. F.,** Classification of non-Hodgkin's lymphomas, *Lancet,* 1, 1295-1296, 1974.
24. **Bennett, M. H., Farrer-Brown, G., Henry, K., and Jellife, A. M.,** Classification of non-Hodgkin's lymphomas, *Lancet,* 2, 405-406, 1974.

25. **Gerrard-Marchant, R., Hamlin, I., Lennert, K. Rilke, F., Stansfeld, G. A., and van Unnik, M. A. J.,** Classification of non-Hodgkin's lymphomas, *Lancet,* 2, 406-408, 1974.
26. **Lennert, K., Mohri, N., Stein, H., and Kaiserling, E.,** The histopathology of malignant lymphomas, *Br. J. Hematol.,* Suppl. 31, 193-203, 1975.
27. **Mathe, G., Rappaport, H., O'Conor, G. T. and Torloni, H.,** Histological and cytological typing of neoplastic diseases of hematopoietic and lymphoid tissues, WHO International Histological Classification of Tumors No. 14, World Health Organization, Geneva, 1976.
28. **Multiple Participants — Workshop,** National Cancer Institute sponsored study of classifications of non-Hodgkin's lymphomas, *(the NHLPC project), Cancer,* 49, 2112-2135, 1982.
29. **Lukes, R. J. and Butler, J. J.,** Spontaneous nodular sclerosis in Hodgkin's disease (abstr.), in *Jt. Annu. Meet. C.A.P. and A.S.C.P.,* Physicians' Record Press, Chicago, 1958, 56.
30. **Butler, J. J.,** Histopathology of malignant lymphomas and Hodgkin's disease, in *Leukemia—Lymphoma,* Year Book Medical Publishers, Chicago, 1970, 123-142.
31. **Neiman, R. S.,** Current problems in the histopathologic diagnosis and classification of Hodgkin's disease, *Pathol. Annu.,* 13(2), 289-328, 1978.
32. **Aisenberg, A. C.,** Current concepts in immunology: cell surface markers in lymphoproliferative disease, *N. Engl. J. Med.,* 304, 321-336, 1981.
33. **Detrick-Hooks, B. and Bernard, A.,** T and B lymphocytes: Assays in malignant and non-malignant disease, *Lab. Manage.,* 41-52, May 1981.
34. **Berard, C. W. Greene, M. H., Jaffe, E. S., Magrat, I., and Ziegler, J.,** A multidisciplinary approach to non-Hodgkins lymphoma — NIH conference, *Ann. Int. Med.,* 94, 218-235, 1981.
35. **Braylan, R. C., Jaffe, E. S., and Berard, C. W.,** Malignant lymphomas: current classification and new observations, *Pathol. Annu.,* 10, 213-270, 1975.
36. **Said, J. W., Hargreaves, H. K., and Pinkus, G. S.,** Non-Hodgkin's lymphomas: an ultrastructural study correlating morphology with immunologic cell type, *Cancer,* 44, 504-528, 1979.
37. **Burke, J. S. and Butler, J. J.,** Malignant lymphoma with high content of epithelioid histiocytes (Lennert's lymphoma), *Am. J. Clin. Pathol.,* 66, 1-9, 1976.

Chapter 8

NATURAL HISTORY OF THE LYMPHOMAS AND STAGING OF DISEASE

Malignant lymphomas form a clinical and histopathological spectrum of neoplastic disorders arising from lymph nodes and the lymphoid components of other tissues. These malignancies are characterized by certain hallmarks and characteristics which bear directly upon diagnosis, histologic patterns, clinical behavior, prognosis, and selection of therapy.[1-32]

The etiology of malignant lymphoma is not known. Viruses,[33-38] radiation,[39-41] genetic,[42,43] familial,[43-45] environmental,[45,46] occupational factors,[47] immunosuppressive[48-50] and anticonvulsant drugs,[51] and associated diseases[52-57] have been implicated as possible causative factors.

With regard to the cellular origins of the lymphomas, in most cases they can be classified by cell of origin.[3,4,58-63] The majority of the non-Hodgkin's lymphomas appear to derive from monoclonal populations of B cells, or have no distinctive cell markers. Undifferentiated and lymphoblastic lymphomas appear to derive from T cells. The cell of origin of Hodgkin's disease is still uncertain, but it may be derived from either T-lymphocytes or from the macrophage line.[3,4,15,56-63]

Initially, correct tissue diagnosis and classification of lymphoma is of utmost importance in the subsequent proper management of these malignancies.[64] A diagnosis and histological subclassification of lymphoma can be made only by microscopic examination of excised lymph nodes or tumor mass. Needle aspiration cytology and/or Giemsa-stained imprints or smears of fresh lymph node biopsy material may suggest a diagnosis, but do not yield sufficient tissue for viewing the topographic relationship of the different cell forms in an organized tissue setting, in order to subclassify the type of lymphoma.[3,4] Reliance on cytological examination of lymphoma tissue is not advisable and the error rate is high.[4] Even experienced pathologists, using fixed sections of lymph nodes, disagree on subclassification of lymphoma in up to 25% of the cases and indeed disagree as to whether the resected tissues show evidence of malignant disease in as many as 6% of the cases.[25]

The patterns of presentation and anatomic distribution (lymph node or organ involvement) are strikingly different in Hodgkin's disease as compared to non-Hodgkin's lymphoma. The usual presentation of Hodgkin's disease is a painless enlargement of a superficial lymph node or nodal group, usually without associated symptoms, incidentally found while bathing, shaving, or on routine physical examination. The cervical rubbery, discrete nodes on both sides of the neck are found in approximately 60 to 80% of the patients.[3-6,12-14,16-23,26,27,65] Axillary lymphadenopathy as initial presentation is found in approximately 6 to 20% of Hodgkin's disease cases and enlarged inguinal nodes in 6 to 12% of cases.[3-6] On the contrary, in non-Hodgkin's lymphoma extranodal presentation is common. Peters et al.[66] found the incidence to be as high as 61% for reticulum cell sarcoma and 40% for lymphosarcoma. Of 100 consecutive laparotomy staged patients in the series of Johnson et al.,[67] 64% were found to have extranodal disease. In the University of Texas M.D. Anderson Hospital experience, primary extranodal presentations were found in 58% of localized disease (stage I and II) patients with diffuse non-Hodgkin's lymphomas treated between 1961 and 1969.[68] In 61% of these cases, the extranodal sites were limited to the head and neck. According to DeVita and Hellman,[4] localized nodal disease in non-Hodgkin's lymphoma is present in less than 10% of cases at the time of initial diagnosis. Another feature of Hodgkin's disease consists of "centripetal" lymph node disease which tends to be in axial lymph nodes, frequently localized at the time of presentation, and characteristically showing contiguous spread via lymphatics to adjacent lymph nodes.[3,4] On the other hand, in non-Hodgkin's lymphoma, lymph node disease is often "centrifugal" and noncontiguous. Involvement of Waldeyer's ring, epitrochlear, popliteal and mesenteric nodes, testes, and gastrointestinal tract is more commonly found in patients with non-Hodgkin's lymphoma than in Hodgkin's disease patients, and is quite rarely localized at the time of diagnosis. Mediastinal presentation of Hodgkin's disease is seen in approximately 50% of patients,[3] whereas in non-Hodgkin's lymphoma the occurrence is about 20%.[69] Organ involvement (bone marrow, liver, and lung) is uncommon in patients with Hodgkin's disease at the time of diagnosis, whereas it is not infrequently found in non-Hodgkin's lymphoma.[3,4,70-75]

Systemic symptoms consisting of fever, night sweats, weight loss in excess of 10% of body weight, and perhaps pruritus during the 6 months preceding the diagnosis, occur in 40% of patients with Hodgkin's disease; approximately 20% with non-Hodgkin's lymphoma may have some combination of the above symptoms, with fever being the commonest.[2-4,24,55] Symptomatic patients, even with seemingly isolated lymphadenopathy, should be thoroughly investigated for intrathoracic or intraabdominal disease, or occult extranodal sites of disease such as the liver and bone marrow. Additional symptoms include weakness, fatigue, anorexia, and alcohol-induced pain in small minority of cases.[3,4,55,76,77] Enlarged lymph nodes can produce local symptoms depending on primary sites or organ involvement. Mediastinal lymphadenopathy may produce nonproductive cough and dyspnea, whereas enlarged lymph node masses in peripheral sites can cause pain, swelling, and specific neuropathy of the involved extremity.[3-6,12,17-27,55] Other symptoms and signs of malignant lymphomas, although rare at initial presentation include superior or inferior vena cava syndrome, spinal cord compression, obstructive jaundice, urethral obstruction, steatorrhea, chronic malabsorption syndrome, pleural and pericardial effusions, specific and nonspecific skin lesions, bone lesions, hematological, immunological, metabolic, and endocrine abnormalities.[2-6,16-23,26,65]

Hodgkin's disease is now considered by most experts to be unifocal in origin and to spread contiguously by involving adjacent lymph node structures first. If tumor growth is left untreated, either direct extension into adjacent organs occurs or blood vessel invasion results in dissemination of the disease to the viscera (spleen, liver, lung, bone marrow, or other organs). Staging laparotomy, introduced by the Stanford group in patients with Hodgkin's disease, revealed that the spleen, although not palpable, is often involved with disease, suggesting hematogenous dissemination from the primary tumor site with subsequent spread to the splenic hilar and retroperitoneal nodes and the liver.[3,4,78-83] More rapid spread and incidence of spleen involvement are more marked in mixed cellularity and lymphocytic depletion types than in lymphocytic predominance and nodular sclerosing forms, in which disease tends to remain localized longer.[3,4,7-12,66,67] The histologic subtypes, as discussed previously, have a bearing on the natural history of Hodgkin's disease and appear to exhibit significantly different patterns of lymph node and extra lymphatic distribution.[3,4,7-12,66] The lymphocytic predominance form is usually localized disease involving peripheral superficial lymph nodes and seldom involves mediastinal or hilar nodes. The nodular sclerosing type tends to involve the lower cervical-supraclavicular and the anterior or middle mediastinal nodes, with or without associated hilar node involvement. On the other hand, the mixed cellularity and lymphocytic depletion types are more often seen in generalized disease, tend to spare the mediastinal and hilar nodes, and have a greater propensity for extralymphatic involvement and noncontiguous pattern of spread. In general terms, some of these major relationships between four types of Hodgkin's disease may be summarized as follows.

The histologic pattern of Hodgkin's disease with lymphoid predominance is characteristically associated with (1) localized disease, (2) preservation of immunologic functions, and (3) long survival.

The histologic pattern of Hodgkin's disease with lymphoid depletion is characteristically associated with (1) disseminated disease, (2) loss of immunologic functions, and (3) short survival.

The histologic patterns of nodular sclerosis and mixed cellularity display clinical features occupying a middle ground between the extremes outlined above.[1,3,4]

Fuller et al.,[68] based on their analyses of survival in patients with different histologic types of Hodgkin's disease, have suggested that the type has less prognostic value (at least for lymphocytic predominance, nodular sclerosing, and mixed cellularity) than was thought formerly, due to marked improvement in the results of therapy. Host factors, including immunological functions not yet defined clearly in addition to proper management of disease, certainly play important roles in the longevity of these patients. Somewhat erratic relapse patterns following treatment-induced remissions have been observed rather commonly in these patients. Long-lasting initial remission and longer survival are, of course, interdependent. The course of Hodgkin's disease left untreated, regardless of stage, is short (1 to 2 years) and most patients die within 2 years as a consequence of deterioration of immunological functions and subsequent fatal infections, or organ failures.[84]

The natural history of non-Hodgkin's lymphoma is more erratic and less predictable. The clinical patterns of presentation and spread of non-Hodgkin's lymphoma relate to the issue of unicentric vs. multicentric origin. The disease may be unifocal in origin, but if so it tends to become generalized in a short period of time, so that disseminated non-Hodgkin's lymphoma at the time of diagnosis is the rule rather than the exception.[4] Patients with the nodular (follicular) lymph node pattern have more generalized disease, although often clinically indolent, than those with diffuse types. A substantial number of patients with nodular lymphoma (41%) will progress to a diffuse variety as a part of the natural history of their disease.[4,85] Involvement of bone marrow and liver, frequently seen in patients with non-Hodgkin's lymphoma at the time of diagnosis, does not impart the same degree of adverse prognosis as patients with Hodgkin's disease. Survival rates in patients with generalized non-Hodgkin's lymphoma confined to lymph nodes (stage III) are similar to those with organ involvement (stage IV).[86,87]

Lymphomas of diffuse small cell (well-differentiated lymphocytic) type disseminate widely at an early stage, but may have an indolent course for many years.[88] The associated peripheral blood picture is most commonly that of chronic lymphocytic leukemia. The common causes of death are bone marrow failure and hypogammaglobulinemia resulting in susceptibility to fatal infections.

Other non-Hodgkin's lymphoma patterns with leukemia potentiality include the poorly differentiated lymphocytic type in which the leukemic potentiality is moderate, retaining the basic cell type in its leukemic form; very commonly the peripheral blood pattern is that of "lymphosarcoma cell leukemia". Poorly differentiated lymphocytic lymphoma of the convoluted cell type and the undifferentiated lymphomas often terminate in acute leukemia, lymphocytic or stem cell type.[1,2,4,6] With the histologic pattern of large cell lymphoma, the leukemic potential appears to be minimal.

Regarding grades of malignancy, non-Hodgkin's lymphomas can be divided into three categories:[4,20]

1. Low-grade — comprising diffuse well-differentiated lymphocytic lymphoma and/or chronic lymphocytic leukemia, nodular poorly differentiated lymphocytic lymphoma, and nodular mixed lymphomas
2. Intermediate-grade — comprising nodular large cell lymphoma, diffuse poorly differentiated lymphocytic, and diffuse mixed cell lymphomas
3. High-grade — comprising diffuse large cell lymphoma, diffuse lymphoblastic, and diffuse undifferentiated lymphomas

In untreated non-Hodgkin's lymphomas (except for those of lowest grade) or treatment failures, progressive involvement of other lymph nodes and/or organs occurs with associated symptoms, deterioration of immunological functions, infections, and fatal outcome in less than 2 years in most instances. In general, the nodular lymphoma patients fare better than the diffuse, and well-differentiated forms have a better prognosis than the poorly differentiated or undifferentiated forms.

After the diagnosis and tissue subclassification of Hodgkin's disease or non-Hodgkin's lymphoma is made, staging of the disease is a mandatory procedure, contributing to selection of therapy and prognosis. Staging procedures include CBC, SMA-18 (simultaneous determination of 18 laboratory tests including liver and renal function), urinalysis, chest X-ray, intravenous pyelogram, bilateral lower extremity lymphangiogram, EKG, bone marrow aspiration, and bilateral iliac crest biopsy. Other studies may be indicated based on disease presentation, such as bone survey, whole chest tomography, abdominal CAT scan, and ultrasonogram to supplement lymphangiographic findings and further delineate abdominal lymphomatous disease. Liver biopsy should be obtained if there is a strong clinical suspicion of liver involvement. Staging laparotomy should be performed, if management decisions will depend on the identification of abdominal disease and if contraindications do not exist.[78-83,89-94] Since significant morbidity and mortality associated with staging laparotomy have been reported,[95-100] Sweet et al.[101] have proposed that the accuracy of assessment of the retroperitoneal lymph nodes by CAT scan[102-104] and sonography,[105,106] in conjunction with lymphangiography, is sufficient to replace laparotomy, especially in non-Hodgkin's lymphoma. As a result of thorough staging procedures, including laparotomy, up to 40% or more of patients will be restaged (up- or downstaged).[3,4,98] The importance of staging procedures to determine the extent of disease cannot be overemphasized; this information is crucial to making proper decisions concerning therapy (radiotherapy, chemotherapy, or combined modality treatment).

A distinction should be made between the clinical stage (CS) which is based on preoperative assessment (history and physical findings, laboratory tests, radiographic examinations, and scanning procedures), and pathological stage (PS) based on the findings at the time of laparotomy or other invasive procedures such as laparoscopy, bone marrow, and/or additional lymph node or tissue biopsies.[3,4] Once staging procedures are completed and extent of disease known, patient is assigned into one of four stages of lymphomatous disease according to Ann Arbor classification.[107] All stages are subclassified as "A" or "B" to indicate the absence or presence of systemic symptoms already discussed.

In view of the different clinical behaviors, presentations, and frequent noncontiguous (perhaps hematogeneous) spread of non-Hodgkin's lymphoma, no rigid staging plan is appropriate for all patients.

Since our studies of serum copper changes in malignant lymphomas started in 1966, Rye staging[108] for Hodgkin's disease and Peters[66] for non-Hodgkin's lymphoma was used until 1971; Ann Arbor classification[107] has been used since in both types of malignant lymphomas.

Following the tissue classification and careful staging of lymphomatous disease, rational decisions in regard to treatment can be made. Radiotherapy alone, ranging from involved fields to mantle technique and inverted Y or total abdomen,[3,4,109-127] has been widely employed. In recent years combined treatment modalities (irradiation plus chemotherapy), depending on the stage at presentation and the histological pattern of lymphoma, have been utilized. Radiotherapy has been most useful in early stages.[2-4,17-23,109-127] Recently a great deal of interest has been concentrated on adjuvant chemotherapy for patients believed to have a high risk of relapse.[128-131]

As far as the criteria for response are concerned, complete response or remission is defined as disappearance of all measurable disease by physical examination and laboratory parameters including all radiographs which were originally abnormal. The only exception is mediastinal disease which may never reduce to predisease status, even though disease is not present. Partial response or remission is defined as greater than 50% reduction in the sum of the diameters of the tumor for more than 3 months duration.

Consistently good results in using radiotherapy alone (70 to 90% 10-year survival rates) encouraged Kaplan[3] and DeVita and Hellman[4] to support irradiation as primary treatment for stages I and II Hodgkin's disease. Kaplan's group[128] recently reported the Stanford experience in the management of 230 patients with pathologic stages I to II Hodgkin's disease treated with irradiation alone or irradiation followed by six cycles of adjuvant combination chemotherapy consisting of nitrogen mustard (Mustargen ®), vincristine (Oncovin ®), procarbazine (Matulane ®), and prednisone (MOPP). The actuarial survival at 10 years was 84% for patients in either treatment group. Freedom from relapse at 10 years was 77% among patients treated with irradiation alone and 84% after treatment with combined modality therapy. Freedom from second relapse at 10 years was 89 and 94%, respectively. Patients with large mediastinal masses (mediastinal mass ratio greater than or equal to one third) had a significantly poorer record of freedom from relapse when treated with irradiation alone than when treated initially with combined modality therapy (45 vs. 81% at 10 years). The 10-year survival of these patients, however, was not significantly different (84 vs. 74%).

Chemotherapy alone, radiotherapy alone, high-dose irradiation combined with chemotherapy, and chemotherapy with addition of low-dose irradiation have all been advocated as primary therapy for stage III Hodgkin's disease. A recent analysis of 99 Hodgkin's disease patients with stage IIIA and IIIB treated at the University of Texas M.D. Anderson Hospital with a combined chemotherapy (MOPP)-radiotherapy program showed a complete remission rate of 89% and a 5-year relapse-free and overall survival of 67 and 76%, respectively.[132,133] The Stanford group,[134] using a combination program with chemotherapy preceding radiotherapy, reported improved results with 79% disease-free survival and 84% actuarial survival at 45 months. Bonadonna's group[135,136] showed that patients with advanced Hodgkin's disease could be treated with ABVD [Adriamycin® (Doxorubicin), bleomycin (Blenoxane ®), vinblastine (Velban ®), and dacarbazine (DTIC®)] or MOPP plus low-dose radiation with improved results; 86.7% of all complete responders were alive and disease-free at 4 years, an improvement over the use of chemotherapy alone.

Hellman's group[137] showed that a combined treatment approach, with chemotherapy given preceding and following radiotherapy, is better for IIIB Hodgkin's disease than is radiotherapy alone. Disease-free and overall survivals were 66 and 84%, respectively, with a median follow-up of 4 years.

Patients with stage IV Hodgkin's disease are treated with combination chemotherapy programs such as MOPP or ABVD, and rather extensive reviews are in print.[3,4,138-140]

Non-Hodgkin's lymphomas are also sensitive to radiotherapy. However, since localized non-Hodgkin's lymphoma at the time of the diagnosis is rare, and in view of their histopathologic and clinical heterogeneity, only a small proportion of these malignancies are amenable to cure with local or regional irradiation. Pathologic stages I, and contiguous stage II, of nodular poorly differentiated lymphocytic, nodular mixed, nodular histocytic, and diffuse lymphomas may be treated with radiotherapy alone, but combined treatment modalities (irradiation and chemotherapy) are also employed.[2-4,17,19-23] Total nodal irradiation has been utilized as treatment for patients with nodular lymphoma stage III.[3,4] For more advanced disease in nodular or diffuse lymphoma (stages III and IV) combination chemotherapy with CHOP-Bleo [Cyclophosphamide (Cytoxan ®), Adriamycin®, Oncovin®, prednisone, and bleomycin] or other drug combinations, with or without regional irradiation to involved areas, is generally utilized.[141-146] Response rate and 5-year survival (70%) is considerably better in nodular lymphoma than in diffuse forms (65 and 50%, respectively). The intermediate and well-differentiated lymphocytic lymphomas, diffuse poorly differentiated lymphocytic lymphomas of the convoluted cell type, Burkitt's and non-Burkitt's lymphoma which generalize very rapidly, are usually treated with chemotherapy. The major long-term complications of chemotherapy in patients with lymphomas are sterility and the risk of second cancers.[147-151]

A full discussion of the natural history and management of Hodgkin's disease and non-Hodgkin's lymphoma are beyond the scope of this volume and readers should refer to other sources, many of them listed in the bibliography.

REFERENCES

1. **Shullenberger, C. C.,** Natural history patterns in leukemia and lymphoma as related to clinicopathologic classification, in *Leukemia Lymphoma,* Year Book Medical Publishers, Chicago, 1970, 143-148.
2. **Gamble, J. F., Fuller, L. M., Loh, K. K., Martin, G. R., Jing, Bas-Shan, Sinkovics, J. G., Shullenberger, C. C., and Butler, J. J.,** Hodgkin's disease and non-Hodgkin's lymphomas, in *Cancer Patient Care* at M.D. Anderson Hospital and Tumor Institute, Clark, L. E. and Howe, C. D., Eds., Year Book Medical Publishers, Chicago, 1976, 125-163.
3. **Kaplan, H. S.,** *Hodgkin's Disease,* 2nd ed., Harvard University Press, Cambridge, Mass., 1980.
4. **DeVita, V. T., Jr. and Hellman, S.,** *Hodgkin's Disease and the Non-Hodgkin's Lymphomas,* DeVita, V. T., Jr., Hellman, S., and Rosenberg, S. A., Eds., Lippincott, New York, 1982, chap. 35.
5. **Rosenberg, S. A.,** *Hodgkin's Disease in Cancer Medicine,* Holland, J. F. and Frei, E., Eds., Lea & Febiger, Philadelphia, 1973, 1276-1301.
6. **Carbone, P. P. and DeVita, V. T.,** Malignant lymphoma, in *Cancer Medicine,* Holland, J. F. and Frei, E., Eds., Lea & Febiger, Philadelphia, 1973, 1302-1330.
7. **Rappaport, H.,** Tumors of the hematopoietic system, in *Atlas of Tumor Pathology,* Armed Forces Institute of Pathology, Washington, D.C., 1966, 442.
8. **Lukes, R. J., Butler, J. J., and Hicks, E. B.,** Natural history of Hodgkin's disease as related to its pathologic picture, *Cancer,* 19, 317, 1966.
9. **Lukes, R. J.,** The pathologic picture of the malignant lymphomas, in *Proc. Int. Conf. on Leukemia-Lymphoma,* Zarafonetis, C. J. D., Ed., Philadelphia, Lea & Febiger, Philadelphia, 1968, 333-356.
10. **Butler, J. J.,** The natural history of Hodgkin's disease and its classification, in *The Reticuloendothelial System,* Rebuck, J. W., Berard, C. W., and Abell, M. R., Eds., Williams & Wilkins, Baltimore, 1975, 184-212.
11. **Butler, J. J., Stryker, J. A., and Shullenberger, C. C.,** A clinicopathological study of stages I and II non-Hodgkin's lymphomas using the Lukes-Collins classification, *Br. J. Cancer,* 31(Suppl. 2), 208, 1975.
12. **Symp. Contemporary Issues in Hodgkin's Disease,** Biology, staging, and treatment, *Cancer Treat. Rep.,* 66, 601, 1982.
13. **Diehl, V., Kirchner, H. H., Schaadt, M., Fonatsch, C. H. R., Stein, H., Gerdes, J., and Boie, C.,** Hodgkin's disease: establishment and characterization of four in vitro cell lines, *J. Cancer Res. Clin. Oncol.,* 101, 111, 1981; *Cancer Treat. Rep.,* 66, 615, 1982.
14. **Kaplan, H. S.,** On the natural history, treatment and prognosis of Hodgkin's disease, *Harvey Lectures, 1968-1969,* Academic Press, New York, 1970, 215-259.
15. **Order, S. E. and Hellman, S.,** Pathogenesis of Hodgkin's disease, *Lancet,* 1, 571, 1972.
16. **Bender, R. A. and DeVita, V. T.,** Non-Hodgkin's lymphoma, in *Randomized Clinical Trials in Cancer: a Critical Review by Sites,* Staquet, M. J., Ed., Raven Press, New York, 1978, 77-102.
17. **Golomb, H. M., Ed.,** Non-Hodgkin's lymphomas, *Semin. Oncol.,* 7, 221, 1980.
18. **Coltman, C. A., Jr., Ed.,** Hodgkin's disease, *Semin. Oncol.,* 7, 91, 1980.
19. **Stuart, A. E., Stansfeld, A. G., and Lander, I., Eds.,** *Lymphomas Other Than Hodgkin's Disease,* Oxford University Press, New York, 1981.
20. **Hoppe, R. T., Rosenberg, S. A., Berard, C. W., Bloomfield, C. D., Bonadonna, G., Glatstein, E., and Rudders, R. A.,** A working formulation of non-Hodgkin's lymphoma for clinical usage — clinicopathologic correlation of low grade and follicular lymphomas (abstr.), *Proc. Am. Soc. Clin. Oncol.,* 22, 519, 1981.
21. **Lukes, R. J. and Butler, J. J.,** The pathology and nomenclature of Hodgkin's disease, *Cancer Res.,* 26, 1063, 1966.
22. **Butler, J. J.,** Histopathology of malignant lymphomas and Hodgkin's disease, in *Leukemia-Lymphoma,* Year Book Medical Publishers, Chicago, 1970, 123-142.
23. **Rosenberg, S. and Kaplan, H., Eds.,** Malignant lymphomas, *Bristol-Myers Cancer Symp. Ser.,* Grune & Stratton, New York, 1982.
24. **Keller, A. R., Kaplan, H. S., Lukes, R. J., and Rappaport, H.,** Correlations of histopathology with other prognostic indicators in Hodgkin's disease, *Cancer,* 22, 487, 1968.

25. **Jones, S. E., Butler, J. J., Byrne, G. E., Jr., Coltman, C. A., and Moon, T. E.,** Histopathologic review of lymphoma cases from the Southwest Oncology Group, *Cancer,* 39(3), 1071, 1977.
26. **Jenkin, R. R. T. and Berry, M. P.,** Hodgkin's disease in children, *Semin. Oncol.,* 7, 202, 1980.
27. **Jones, S. E., Fuks, Z., Kaplan, H. S., and Rosenberg, S. A.,** Non-Hodgkin's lymphomas. Results of radiotherapy, *Cancer,* 32, 682, 1973.
28. **Jones, S. E.,** Clinical features and course of the non-Hodgkin's lymphoma, *Clin. Haematol.,* 3, 131, 1974.
29. **Fuller, L. M.,** Results of large volume irradiation in the management of Hodgkin's disease and malignant lymphomas originating in the abdomen, *Radiology,* 87, 1058, 1966.
30. **DeVita, V. T.,** Hodgkin's disease: the salvage of radiation treatment failures, *Int. J. Radiol. Biol. Phys. Oncol.,* 17, 252, 1976.
31. **DeVita, V. T., Lewis, B. J., Rozencweig, M., and Muggia, F. M.,** The chemotherapy of Hodgkin's disease, *Cancer,* 42, 979, 1978.
32. **DeVita, V. T., Simon, R. M., Hubbard, S. M., Young, R. C., Berard, C. W.. and Moxley, J. H.,** Curability of advanced Hodgkin's disease, *Ann. Int. Med.,* 92, 587, 1980.
33. **Epstein, M. A., Achong, B. G., and Barr, Y. M.,** Virus particles in cultivated lymphoblasts from Burkitt's lymphoma, *Lancet,* 1, 702, 1964.
34. **Dmochowski, L.,** Viral studies in human leukemia and lymphoma, in *Proc. Int. Conf. on Leukemia-Lymphoma,* Zarafonetis, C. J. D., Ed., Lea & Febiger, Philadelphia, 1968, 97-113.
35. **Epstein, A. L. and Kaplan, H. S.,** Biology of the human malignant lymphomas. I. Establishment in continuous culture and heterotransplantation of diffuse histocytic lymphomas, *Cancer,* 34, 1851, 1974.
36. **Gallo, R. C., Yong, S. S., and Ting, R. C.,** RNA dependent DNA polymerase of human acute leukemia cells, *Nature (London),* 28, 927, 1970.
37. **Kaplan, H. S.,** Etiology of lymphomas and leukemia: role of C-type RNA viruses, *Leukemia Res.,* 2, 253, 1978.
38. **Theilen, G. H.,** Comparative aspects of naturally occurring leukemias, and lymphomas in man and animals in relation to causation, control and prevention, in *Leukemias, Lymphomas and Papillomas — Comparative Aspects,* Bachman, P. A., Ed., Taylor and Francis, London, 1980.
39. **Anderson, R. E. and Kshida, K.,** Malignant lymphomas in survivors of the atomic bomb in Hiroshima, *Ann. Int. Med.,* 61, 853, 1964.
40. **Miller, R. W.,** Delayed radiation effects in atomic bomb survivors, *Science,* 166, 569, 1969.
41. **Anderson, R. E., Nishiyama, H., Yohei, I., Kenzo, T., and Nobukazo, O.,** Pathogenesis of radiation related leukemia and lymphoma. Speculations based primarily on experience of Hiroshima and Nagasaki, *Lancet,* 1, 1060, 1972.
42. **Fraumeni, J. F.,** Constitutional disorders of man predisposing to leukemia and lymphoma, in *Comparative Morphology of Hematopoietic Neoplasms (Natl. Cancer Inst. Monogr.),* No. 32, Lingeman, C. H. and Garner, F. M., Eds., Bethesda, Md., 1969, 221-232.
43. **Snyder, A. L., Li, F. P., Henderson, E. S., and Todaro, G. J.,** Possible inherited leukaemogenic factors in familial acute leukemia, *Lancet,* 1, 586, 1970.
44. **Grufferman, S., Cole, P., Smith, P. G., and Lukes, R. J.,** Hodgkin's disease in siblings, *N. Engl. J. Med.,* 296, 248, 1977.
45. **Vianna, N. J., Davies, J. N. P., Polan, A. K., and Wolfgang, P.,** Familial Hodgkin's disease: an environmental and genetic disorder, *Lancet,* 2, 854, 1974.
46. **Vianna, N. J., Greenwald, P., Davies, J. N. P.,** Extended epidemic of Hodgkin's disease in high school students, *Lancet,* 1, 1209, 1971.
47. **Milham, S., Jr. and Hesser, J.,** Hodgkin's disease in woodworkers, *Lancet,* 2, 136, 1967.
48. **McKhann, C. F.,** Primary malignancy in patients undergoing immunosuppression for renal transplantation, *Transplantation,* 8, 209, 1969.
49. **Penn, I.,** The incidence of malignancies in transplant recipients, *Transplant. Proc.,* 7, 323, 1975.
50. **Matas, A. J., Hertel, B. F., Rosai, J., and Simmons, R.,** Post-transplant malignant lymphoma. Distinctive morphologicic features related to its pathogenesis, *Am. J. Med.,* 61, 716, 1976.
51. **Hyman, G. and Sommer, S.,** The development of Hodgkin's disease and other lymphomas during anti-convulsant therapy, *Blood,* 28, 416, 1966.
52. **Zulman, J., Jaffe, R., and Talal, N.,** Evidence that the malignant lymphoma of Sjogrens syndrome is a monoclonal B-cell neoplasm, *N. Engl. J. Med.,* 299, 1215, 1978.
53. **Kassan, J. S., Thomas, T. L., Haralampos, M., Montsopoulos, H. M., Hoover, R., Kimberlt, R. P., Budman, D. R., Costa, J., Decker, J. L., and Chused, M.,** Increased risk of lymphoma in Sicca syndrome, *Ann. Int. Med.,* 89, 888, 1978.
54. **Talal, N., Sokoloff, L., and Barth, W.,** Extra salivary lymphoid abnormalities in Sjogrens syndrome (reticulum cell sarcoma, "Pseudolymphoma," Macroglobulinemia), *Am. J. Med.,* 43, 50, 1967.
55. **MacSween, R. N. M.,** Reticulum cell sarcoma and rheumatoid arthritis in a patient with XY/XXY/XXX/Y Klinefelter's syndrome and normal intelligence, *Lancet,* 1, 460, 1965.

56. **Tan, C., Etcubanas, E., Lieberman, P., Isenberg, H., King, O., and Murphy, M. L.,** Chediak-Higashi syndrome in a child with Hodgkin's disease, *Am. J. Dis. Child.,* 121, 135, 1971.
57. **Miller, D. G.,** The association of immune disease and malignant lymphoma, *Ann. Int. Med.,* 66, 507, 1967.
58. **Mann, R. B., Jaffe, E. S., and Berard, C. W.,** Malignant Lymphomas — a conceptual understanding of morphologic diversity. A review, *Am. J. Pathol.,* 94(1), 105, 1979.
59. **Berard, C. W.,** Reticuloendothelial system. An overview of neoplasia, in *The Reticuloendothelial System,* Williams & Wilkins, Baltimore, 1975, 310-317.
60. **Lukes, R. J. and Collin, R. D.,** Immunologic characterization of human malignant lymphomas, *Cancer,* 34, 1488, 1974.
61. **Braylan, R. C., Jaffee, E. S., and Berard, C. W.,** Malignant lymphomas: current classification and new observations, *Pathology Annual,* Sommers, S. C., Ed., Appleton-Century-Crofts, New York, 1975, 213-270.
62. **Berard, C. W., Gallo, R. C., Jaffee, E. S., Green, I., and DeVita, V.,** Current concepts of leukemia and lymphoma: Etiology, pathogenesis and therapy, *Ann. Int. Med.,* 85, 351, 1976.
63. **DeVita, V. T.,** Lymphocyte reactivity in Hodgkin's disease: a lymphocyte civil war, *N. Engl. J. Med.,* 289, 801, 1973.
64. **Miller, T. P., Byrne, G. E., and Jones, S. E.,** Mistaken clinical and pathologic diagnoses of Hodgkin's disease, *Cancer Treat. Rep.,* 66, 645, 1982.
65. **Moran, E. M. and Ultman, J. E.,** Clinical features and course of Hodgkin's disease, *Clin. Hematol.,* 3, 91, 1974.
66. **Peters, M. V., Hasselbach, R., and Brown, T. C.,** The natural history of the lymphomas related to the clinical classification, in *Proc. Int. Conf. Leukemia-Lymphoma,* Zarafonetis, C. J. D., Ed., Lea & Febiger, Philadelphia, 1968, 357-370.
67. **Johnson, R. E., Thomas, L. B., and Chretien, P.,** Correlation between clinico-histologic staging and extranodal relapse in Hodgkin's disease, *Cancer,* 25, 1071, 1970.
68. **Fuller, L. M., Madoc-Jones, H., Gamble, J. F., Butler, J. J., Sullivan, M. P., Fernandez, C. F., and Gehan, E. A.,** New assessment of the prognostic significance of histopathology in Hodgkin's disease for laparotomy-negative stage I and stage II patients, *Cancer,* 39, 2174, 1977.
69. **Filly, R., Blank, N., and Castellino, R. A.,** Radiographic distribution of intrathoracic disease in previously untreated patients with Hodgkin's disease and non-Hodgkin's lymphoma, *Radiology,* 120, 277, 1976.
70. **Rosenberg, S. A.,** Hodgkin's disease of the bone marrow, *Cancer Res.,* 31, 1733, 1971.
71. **Ferraut, A., Rodhain, J. L., Michaux, L., Piret, L., Maldague, B., and Sokol, G.,** Detection of skeletal involvement in Hodgkin's disease: a comparison of radiography bone scanning and bone marrow biopsy in 38 patients, *Cancer,* 35, 1346, 1975.
72. **McKenna, R. W., Bloomfield, C. D., and Brunning, R. D.,** Nodular lymphoma: bone marrow and blood manifestations, *Cancer,* 36, 428, 1975.
73. **Reimer, R. R., Chabner, B. A., Young, R. C., Reddick, R., and Johnson, R. E.,** Lymphoma presenting in bone: results of histopathology, staging and therapy, *Ann. Intern. Med.,* 87(1), 50, 1977.
74. **Shoji, H. and Miller, T.,** Primary reticulum cell sarcoma of bone. Significance of clinical features upon prognosis, *Cancer,* 28, 1234, 1971.
75. **Shari, M., Modan, B., Goldman, P. H., Bradstaeter, S., and Ramot, B.,** Primary gastrointestinal lymphoma, *Isr. J. Med. Sci.,* 5, 1173, 1969.
76. **Atkinson, M. K., Austin, D. E., McElwain, T. J., and Peckham, M. J.,** Alcohol pain in Hodgkin's disease, *Cancer,* 37, 895, 1976.
77. **James, A. H.,** Hodgkin's disease with or without alcohol-induced pain, *Q. J. Med.,* 29, 47, 1960.
78. **Glatstein, E., Guernsey, J. M., Rosenberg, S. A., and Kaplan, H. S.,** The value of laparotomy and splenectomy in the staging of Hodgkin's disease, *Cancer,* 24, 709, 1969.
79. **Glatstein, E., Trueblood, H. W., Enright, L. P., Rosenberg, S., and Kaplan, H. S.,** Surgical staging of abdominal involvement in unselected patients with Hodgkin's disease, *Radiology,* 97, 425, 1970.
80. **Desser, R. K., Moran, E. M., and Ultmann, J. E.,** Staging of Hodgkin's disease and lymphoma, *Med. Clin. North Am.,* 57, 479, 1973.
81. **Rosenberg, S. A.,** A critique of the value of laparotomy and splenectomy in the evaluation of patients with Hodgkin's disease, *Cancer Res.,* 31, 1737, 1971.
82. **Piro, A. J., Hellman, S., and Moloney, W. C.,** The influence of laparotomy on management decisions in Hodgkin's disease, *Arch. Intern. Med.,* 130, 844, 1972.
83. **DeVita, V. T.,** The role of staging laparotomy in combined modality therapy of Hodgkin's disease, *World J. Surg.,* 2(1), 105, 1978.
84. **Craft, C. B.,** Results with Roentgen ray therapy in Hodgkin's disease, *Bull. Univ. Minn. Hosp.,* 11, 391, 1940.

85. **Jones, R., Hubbard, S. M., Osborne, C., Merrill, J., Garvin, J., Young, R., and DeVita, V.,** Histologic conversions in non-Hodgkin's lymphoma: evolution of nodular lymphomas to diffuse lymphomas, *Clin. Res.,* 26, 437, 1978.
86. **Rosenberg, S. A. and Kaplan, H. S.,** Clinical trials in the non-Hodgkin's lymphomata at Stanford University: experimental design and repliminary results, *Br. J. Cancer,* 31(Suppl. 2), 456, 1975.
87. **Roeser, H. P., Hocker, G. K., Kynaston, B., Roberts, S. J., and Whitaker, S. V.,** Advanced non-Hodgkin's lymphomas: response to treatment with combination chemotherapy and factors influencing prognosis, *Br. J. Haematol.,* 30, 323, 1975.
88. **Wintrobe, M. M.,** Chronic lymphocytic leukemia, in *Clinical Hematology,* 7th ed., Lea & Febiger, Philadelphia, 1974, 1520.
89. **Chabner, B. A., Johnson, R. E., Young, R. C., Canellos, P. G., Hubbard, S. P., Johnson, S. K., and DeVita, V.,** Sequential non-surgical and surgical staging of non-Hodgkin's lymphoma, *Ann. Int. Med.,* 85(2), 149, 1976.
90. **Chabner, B. A., Fisher, R. I., Young, R. C., and De Vita, V.,** Staging of non-Hodgkin's lymphoma, *Semin. Oncol.,* 7(3), 285, 1980.
91. **Begent, R. H. J. and Wiltshaw, E.,** The effect of splenectomy in the haematological response to radiotherapy in Hodgkin's disease, *Br. J. Haematol.,* 27, 331, 1974.
92. **Salzman, J. R. and Kaplan, H. S.,** Effect of splenectomy on hematologic tolerance during total lymphoid radiotherapy of patients with Hodgkin's disease, *Cancer,* 27, 471, 1971.
93. **Panattiere, F. J. and Coltman, C. A.,** Splenectomy effects on chemotherapy in Hodgkin's disease, *Arch. Int. Med.,* 131, 363, 1973.
94. **Panattiere, F. J., Coltman, C. A., and Delaney, F. C.,** Splenectomy, chemotherapy, and survival in Hodgkin's disease, *Arch. Int. Med.,* 137, 341, 1977.
95. **Brogadir, S., Fialk, M. A., Coleman, M., Vinciguerra, V. P., Degnan, T., Pasmantier, M., and Silver, R. T.,** Morbidity of staging laparotomy in Hodgkin's disease, *Am. J. Med.,* 64(3), 429, 1978.
96. **Meeker, W. R., Richardson, J. D., West, W., and Parker, J. C.,** Critical evaluation of laparotomy and splenectomy in Hodgkin's disease, *Arch. Surg.,* 105, 222, 1972.
97. **Desser, R. L. and Ultmann, J. E.,** Risk of severe infection in patients with Hodgkin's disease of lymphoma after diagnostic laparotomy and splenectomy, *Ann. Int. Med.,* 77, 143, 1972.
98. **Young, R. C., Anderson, T., and DeVita, V. T.,** The treatment of Hodgkin's disease: emphasizing programs at the Clinical Center, National Institutes of Health, *Curr. Probl. Cancer,* 1(7), 1, 1977.
99. **Chilcote, R. R., Baehner, R. H., and Hammond, D.,** Septicemia and meningitis in children splenectomized for Hodgkin's disease, *N. Engl. J. Med.,* 295, 798, 1976.
100. **Slaven, R. and Nelson, T. S.,** Complications from staging laparotomy for Hodgkin's disease, *Natl. Cancer Inst. Monogr.,* 36, 457, 1973.
101. **Sweet, D. L., Jr., Kinnealey, A., and Ultmann, J. E.,** Hodgkin's disease: Problems of staging, *Cancer,* 42(2), 957, 1978.
102. **Redman, H. C., Glatstein, E., Castellino, R. A., and Federall, W. A.,** Computed tomography as an adjunct in the staging of Hodgkin's disease and non-Hodgkin's lymphomas, *Radiology,* 124(2), 381, 1977.
103. **Breeman, R. S., Castellino, R. A., Harell, G. S., Marshall, W. H., Glatstein, E., and Kaplan, H. S.,** CT-Pathologic correlations in Hodgkin's disease and non-Hodgkin's lymphoma, *Radiology,* 126, 159, 1978.
104. **Jones, S. E., Tobias, D. A., and Waldman, R. S.,** Complete tomographic scanning in patients with lymphoma, *Cancer,* 41, 480, 1978.
105. **Rochester, D., Bowie, J. D., Kunzmann, A., and Erichester, R.,** Ultrasound in the staging of lymphoma, *Radiology,* 124(2), 483, 1977.
106. **Filly, R., Marglin, S., and Castellino, R. A.,** The ultrasonographic spectrum of abdominal and pelvic Hodgkin's disease and non-Hodgkin's lymphoma, *Cancer,* 38, 2143, 1976.
107. **Carbone, P. P., Kaplan, H. S., Mushoff, K., Smithers, D. V., and Tubiana, M.,** Report of the committee on Hodgkin's disease staging, *Cancer Res.,* 31, 1860, 1971.
108. **Rosenberg, S. A.,** Report of the committee on the staging of Hodgkin's disease in symposium: obstacles to the control of Hodgkin's disease, *Cancer Res.,* 26, 1310, 1966.
109. **Hellman, S., Mauch, P., Goodman, R. L., Rosenthal, D. S., and Moloney, W. C.,** The place of radiation therapy in the treatment of Hodgkin's disease, *Cancer,* 42, 971, 1978.
110. **Fuller, L. M., Gamble, J. F., Ibrahim, E., Jing, B. S., Butler, J. J., and Shullenberger, C. C.,** Stage II Hodgkin's disease — significance of mediastinal and non-mediastinal presentations, *Radiology,* 109, 429, 1973.
111. **Fuller, L. M., Gamble, J. F., Shullenberger, C. C., Butler, J. J., and Gehan, E. A.,** Prognostic factors in localized Hodgkin's disease treated with regional radiotherapy: clinical presentation and specific histology, *Radiology,* 98, 641, 1971.

112. **Fuller, L. M., Gamble, J. F., Sullivan, M. P., Shullenberger, C. C., and Banker, F. L.,** Management of stage I and stage II Hodgkin's disease, in *I Linfomi Maligni,* Casa Editrice Ambrosiana, Milan, 1974, 185-187.
113. **Fuller, L. M., Jing, B. S., Shullenberger, C. C., and Butler, J. J.,** Radiotherapeutic management of localized Hodgkin's disease involving the mediastinum, *Br. J. Radiol.,* 40, 913, 1967.
114. **Peters, M. V.,** A study of survival in Hodgkin's disease treated radiologically, *Am. J. Roentgenol.,* 63, 299, 1950.
115. **Peters, M. V. and Middlemiss, K. C. H.,** A study of Hodgkin's disease treated by irradiation, *Am. J. Roentgenol.,* 79, 114, 1958.
116. **Peters, M. V.,** Prophylactic treatment of adjacent areas in Hodgkin's disease, *Cancer Res.,* 26, 1232, 1966.
117. **Kaplan, H. S.,** The radical radiotherapy of regionally localized Hodgkin's disease, *Radiology,* 78, 553, 1962.
118. **Carmel, R. J. and Kaplan, H. S.,** Mantle irradiation in Hodgkin's disease. An analysis of technique, tumor irradiation, and complications, *Cancer,* 37, 2812, 1976.
119. **Kaplan, H. S.,** Long-term results of palliative and radical radiotherapy of Hodgkin's disease, *Cancer Res.,* 26, 1250, 1966.
120. **Kaplan, H. S.,** Role of intensive radiotherapy in the management of Hodgkin's disease, *Cancer,* 19, 356, 1966.
121. **Kaplan, H. S. and Rosenberg, S. A.,** The treatment of Hodgkin's disease, *Med. Clin. North Am.,* 50, 1591, 1966.
122. **Kaplan, H. S.,** Clinical evaluation and radiotherapeutic management of Hodgkin's disease and the malignant lymphomas, *N. Engl. J. Med.,* 278, 892, 1968.
123. **Rosenberg, S. and Kaplan, H. S.,** The results of radical radiotherapy in Hodgkin's disease and other lymphomas, in *Proc. Int. Conf. Leukemia-Lymphoma,* Zarafonetis, C. J. D., Ed., Lea & Febiger, Philadelphia, 1968, 403-408.
124. **Kaplan, H. S. and Stewart, J. R.,** Complications of intensive megavoltage radiotherapy for Hodgkin's disease, *Natl. Cancer Inst. Monogr.,* 36, 439, 1973.
125. **Hellman, S., Chaffey, J. T., Rosenthal, D. S., Maloney, C. W., Canellos, G. P., and Skarin, A. T.,** The place of radiation therapy in the treatment of non-Hodgkin's lymphomas, *Cancer,* 39, 843, 1977.
126. **Bush, R. S., Gaspodarowicz, M., Sturgeon, J., and Alison, R.,** Radiation therapy of localized non-Hodgkin's lymphoma, *Cancer Treat. Rep.,* 61, 1129, 1977.
127. **Peters, M. V., Bush, R. S., Brown, T. C., and Reid, J.,** The place of radiotherapy in the control of non-Hodgkin's lymphoma, *Br. J. Cancer,* 31(Suppl. 2), 386, 1975.
128. **Hoppe, R. T., Coleman, N. N., Cox, R. S., Rosenberg, S. A., and Kaplan, H. S.,** The management of Stage I-II Hodgkin's disease with irradiation alone or combined modality therapy: the Stanford experience, *Blood,* 59, 455, 1982.
129. **Wiernik, P. H., Slawson, R. G., Burks, L. C., and Diggs, C. H.,** A randomized trial of radiotherapy (RT) and MOPP (C) vs. MOPP alone for stages IB-IIIA Hodgkin's disease, *Proc. Am. Soc. Clin. Oncol.,* 20, 315, 1979.
130. **Coltman, C. A., Jr., Fuller, L. M., Fisher, R., and Frei, E., III,** Extended field radiotherapy versus involved field radiotherapy plus MOPP in stage I and II Hodgkin's disease, in *Adjuvant Therapy of Cancer,* Jones, S. E. and Salmon, S. E., Eds., Grune & Stratton, New York, 1979, 129-136.
131. **Coltman, C. A. and Fuller, L. M.,** Patterns of relapse in localized (Stage I — II) Hodgkin's disease (H.D.) following extended field radiotherapy (EFXRT) vs. involved field radiotherapy (IFXRT) plus MOPP, *Blood,* Suppl. 50, 188, 1977.
132. **Fuller, L. M., Gamble, J. F., Velasquez, W., Rodgers, R., Butler, J. J., North, L., Martin, G. R., Behan, E. A., and Shullenberger, C. C.,** Evaluation of the significance of prognostic factors in stage III Hodgkin's disease treated with MOPP and radiotherapy, *Cancer,* 45, 1352, 1980.
133. **Prosnitz, L. R., Farber, L. R., Fischer, J. J., Bertino, J. R., and Fischer, D. B.,** Long term remissions with combined modality therapy for advanced Hodgkin's disease, *Cancer,* 37, 2826, 1976.
134. **Hoppe, R. T., Portlock, C. S., Glatstein, E., Rosenberg, S. A., and Kaplan, H. S.,** Alternating chemotherapy and irradiation in the treatment of advanced Hodgkin's disease, *Cancer,* 43, 472, 1979.
135. **Santoro, A., Sucali, R., Bonfante, V., Volterrani, F., Bajetta, E., and Bonadonna, G.,** Comparative therapeutic effects and morbidity of two combined treatment modalities in PS IIB-IM (A + B) Hodgkin's disease, *Proc. Am. Soc. Clin. Oncol.,* 20, 359, 1979.
136. **Bonadonna, G., Santoro, A., Zucali, R., and Valagussa, P.,** Improved 5-year survival in advanced Hodgkin's disease by combined modality approach, *Cancer Clin. Trials,* 2, 217, 1979.
137. **Goodman, R., Mauch, P., Piro, A., Rosenthal, D., Goldstein, M., Tullis, J., and Hellman, S.,** Stages IIB and IIIB Hodgkin's disease. Results of combined modality treatment, *Cancer,* 40, 84, 1977.

138. **Mubashir, B. A., Shullenberger, C. C., and Gamble, J. F.,** Treatment of advanced Hodgkin's disease with combination chemotherapy, *South. Med. J.*, 66, 779, 1973.
139. **Bonadonna, G., Santoro, A., and Bonfante, V.,** Salvage chemotherapy with ABVD in MOPP-resistant Hodgkin's disease, *Proc. Am. Soc. Clin. Oncol.*, 22, 522, 1981.
140. **Rodgers, R. W., Gamble, J. F., Loh, K. K., and Shullenberger, C. C.,** Adriamycin, Bleomycin, DIC, CCNU, and Prednisone (ABCIC) chemotherapy in MOPP-resistant Hodgkin's disease, *Cancer,* 46, 2349, 1980.
141. **Rodriguez, V., Cabanillas, F., Burgess, M. A., McKelvey, E. M., Valdivieso, M., Bodey, G. P., and Freireich, E. G.,** Combination chemotherapy (''CHOP-Bleo'') in advanced (non-Hodgkin) malignant lymphoma, *Blood,* 49, 325, 1977.
142. **Schein, P. S., DeVita, V. T., Hubbard, S., Chabner, B. A., Canellos, G. P., Berard, C., and Young, R. C.,** Bleomycin, adriamycin, cyclophosphamide, and prednisone (BACOP) combination chemotherapy in the treatment of advanced diffuse histiocytic lymphoma, *Ann. Intern. Med.*, 85, 417, 1976.
143. **Luce, J. K., Gamble, J. F., Wilson, H. E., Monto, R. W., Isaacs, B. L., Palmer, R. L., Coltman, C. A., Hewlett, J. S., Gehan, E. A., and Frei, E.,** Combined cyclophosphamide, vincristine and prednisone therapy of malignant lymphoma, *Cancer,* 28, 306, 1971.
144. **Sweet, D. L., Golomb, H. M., Ultmann, J. E., Miller, J. B., Stein, R. S., Lester, E. P., Mintz, U. R., Bitram, J. D., Streuli, R., Daly, K., and Roth, O.,** Cyclophosphamide, vincristine, methotrexate with leucovorin rescue, and cytosine arabinoside (COMLA) combination sequential chemotherapy in the treatment of advanced diffuse histiocytic lymphoma, *Ann. Intern. Med.*, 92, 785, 1980.
145. **Ginsberg, S. J., Crooke, S. T., Bloomfield, C. D., Peterson, B., Kennedy, B. J., Blom, J., Ellison, R. R., Pajak, T. F., and Gottlieb, A. J.,** Cyclophosphamide, doxorubicin, vincristine, and low-dose continuous infusion bleomycin in non-Hodgkin's lymphoma: cancer and leukemia Group B Study #7804, *Cancer,* 49, 1346, 1982.
146. **Lawrence, J., Coleman, M., Allen, L. S., Silver, R. T., and Pasmantier, M.,** Combination chemotherapy of advanced diffuse histiocytic lymphoma with the six drug COP-BLAM regimen, *Ann. Intern. Med.*, 97, 190, 1982.
147. **DeVita, V. T., Jr.,** The consequences of the chemotherapy of Hodgkin's disease: the 10th David A. Karnofsky Memorial Lecture, *Cancer,* 47, 1, 1981.
148. **Sherins, R. J. and DeVita, V. T.,** Effect of drug treatment for lymphoma on male reproduction capacity, *Ann. Int. Med.*, 79, 216, 1973.
149. **Sherins, R., Winokur, S., DeVita, V. T., and Vaitukaitus, J.,** Surprisingly high risk of functional castration in women receiving chemotherapy for lymphoma, *Clin. Res.*, 23, 343A, 1975.
150. **Brody, R. S., Schottenfeld, D., and Reid, A.,** Multiple primary cancer risk after therapy for Hodgkin's disease, *Cancer,* 40, 1917, 1977.
151. **Baccarani, M., Bosi, A., and Papa, G.,** Second malignancy in patients treated for Hodgkin's disease, *Cancer,* 46, 1735, 1980.

Chapter 9

COPPER STUDIES IN LYMPHOMAS AND LEUKEMIAS

I. ANALYSIS OF THE AUTHORS' CLINICAL MATERIAL, METHODS, AND NORMAL RANGE OF SERUM COPPER

A group of 257 patients with Hodgkin's disease (214 adults and 43 children), 241 with non-Hodgkin's lymphoma (236 adults and 5 children), and 165 patients with acute leukemias with over 3000 serum copper level (SCL) determinations during the course of the disease are reviewed here. These patients were treated and followed in the University of Texas M.D. Anderson Hospital and Tumor Institute and the Diagnostic Clinic of Houston from 1966 to 1973 with standardized methods of pathological tissue classifications, staging procedures, and treatment management. Those patients who did not fulfill these criteria or were known to be pregnant, on estrogen medications, or having infections have been excluded in keeping with the recognized relationship between serum copper levels and tissue type, and extent of lymphoma, inflammations, and hormonal imbalances. Our continued observations of serum copper alterations in the course of malignant lymphomas confirmed the views expressed in our previous publications concerning the clinical usefulness of serum copper level determinations in daily oncological practice.[1-17] Since most of these studies on serum copper levels in lymphomas were done in the late 1960s and early 1970s, the terminology and classification of non-Hodgkin's lymphomas used in this chapter is a modification of Rappaport's nomenclature, which was in common use at that time. The details of this nomenclature and its relationship (synonyms) to the current terminology are discussed in Chapter 7.

In a pilot study published in 1968,[2] 433 serum copper observations from 70 patients (37 pediatric and 33 adult) with malignant lymphoma or acute leukemia were reviewed. Many of these patients were followed throughout their entire clinical course, from the diagnosis of disease to death. This pilot study was undertaken to determine the general relationship between serum copper level and disease activity and response to therapy in patients with malignant lymphoproliferative diseases. Pertinent clinical information of this study group is shown in Table 1. As can be seen from Table 1 the largest groups in this clinical material consisted of 28 patients with Hodgkin's disease (22 adults and 6 children) and 28 cases with acute childhood leukemia. The results will be discussed in conjunction with other serum copper level studies in lymphoma (Hodgkin's or non-Hodgkin's type) and acute leukemia, respectively.

In a subsequent study published in 1972,[6] and in an effort to examine certain quantitative aspects of the serum copper relationship with the bone marrow blast percentage noted in a pilot study,[1,2,4] a statistical analysis was made using 346 serum copper level observations on 112 patients (64 pediatric and 48 adults) with acute leukemia. The patients meeting the criteria of routine bone marrow examinations and serum copper levels done on the same day were included in this study. Pertinent clinical data (distribution of the several types of leukemia in pediatric and adult groups) are presented in Table 2. These findings of relationship between serum copper level and percentage of blast cells will be discussed later.

An inventory of these data defined a subgroup of 25 pediatric patients with acute leukemia who had at least 3 paired serum copper level and bone marrow estimations at monthly (or lesser) intervals. This study, based on sequential observations of longitudinal relationship of serum copper level to bone marrow blast cells in these patients followed throughout various phases of disease activity (many of them throughout the entire course of disease), provided additional important clinical information which was reported in 1972[7] and will be reviewed here.

Table 1
CLINICAL MATERIAL STUDIED (PILOT STUDY)

Diagnosis	No. of patients	No. of serum copper determinations
Malignant lymphoma		
Hodgkin's disease	28	183
Reticulum cell sarcoma	9	29
Lymphocytic lymphoma	3	12
Nodular lymphoma	2	5
Subtotal	42	229
Acute leukemia		
Lymphocytic	23	186
Granulocytic	4	13
Monocytic	1	5
Subtotal	28	204
Total	70	433

From Hrgovcic, M., Tessmer, C. F., Minckler, M. T., Mosier, B., and Taylor, H. G., *Cancer,* 21, 743, 1968. With permission.

Table 2
CLINICAL MATERIAL STUDIED (SERUM COPPER AND BONE MARROW)

Diagnosis	No. of patients		Total no. of patients[a]	No. of serum copper determinations & BM		Total no. of SCL & BM
	Pediatric	Adult		Pediatric	Adult	
Acute lymphocytic leukemia (ALL)	47	10	57	190	29	219
Acute granulocytic leukemia (AGL)	10	30	40	24	44	68
Acute monocytic leukemia (AMonL)	1	0	1	11	0	11
Acute unclassified leukemia (AUL)	6	12	18	40	16	56
Total	64	52	116	265	89	354

[a] Ages range from 1 to 79 years.

From Tessmer, C. F., Hrgovcic, M., Brown, B., Wilbur, J., and Thomas, F. B., *Cancer,* 29, 173, 1972. With permission.

In 1973, we reported the significance of the serum copper level in 191 adult patients with Hodgkin's disease.[13] The distribution of clinical material according to the histological subclassifications of Hodgkin's disease, age groups, sex, and number of serum copper level determinations is shown in Table 3. The youngest patient was 16 years old and the oldest was 76.

Table 3
CLINICAL DATA FROM PATIENTS WITH HODGKIN'S DISEASE (ADULT PATIENTS)

Histologic type	No. of patients	Age groups			Sex		SCL determinations
		16—25	26—50	>50	M	F	
Lymphocytic predominance	9	2	5	2	6	3	11
Nodular sclerosing	100	28	56	16	57	43	432
Mixed cellularity	72	12	39	21	56	16	386
Lymphocytic depletion	10		4	6	7	3	67
Total	191	42	104	45	126	65	896

From Hrgovcic, M., in *Zinc and Copper in Medicine,* Karcioglu, Z. A. and Sarper, R. M., Eds., 1980, 481. Courtesy of Charles C Thomas, Publisher, Springfield, Illinois.

Table 4
DISTRIBUTION OF CLINICAL MATERIAL (191 PATIENTS) ACCORDING TO STAGE, DISEASE ACTIVITY, AND PREVIOUS THERAPY AT TIME OF FIRST SERUM COPPER LEVEL DETERMINATION

Histologic type	Total patients	Stage								Status of disease activity		Previous therapy	
		I		II		III		IV					
		A	B	A	B	A	B	A	B	Active	Remission	Untreated	Treated
Lymphocytic predominance	9	6		1	1	1				6	3	6	3
Nodular sclerosing	100	4		32	12	10	19	4	19	76	24	52	48
Mixed cellularity	72	7		8	3	14	17	2	21	52	20	18	54
Lymphocytic depletion	10	1		3			1		5	8	2	4	6
Total	191	18		44	16	25	37	6	45	142	49	80	111

From Hrgovcic, M., in *Zinc and Copper in Medicine,* Karcioglu, Z. A. and Sarper, R. M., Eds., 1980, 481. Courtesy of Charles C Thomas, Publisher, Springfield, Illinois.

Table 4 shows a distribution of the patients studied according to the extent and activity of disease, as well as previous therapy, at the time of the first serum copper level determinations. The methods and findings will be discussed later.

In the same year, we reported serum copper studies of Hodgkin's disease in children.[8] The study material consisted of 37 pediatric cases of Hodgkin's disease in whom 669 serum copper levels were determined. Age ranged from 4 to 18 years with a mean of 12.2 years. Pertinent clinical data are shown in Table 5 and results will be discussed in the section on Hodgkin's disease cases.

In the same year we reported the serum copper level as an index of tumor response to radiotherapy, which will be alluded to later.[8]

The relationship between the serum copper level and disease activity in non-Hodgkin's lymphoma was also noted in our pilot studies, but in view of a limited number of cases this was not clear, particularly in patients with reticulum cell sarcoma. Therefore, the study of

Table 5
CLINICAL DATA FROM PATIENTS WITH HODGKIN'S DISEASE IN CHILDREN

Case	Unit no.	D.O.B.	Sex	Period of SCLs	Total no. of SCLs	Range of SCLs	Initial histology Hodgkin's disease	Stage
MHB	6 97 41	8/5/53	M	1/15/68 to 12/4/68	17	164—242	Nodular sclerosing	IVB
RLB	5 58 23	9/15/55	M	3/18/66 to 12/20/71	15	94.5—281	Mixed	IVB
LMB	7 65 21	11/14/55	M	7/29/69 to 10/13/71	15	86.5—114.5	Nodular sclerosing	IA
RDB	6 59 69	9/7/54	M	5/10/67 to 11/12/71	8	109—137	Lymphocytic predominance	IIA
KHC	5 06 54	10/11/51	M	7/18/66 to 7/26/71	3	71—94	Nodular lympho-cytic, histiocytic	IA
MEC	5 10 38	8/20/49	M	3/11/66 to 5/5/67	19	108—172	Mixed	IIA
RAC	3 10 41	9/19/51	M	2/11/70 to 6/23/71	2	95—105.6	Lymphocytic predominance	IA
LEC	7 12 69	4/20/56	F	5/21/68 to 12/22/71	20	102.5—187	Nodular sclerosing	IIA
DWD	7 17 51	2/8/63	M	6/25/68 to 12/17/71	29	97.5—261	Mixed	IIIB
MWD	6 65 13	3/25/53	M	6/12/70 to 12/20/71	19	106—231	Mixed	IIIA-IVB
VME	7 57 58	1/23/60	F	5/6/69 to 8/23/71	4	91—122.5	Lymphocytic predominance	IIA
RAE	5 40 41	1/9/55	M	2/20/68 to 12/17/71	43	90.7—203	Mixed	IIIB
DGG	7 26 94	3/30/60	F	8/23/68 to 11/10/71	20	81.4—187.5	Nodular sclerosing	IIA
SMH	7 05 93	7/24/56	M	3/20/68 to 7/19/71	7	80—250	Unclassified	IA
AH	6 14 63	1/10/52	M	4/26/66 to 5/23/70	40	96—250	Nodular sclerosing	IVB

KLK	7 53 02	3/16/59	M	4/1/69 to 12/9/71	25	101.8—201	Nodular sclerosing	IIA-IIB
EK	7 73 28	10/3/56	M	9/4/69 to 11/2/71	10	91.5—152.2	Nodular sclerosing	IIIA
TDK	8 16 88	2/22/59	M	8/3/70 to 12/14/71	17	97.5—193	Nodular sclerosing	IIIB
RLB	6 73 74	1/1/57	M	8/6/69 to 12/20/71	9	135.5—270	Mixed	IA-IIIB
DLL	7 04 54	1/13/64	M	3/7/68 to 10/25/71	38	78—252	Lymphocytic predominance, mixed	IIA-IVB
EDL	7 81 32	11/15/56	M	11/20/69 to 10/20/71	13	89.5—203.3	Nodular sclerosing	IIIB
JM	7 44 82	1/1/63	M	2/18/69 to 10/18/71	26	106—207.5	Nodular sclerosing	IIA-IIIA
MHM	6 42 15	1/12/54	M	8/21/70 to 8/13/71	2	89—94	Lymphocytic predominance	IIIA
TLM	6 73 77	8/16/61	F	3/5/68 to 5/3/71	19	114—266	Nodular sclerosing	IVB
RWM	4 11 75	11/13/50	M	4/15/66 to 6/7/71	26	85—220	Mixed	IIIB
BN	5 93 95	9/21/55	M	3/11/66 to 12/24/71	65	84—173	Mixed	IIB-IIIB
TP	7 96 05	8/6/55	M	3/18/70 to 12/13/71	13	95—132	Mixed	IIIA
RWP	7 36 94	2/11/53	M	12/18/69 to 12/3/71	19	90—170	Nodular sclerosing	IIIB
RAP	7 62 28	9/17/60	F	6/10/69 to 6/10/71	2	127—172	Nodular sclerosing	IIIB
DR	6 03 31	6/6/57	M	1/31/66 to 12/17/71	21	80—162	Mixed	IIIA
ADR	7 08 59	12/24/57	M	4/1/68 to 12/4/70	24	134—287	Nodular sclerosing	IVB
RS	6 25 56	11/21/61	M	7/25/66 to 5/3/67	13	115—238	Mixed	IVB
JS	7 61 12	7/22/56	M	5/30/69 to 11/5/71	25	122—295	Nodular sclerosing	IIB-IIIB

Table 5 (continued)
CLINICAL DATA FROM PATIENTS WITH HODGKIN'S DISEASE IN CHILDREN

Case	Unit no.	D.O.B.	Sex	Period of SCLs	Total no. of SCLs	Range of SCLs	Initial histology Hodgkin's disease	Stage
JLT	3 97 59	12/5/48	M	6/17/70 to 12/2/71	9	85—110	Unclassified	IIIA
DLW	6 99 63	6/29/53	F	11/24/69 to 8/16/71	5	85—122.3	Nodular sclerosing	IIA
KLW	8 07 25	5/20/57	F	5/22/70 to 12/27/71	12	71—139	Nodular sclerosing	IIIA
RY	7 41 33	5/2/66	M	12/19/68 to 3/18/71	16	107—224	Nodular sclerosing	IIA-IVB

From Tessmer, C. F., Hrgovcic, M., and Wilbur, J. R., *Cancer*, 31, 303, 1973. With permission.

serum copper levels in non-Hodgkin's lymphoma patients was undertaken consisting of 353 observations of serum copper level in 236 patients with various types of non-Hodgkin's lymphoma.[14] The pertinent clinical data in relation to histologic groups, cell type, clinical behavior, and the serum copper level will be discussed later.

In 1975, we reported our interpretations of the serum copper level in Hodgkin's disease and have also stressed that the serum copper is nonspecific, and that false-positive levels may be observed secondary to infection or inflammation, in pregnancy, and in patients taking certain medications, particularly contraceptive drugs containing estrogen, or certain dyes such as phenazopyridine hydrochloride (Pyridium®).[15] This study material was based on 318 observations of serum copper levels in 39 adult patients with Hodgkin's disease. These data will be discussed in detail in the section on serum copper interpretations.

As a control group, 45 patients with carcinoma of the prostate were studied. This control group was studied from two aspects: (1) as a solid tumor group having no known relation to serum copper levels and (2) as a group undergoing radiation therapy with exposure levels equal or exceeding those of the lymphoma patients.

All of the patients with the tissue diagnosis of Hodgkin's disease or non-Hodgkin's lymphoma seen from 1966 to 1973, in which one or more serum copper level determinations have been made, were included in this study regardless of the extent and activity of disease or previous therapy. Pregnant women and patients on estrogens or those with infections have been excluded as mentioned previously, in keeping with the recognized relationship between estrogens, inflammation, and serum copper levels. The tissue examination findings in malignant lymphomas were reported according to the Lukes and Butler classifications.[18,19] The mandatory staging procedures as discussed in Chapter 8 were performed.[20,21] There were 40 Hodgkin's disease patients in this study group, clinically stages I and II, subjected to staging laparotomy. The Rye/Ann Arbor[22,23] clinical staging categories were adopted for Hodgkin's disease, whereas the Peters et al.[24] classification was used in non-Hodgkin's lymphoma.

The serum copper level in lymphoma and acute leukemia patients was determined usually once weekly throughout the period of hospitalization or X-ray treatment. In the outpatient group, the serum copper level was determined on the day of examination in the outpatient clinic, and thereafter at 1- to 3-week or 2-month intervals, depending upon the treatment schedule, remission induction, and/or maintenance. In the follow-up of the therapy group, the serum copper level was determined at 2- to 6-month intervals, depending upon the date of completion of treatment and the duration of remission — with the exception of several volunteers in long-term remission who have continued to provide samples for serum copper level determinations at intervals of 2 weeks to 1 month during a period of 4 to 6 years. The serum copper level records of 87 patients with Hodgkin's disease and 168 cases with non-Hodgkin's lymphoma consisted of one determination only. These single determinations are from patients seen in various phases of activity and extent of disease, and are included in order to determine an overall basic relationship of the serum copper level to the status of disease activity. Coordination and analysis of these single values is made possible by assigning to them one of the clinical categories described below. In order to correlate the alterations of the serum copper level with the clinical activity of lymphomatous disease, six clinical categories of disease activity were established as shown in Table 6. This classification is more appropriate and precise in the retrospective analysis of serum copper level evaluation in this study than classifying patients into three commonly used (active, partial, and complete remission) groups, particularly in borderline cases (partial remission or response).

The appropriate clinical category is assigned utilizing the date of each copper value and without reference to the serum copper level as an indication of the clinical state of disease activity of that patient at the time the specimen for serum copper level was drawn. The status of disease activity was determined independently by attending staff physicians who were not aware of serum copper values, or involved in these studies.

Table 6
DEFINITION OF CLINICAL CATEGORIES

Category	Abbreviation
1. Pretreatment	Pre Rx
2. Under treatment with evidence of disease	Rx WED
3. Under treatment, activity of disease undetermined	RX Dund
4. Under treatment, no evidence of disease	Rx NED
5. No current treatment, no evidence of disease	No Rx NED
6. Relapse, or recurrent disease, with or without maintenance treatment	Relapse

From Hrgovcic, M., Tessmer, C. F., Thomas, F. B., Fuller, F. M., Gamble, J. F., and Shullenberger, C. C., *Cancer*, 31, 1337, 1973. With permission.

All serum copper values are plotted according to the assigned clinical category in order to identify certain general trends. The same method is applied to correlate the serum copper level with histological classifications, stages, and systemic symptoms in both Hodgkin's disease and non-Hodgkin's lymphoma. Serum copper samples in the control group (carcinoma of prostate) were taken twice weekly during the course of radiotherapy which was administered as 200 rad daily for 5 days of the week, a total of 5000 rad to a pelvic field (12 × 12 cm) continued to 7000 rad in a smaller field (8 × 8 cm).

As discussed in Chapter 2, Methods of Copper Determination, there are several microanalytic techniques for the quantitative measurement of copper. To this end, we have used four different trace element analysis (colorimetric, atomic absorption, X-ray fluorescent, and proton-induced X-ray emission — PIXE) techniques with emphasis on how they can be applied to the measurements of trace elements in large numbers of biological samples under routine analysis conditions. In 1974 we reported the results of trace element analysis of human serum by X-ray fluorescence and by X-ray excitation with protons under routine analysis conditions and compared the results with those obtained from flame atomic absorption on equivalent samples; this will be discussed in detail.[11]

The serum copper levels were determined in the early part of this study by the colorimetric method of Jensen et al.[25] as modified by Hrgovcic et al.[2] Since 1968, the serum copper levels have been determined by an atomic absorption technique, utilizing direct dilution of serum (1:4), in the Perkin-Elmer Model 403 with integrated absorption readout.[26] The latter technique has made possible a minimum sample size of 0.2 mℓ of serum, increased sensitivity and specificity, and reduced laboratory processing and time.

In order to compare the efficiency, precision, advantages, and pitfalls of each technique, direct comparison of techniques was done. We studied 100 serum samples from patients who were admitted to the M.D. Anderson Hospital and Tumor Institute (Hematology Section) with various lymphoreticular disorders: 4 to 5 mℓ of serum from each patient was separated into three aliquots, frozen, and later studied with the three techniques. The precision of each technique was checked from three different batches of pooled blood-donor serum in which the copper, iron, and zinc concentrations were approximately 1 ppm. In each case, many small aliquots were individually prepared and processed. The precision of each technique for measuring the copper content in the blood serum samples indicated that the flame atomic absorption results are the most precise, followed by X-ray fluorescence and X-ray excitation with proton. All three techniques are found capable of measuring iron, copper, and zinc concentrations at the part per million level with a relative error of less than 10%. The copper:zinc ratio was taken as the direct ratio of measured peak counts; thus, the standard deviation of this ratio was a measure of the precision of the X-ray technique less the doping procedures involved. It was apparent that the doping procedures introduce a sizable part of the error to the X-ray technique. Of the three techniques, the flame atomic absorption apparatus produced the most consistent results and was subject to fewer sources of contamination. The disadvantages of atomic absorption include measurements of only one or two (Cu and Zn) trace elements at a time and the fact that the precision depended upon the homogeneity and the viscosity of the sample solution, since both sample and standard solutions are assumed to be nebulized and admitted to the flame at the same rate. An inhomogeneous mixture of the protein-bound iron in serum would appear to be responsible, in addition to other factors, for the abnormally high relative standard deviation of measured iron concentrations.

The primary advantage of the X-ray fluorescence technique is that a number of trace element concentrations can be measured simultaneously. This technique has the added advantage that the apparatus is easy to operate while good signal-to-noise ratios allow simple on-line analysis. At present there are two disadvantages: the amount of sample needed is large and the number of steps and the time involved in preparing samples is great.

The primary advantages of proton-induced X-ray emission techniques are (1) the largest number of elements with concentrations in the parts per million range can be analyzed simultaneously, (2) only milligrams of sample are required; thus, matrix effects are negligible. It should be noted that, of the three systems described, this is the most complex because of the use of a tandem Van de Graaff accelerator. As a practical tool, however, it is assumed that a 2 to 3 MeV single-ended Van de Graaff would be used. Although a 2 to 3 MeV machine would still be classed as sophisticated laboratory equipment, its use and maintenance would not constitute a serious disadvantage. In addition, the use of commercially prepared, thin, heat-resistant low-Z plastic foil backings would cut preparations time to minutes per sample. One major disadvantage cannot be eliminated. Because of *bremsstrahlung* emission from "knock-on" electrons produced in both sample and backing, background radiation levels under many of the peaks require that sophisticated mathematical methods be used to "strip" the spectra. Increasing the sample thickness does not reduce this problem. Ashing the sample would reduce the *bremsstrahlung* emission but would compromise some of the ease of sample preparation. Thus, larger computers capable of processing and analyzing data on-line are a necessary part of such a system.

In conclusion and based on our experience with these three techniques used, we believe that as practical analytic tools, each possesses certain inherent advantages that must be weighed in deciding which technique is the most appropriate for a given study. Specifically, the order of simplicity and precision (flame atomic absorption, X-ray fluorescence, and X-ray excitation with protons) is in inverse proportion to the number of elemental concentrations which can be studied simultaneously. These studies, consisting of a comparison of the three techniques of trace-element determinations (copper, zinc, and iron) were performed in collaboration with Rice University and are reported elsewhere with complete and detailed data analysis.[11]

A. Quality Control

To provide for continuing quality control and to establish the reproducibility of the method, a large amount of serum was collected from a normal man and divided into 2-mℓ aliquots which were then stored and frozen until used. From this material, 19 samples were tested in duplicate during 1 month to establish the reproducibility. These test results ranged from 91.7 to 96.9 μg/100 mℓ. The mean value was 94.3 μg/100 mℓ with standard deviation of ±1.3 μg/100 mℓ. Thereafter, a duplicate sample of this stock control serum was run with each group of patients' samples in addition to the three standard copper solutions.

B. Normal Values

To determine normal values under our technique and conditions, 153 individuals without

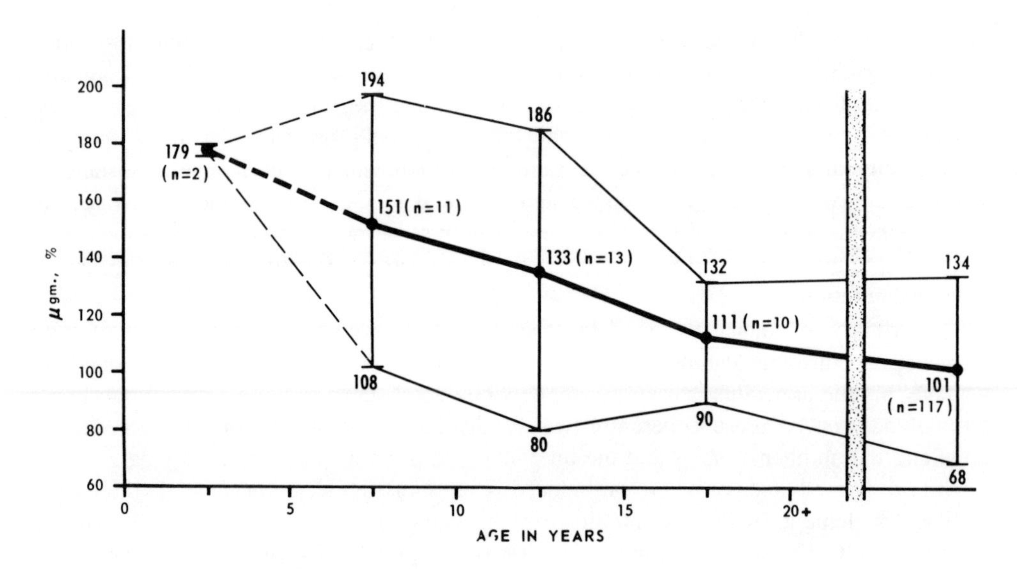

FIGURE 1. Serum copper levels in the control population distributed by ages. (From Hrgovcic, M., Tessmer, C. F., Minckler, M. T., Mosier, B., and Taylor, H. G., *Cancer,* 21, 743, 1968. With permission.)

overt disease were tested for serum copper levels. There were 77 serum samples obtained incidental to blood bank donations from a group of healthy male adult prisoners. An additional 40 serum samples were obtained incidental to blood donations from a mixed population of healthy adults in the blood bank of The University of Texas M.D. Anderson Hospital and Tumor Institute. Evaluated separately, no significant statistical difference was found between the two adult groups. These 117 sera gave serum copper levels ranging from 69 to 133 μg/100 mℓ, an average value of 101 μg/100 mℓ, and a standard deviation of ± 16 μg/100 mℓ.

Serum from 36 children aged 2 to 18 years showed a distinct age-related pattern. Values of about 180 μg/100 mℓ in the 0 to 4.9 age group dropped in linear fashion to the adult levels given above. No sex differences were apparent. These data are summarized in Figure 1 and are in reasonable conformity to those reported in the literature[25-39] (see also Chapter 3, Table 1).

Since the use of the serum copper level in children is somewhat more complex and controversial, and in view of the deviation of the normal values from those of normal adult controls as shown in pilot studies,[2,3] further examination of this variation led to a correction factor, and the demonstration that in the abnormal values encountered in active leukemic disease, discussed below, such a correction factor provided complete coincidence and correlation with the above adult values.[6] The correction factor has also been utilized in the evaluation of Hodgkin's disease in children, providing the basis for a uniform normal range.[8] The study group consisted of 681 samples obtained in conjunction with the laboratory aspect of a health survey in several elementary public schools in Waco, Texas.[10] Distribution of clinical material according to age, race, and sex is shown in Table 7 and means and standard deviations of serum copper levels of Caucasian, Mexican-American, and Black children according to age is shown in Table 8. The serum copper levels of male and female children of each age group within each racial category were not statistically different. On the other hand, a significant racial difference was noted, the serum copper level in Black children being higher than that for Caucasian and Mexican-Americans, whereas the levels for the last two did not differ significantly from each other.

Further analysis of the age relationship on this basis is indicated by the following regression equations for the means and standard deviations:

Table 7
SERUM COPPER IN NORMAL CHILDREN ACCORDING TO AGE, RACE, AND SEX

	White		Mexican-American		Black	
Age (years)	**Males**	**Females**	**Males**	**Females**	**Males**	**Females**
6	16	8	11	16	16	19
7	21	19	25	27	15	9
8	15	25	17	18	1	3
9	26	13	36	30	7	6
10	23	22	28	33	8	7
11	17	20	24	16	7	12
12	10	11	5	26	7	6
N = 681	128	118	146	166	61	62
	36.1%		45.8%		18.1%	

Reproduced with permission from Tessmer, C. F., Krohn, W., Johnston, D., Thomas, F. B., Hrgovcic, M., and Brown, B., *American Journal of Clinical Pathology,* Volume 60, pages 870-878, 1973.

Table 8
MEANS AND STANDARD DEVIATIONS OF SERUM COPPER LEVELS OF CAUCASIAN, MEXICAN-AMERICAN, AND BLACK CHILDREN GROUPS ACCORDING TO AGE

Age (years)	Caucasian-Mexican-American (μg/100 mℓ)	Number	Black (μg/100 mℓ)	Number
6	142.7 ± 24.82	51	153.2 ± 25.69	35
7	135.7 ± 24.12	92	148.1 ± 25.33	24
8	136.5 ± 22.74	75	153.2 ± 3.47	4
9	130.7 ± 22.32	105	147.8 ± 21.92	13
10	134.6 ± 20.95	106	138.7 ± 26.00	15
11	124.9 ± 24.11	77	143.7 ± 32.71	19
12	120.4 ± 19.68	52	136.8 ± 19.38	13
Total		558		123

Regression

	Caucasian-Mexican-American		Black
Mean	159.0 − 2.96 × age	Mean	183.6 − 3.99 × age
SD	27.83 − 0.576 × age	SD	23.74 − 0.143 × age

Reproduced with permission from Tessmer, C. F., Krohn, W., Johnston, D., Thomas, F. B., Hrgovcic, M., and Brown, B., *American Journal of Clinical Pathology,* Volume 60, 870-878, 1973.

a. Caucasian and Mexican-American
   Mean 159.0 − 2.96 × age
   SD 27.83 − 0.576 × age
b. Blacks
   Mean 183.6 − 3.99 × age
   SD 23.74 + 0.143 × age

A plot of the mean serum copper level of each age according to racial grouping and including the regression, is shown in Figure 2. The values demonstrate graphically the clear

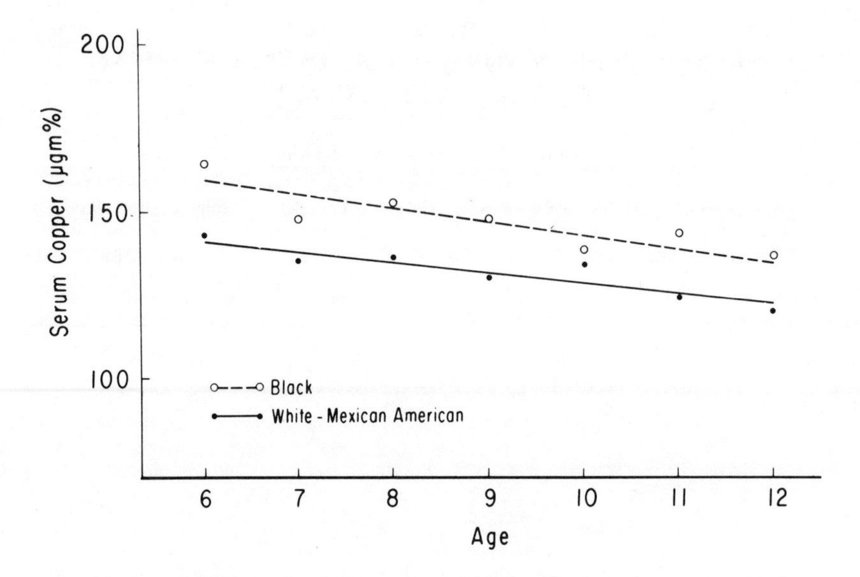

FIGURE 2. Plot of mean serum copper level of White, Mexican-American, and Black groups according to age, with regression. (Reproduced with permission from Tessmer, C. F., Krohn, W., Johnston, D., Thomas, F. B., Hrgovcic, M., and Brown, B., *American Journal of Clinical Pathology,* Volume 60, pages 870-878, 1973.)

relationship of serum copper level to age in this range of 6 to 12 years, as well as the significantly higher values in Black children.

The most useful expression of the relationship of serum copper to age in the range of this study group, however, is provided by a single formula for age correction of observed serum copper levels. As indicated by the foregoing data, this correction is necessarily presented for the two racial groupings.

Caucasian and Mexican-American

$$SCL_{corr} = \frac{SCL_{obs} + 16.7 - 0.7 \times age}{1.739 - 0.036 \times age}$$

Blacks

$$SCL_{corr} = \frac{SCL_{obs} - 33.74 + 4.89 \times age}{1.48 + 0.00894 \times age}$$

where SCL_{corr} is the value of serum copper in μg/100 mℓ to be obtained by correction to adult (20-year) control levels.

These formulas represent corrections for the mean and standard deviation, based on the control adult values for serum copper utilized in this laboratory, 101 ± 16 μg/100 mℓ.

Complete and detailed data analysis of this study concerning normal serum copper levels in children-age correction factor was reported previously.[10]

As a portion of our studies of copper metabolism, and in order to identify and characterize the copper proteins involved in lympho- and myeloproliferative malignant diseases, the chromatographic separation of the copper proteins of a dozen normal human sera was performed, which resulted in identification of two copper-binding proteins.[40] As can be seen from Figure 3 the serum copper has been separated into two distinct components. The major peak represents copper bound to ceruloplasmin (mol wt 130,000) while the minor peak appears to be a nonceruloplasmin copper protein ''ferroxidase II'' reported and described

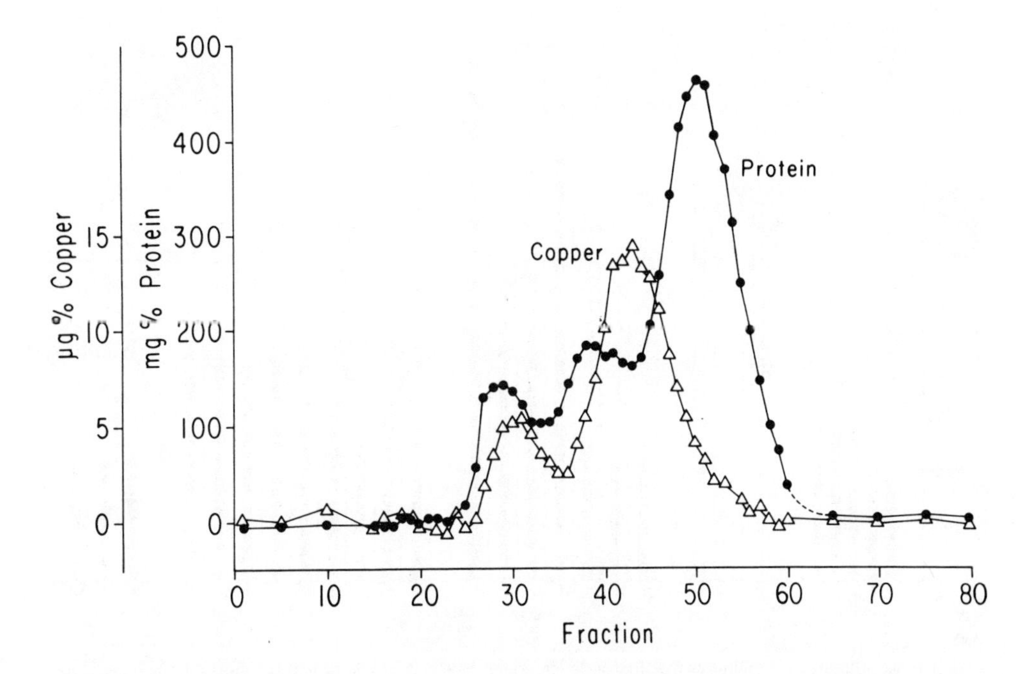

FIGURE 3. Copper-binding protein of normal human plasma: identification and preliminary observations.

by Topham and Frieden.[41] The identification of this and other copper-binding proteins in normal and diseased conditions may be of substantial significance in the study of copper metabolism. The characterization of the elevated copper protein may provide some insight into the index of lymphomatous disease activity including source, degree of specificity, and true relation to the disease process, which is not yet apparent.

II. HODGKIN'S DISEASE PATIENTS

A. Serum Copper in Hodgkin's Disease in Adults

In view of the age-related variation of serum copper levels in children and difficulties in their interpretations we will discuss separately the serum copper alterations in adults and the pediatric age group. The adult Hodgkin's disease group studied is based on 1051 serum copper levels determinations in 214 patients, with pertinent clinical information shown in Tables 1, 3, and 4.

1. Results

In the pilot study[2] we reported serum copper level in 28 patients (22 adults and 6 pediatric cases) with Hodgkin's disease. The serum copper level before and after treatment are shown in Figure 4. Five of these patients were in complete clinical remission and had serum copper values ranging from 71 to 130 μg%. This compares to the 95% confidence interval for normal adult serum copper levels of 69 to 133 μ % (N = 117).[2] The 20 patients with active Hodgkin's disease had serum copper level ranging from 172 to 426 μg%. Thus, it was suggested that the active phase of the disease is characterized by a marked hypercupremia, while the inactive phase is characterized by normal serum copper levels.

Of the 28 patients with Hodgkin's disease, 2 failed to show clear correlation between the serum copper level and disease activity. Their serum copper levels remained within normal limits in spite of active disease. A third patient was of unknown disease status at admission. He had been referred to the University of Texas at Houston M.D. Anderson Hospital and Tumor Institute in March 1966 for additional X-ray therapy following a course of nitrogen

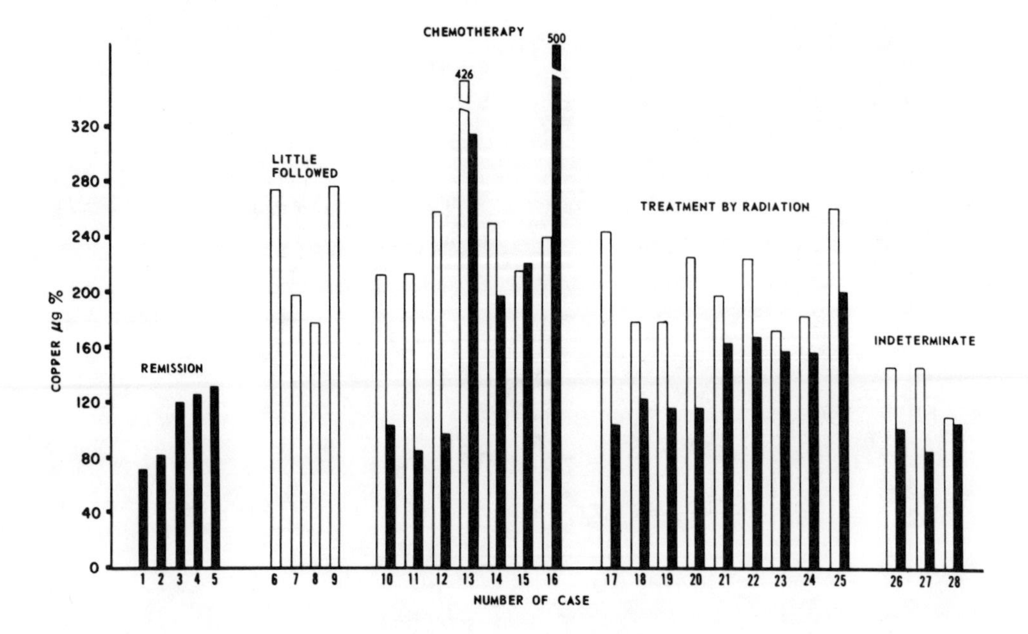

FIGURE 4. Survey of 28 adult and pediatric patients with Hodgkin's disease and serum copper levels before and after treatment. (From Hrgovcic, M., Tessmer, C. F., Minckler, M. T., Mosier, B., and Taylor, H. G., *Cancer*, 21, 743, 1968. With permission.)

mustard treatment at another institution. A positive response to chemotherapy had ensued and constitutional symptoms had disappeared. Possibly the near-normal serum copper level (146 µg%) which was noted at admission relates to the prior therapy and clinical response.

Of the 20 patients with active Hodgkin's disease, 4 had inadequate serum copper level records. Two refused further follow-up and two died without additional determinations. Of the remaining 16 patients, 7 were treated with chemotherapeutic agents cyclophosphamide (Cytoxan®) and vinblastine sulfate (Velban®) and 9 were given radiotherapy. Three of the chemotherapy patients and four of the radiotherapy patients had a favorable response to treatment. The initially high serum copper levels of these patients decreased to normal values preceding or accompanying clinical remission.

Four of the patients who received chemotherapy showed only slight or no improvement, which was reflected by a persistent hypercupremia. Five radiotherapy patients showed signs of some clinical improvement, and although the serum copper levels of these patients decreased, they did not reach normal range. All nine of these patients from both treatment groups had active disease at the time of the final serum copper determinations.

A subsequent study[13] of serum copper levels in 191 adult patients with Hodgkin's disease and pertinent clinical data shown in Tables 3 and 4 was made. All (896) serum copper levels are presented in graphic and tabular form in Figure 5 and Table 9 according to clinical category including distribution, mean, number of samples, and number of patients in each category. The basic overall relation of serum copper level to disease activity is represented by the mean serum copper level of each category. Increased serum copper levels in the active phase of the disease gradually decrease with sequential reduction of the disease activity as response to therapy, with normal serum copper values during remission, and an elevation of the serum copper level in relapse.

Since a number of patients have contributed more than one value to each of the clinical categories, and the mean may be unduly influenced by the values of single cases, the mean values of all serum copper levels of each patient in a given clinical category are shown in Figure 6. The same sequence of the basic relationship of the serum copper level to the

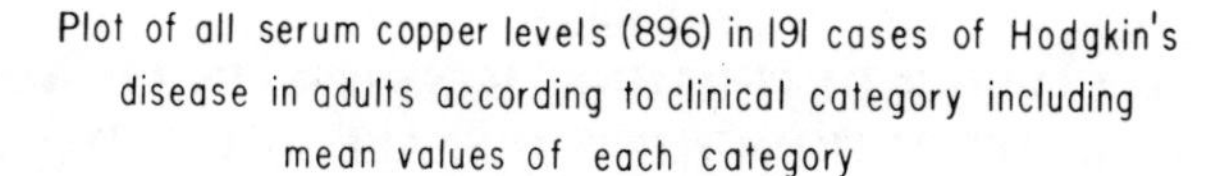

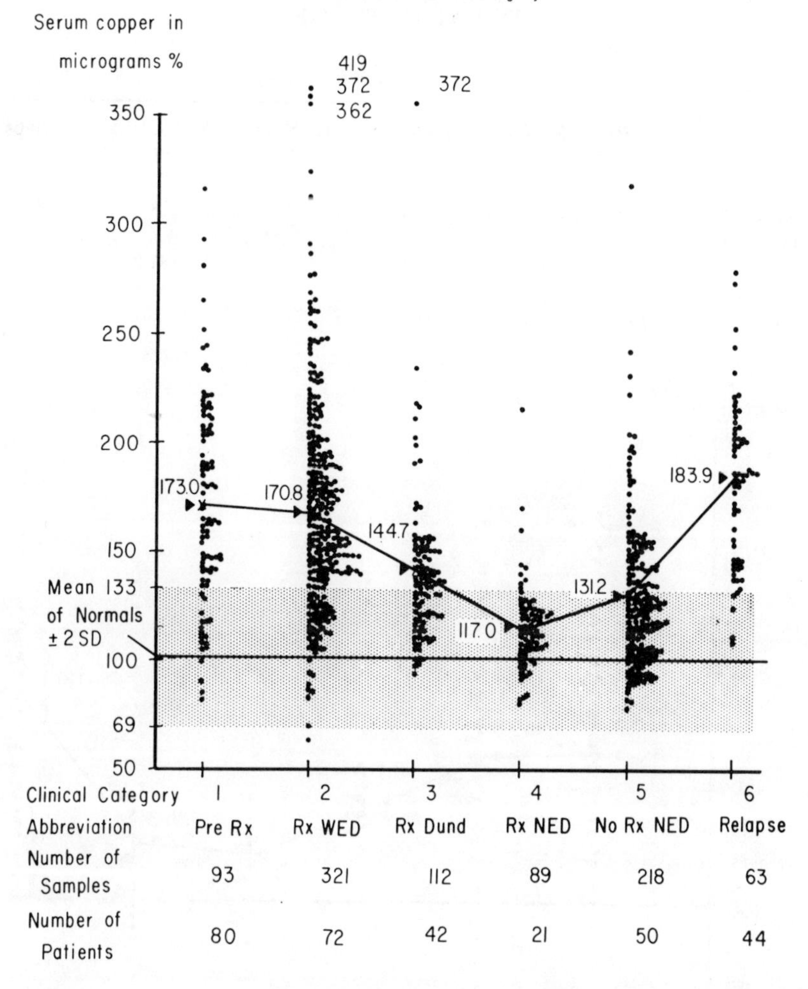

FIGURE 5. All SCL are plotted in the clinical category to which they are assigned (see text). Represented by the mean SCL of each category, it may be seen that increased SLCs in the active phase of disease decrease with response to treatment and rise with relapse. (From Hrgovcic, M., Tessmer, C. F., Thomas, F. B., Fuller, M. L., Gamble, J. F., and Shullenberger, C. C., *Cancer,* 31, 1337, 1973. With permission.)

disease activity process is noted; that is, a high serum copper level in the active phase of the disease, decreasing with response to therapy, normal during remission, and increasing with relapse.

A comparison of Figures 5 and 6 indicates an essentially similar pattern with no significant distortion by single cases. These figures serve to point out wide variations of serum copper level in certain categories and groups, seemingly inconsistent with the hypothesis that serum copper is mirroring the activity of the disease process. These aspects are examined in further detail.

Serum copper values above normal in categories 3, 4, and 5 in the absence of clinically clear evidence of Hodgkin's disease, and thus considered inconsistent, are presented in Table 10. Figures 5 and 6 also indicate inconsistent patterns of serum copper levels in untreated and relapsed cases (clinical categories 1 and 6). These are analyzed in Table 11.

Table 9
SERUM COPPER DETERMINATIONS (896) IN 191 ADULT PATIENTS WITH HODGKIN'S DISEASE ACCORDING TO CLINICAL CATEGORY

	Clinical category					
	PreRx 1	**Rx WED 2**	**Rx Dund 3**	**Rx NED 4**	**No Rx NED 5**	**Relapse 6**
No. of samples (total = 896)	93	321	112	89	218	63
Mean	173.0	170.8	144.7	117.0	131.2	184.5
SD	47.97	47.5	37.6	18.5	31.6	36.2
No. of patients	80	72	42	21	50	44

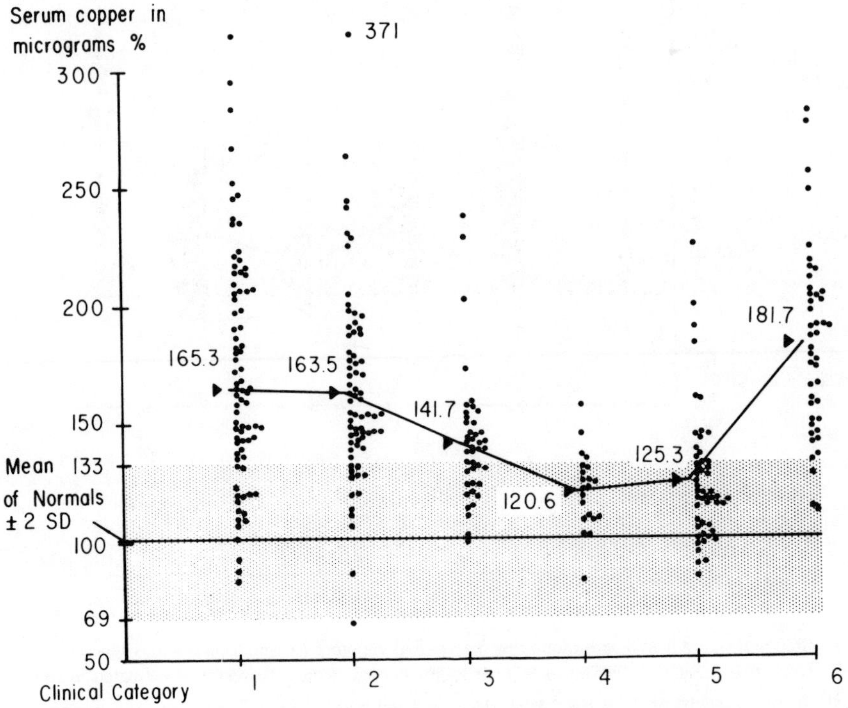

Abbreviation	Pre Rx	Rx WED	Rx Dund	Rx NED	No Rx NED	Relapse
Number of Samples	93	321	112	89	218	63
Number of Patients	80	72	42	21	50	44

FIGURE 6. Each value in this plot represents the mean of all SCLs in a given clinical category from one patient. This is presented to avoid possible distortion of the overall mean by a large number of values from a single patient. The pattern, however, is essentially similar to that in Figure 5 and 6 (see legends for explanation). (From Hrgovcic, M., Tessmer, C. F., Thomas, F. B., Fuller, M. L., Gamble, J. F., and Shullenberger, C. C., *Cancer*, 31, 1337, 1973. With permission.)

Table 10
ANALYSIS OF INCONSISTENT CASES IN CLINICAL CATEGORIES 3, 4, AND 5 (HIGH SERUM COPPER LEVELS IN THE ABSENCE OF CLINICALLY DEMONSTRABLE ACTIVE HODGKIN'S DISEASE)

	Rx Dund 3	Rx NED 4	No Rx NED 5
Total no. of cases	42	21	50
Total inconsistent SCL cases (SCL above normal range)	25	3	12
Prerelapse	7	1	6
Infection	4	1	4
Without explanation	14	1	2

From Hrgovcic, M., in *Zinc and Copper in Medicine,* Karcioglu, Z. A. and Sarper, R. M., Eds., 1980, 481. Courtesy of Charles C Thomas, Publisher, Springfield, Illinois.

Table 11
ANALYSIS OF INCONSISTENT CASES IN CLINICAL CATEGORIES 1 AND 6 (NORMAL SERUM COPPER LEVELS IN UNTREATED AND RELAPSED PATIENTS)

	Pre Rne Rx 1	Relapse 6
Total no. of cases	80	44
Total inconsistent SCL cases (SCL within normal range)	17	4
Stage IA	8	0
Terminal phase	0	2
Without explanation	9	2

From Hrgovcic, M., in *Zinc and Copper in Medicine,* Karcioglu, Z. A. and Sarper, R. M., Eds., 1980, 481. Courtesy of Charles C Thomas, Publisher, Springfield, Illinois.

The mean values of serum copper levels in each of the initial histologic classifications of Hodgkin's disease are presented according to clinical category in Figure 7 and Table 12. Again, similar basic relation of the disease activity was noted in three histologic subgroups (nodular sclerosing, mixed cellularity, and lymphocytic depletion). The exceptions are serum copper levels of patients of the lymphocytic predominance type which appear to show no elevation. The limitations of these data will be alluded to later.

A review of the relationship of the serum copper level (mean values) to clinical category by stage of disease (Figure 8) shows, in general, the same sequence noted in previous figures, that is, reflections of the disease activity process in the serum copper level. In addition, a significant difference in serum copper levels between localized and generalized Hodgkin's disease is noted, being consistently higher in stages representing more extensive disease. These stages include values of patients both with and without systemic symptoms. A comparison of the mean values of the serum copper level of all the cases, in all stages, show a higher serum copper level in patients with systemic symptoms compared to those without (Figure 9).

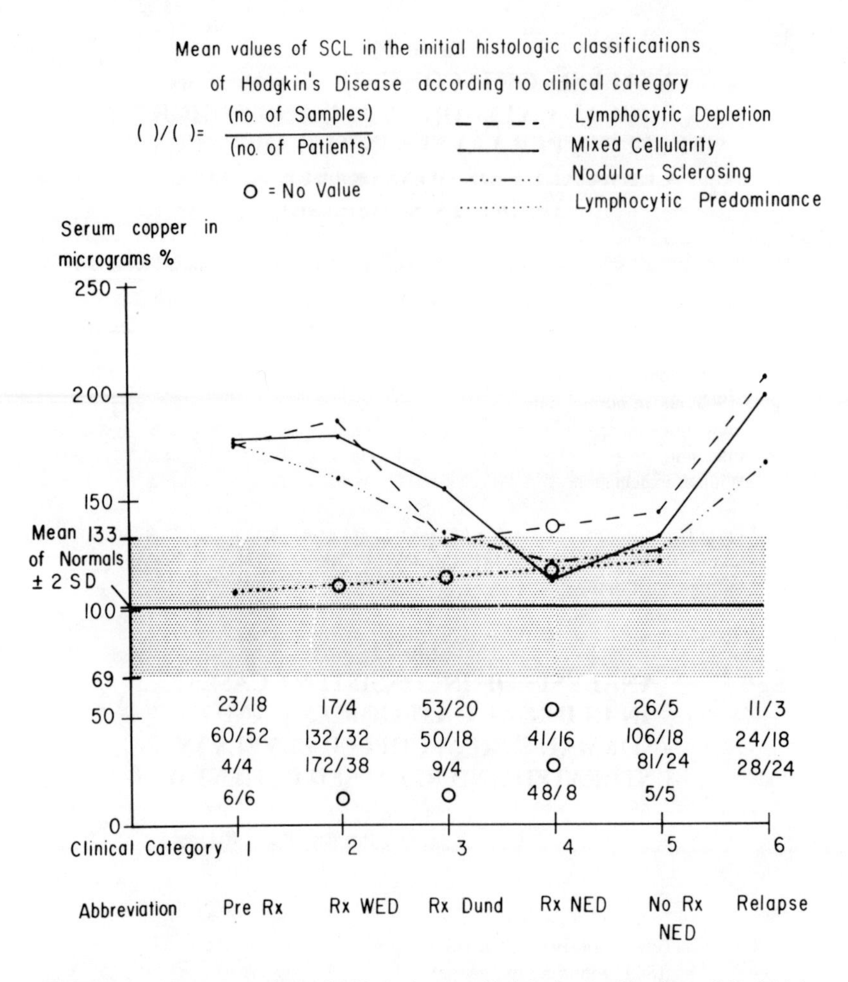

FIGURE 7. The mean values of SCL in the four histiologic classifications of Hodgkin's disease according to the assigned category shows generally similar relation to the clinical evidence of disease activity, that is, high SCL in the active phase of the disease, falling with response to treatment, and rising with relapse. The exceptions are SCL of patients with the lymphocytic predominance type which appear to show no elevations (see text). (From Hrgovcic, M., Tessmer, C. F., Thomas, F. B., Fuller, M. L., Gamble, J. F., and Shullenberger, C. C., *Cancer,* 31, 1337, 1973. With permission.)

The variability of the serum copper level in patients with extended periods in a category merits special interest. This factor has been analyzed in cases of complete remission (clinical categories 4 and 5). The total number of cases in category 4 and 5 (40 patients) represent those with two or more serum copper levels over a period of 2 months or longer while remaining in a single category.

Of this group, category 4 (Rx, NED) constitutes 12 cases and is represented in Figure 10. It may be seen that there are no significant variations in the serum copper levels with the passage of time in the same individual.

Those cases of category 5 (No Rx, NED), falling in the above group of 40 patients, are shown in two subgroups. The first of these subgroups (20 patients) represents those without evidence of disease activity *or* complications and are seen in Figure 11. There are no significant variations in the same individual. The data in this subgroup are limited in certain cases. The second subgroup (Figure 12) presents serum copper levels in eight cases, two with known complications and what may be termed inconsistent hypercupremia. Six of these

Table 12
SERUM COPPER LEVELS IN 191 PATIENTS WITH HODGKIN'S DISEASE ACCORDING TO CLINICAL CATEGORY AND HISTOLOGIC TYPE

	Clinical category						
	PreRx 1	**Rx WED 2**	**Rx Dund 3**	**Rx NED 4**	**No Rx NED 5**	**Relapse 6**	**Total**
Lymphocytic predominance							
No. of samples	6	0	0	0	5	0	11
Mean	106.8	—	—	—	114.4	—	
SD	19.2	—	—	—	18.5	—	
No. of patients	6	—	—	—	5	—	
Nodular sclerosing							
No. of samples	60	172	50	41	81	28	432
Mean	177.4	162.7	135.5	121.1	126.8	166.7	
SD	46.6	36.1	24.3	22.1	30.1	33.5	
No. of patients	52	38	18	16	24	24	
Mixed cellularity							
No. of samples	23	132	53	48	106	24	386
Mean	177.8	180.9	155.5	113.4	132.3	195.9	
SD	45.6	59.0	46.9	15.0	33.3	37.5	
No. of patients	18	32	20	8	18	18	
Lymphocytic depletion							
No. of samples	4	17	9	0	26	11	67
Mean	176.8	187.7	131.2	—	143.3	205.7	
SD	35.3	56.3	6.9	—	31.3	17.3	
No. of patients	4	4	4	—	5	3	
Total samples	93	321	112	89	218	63	896

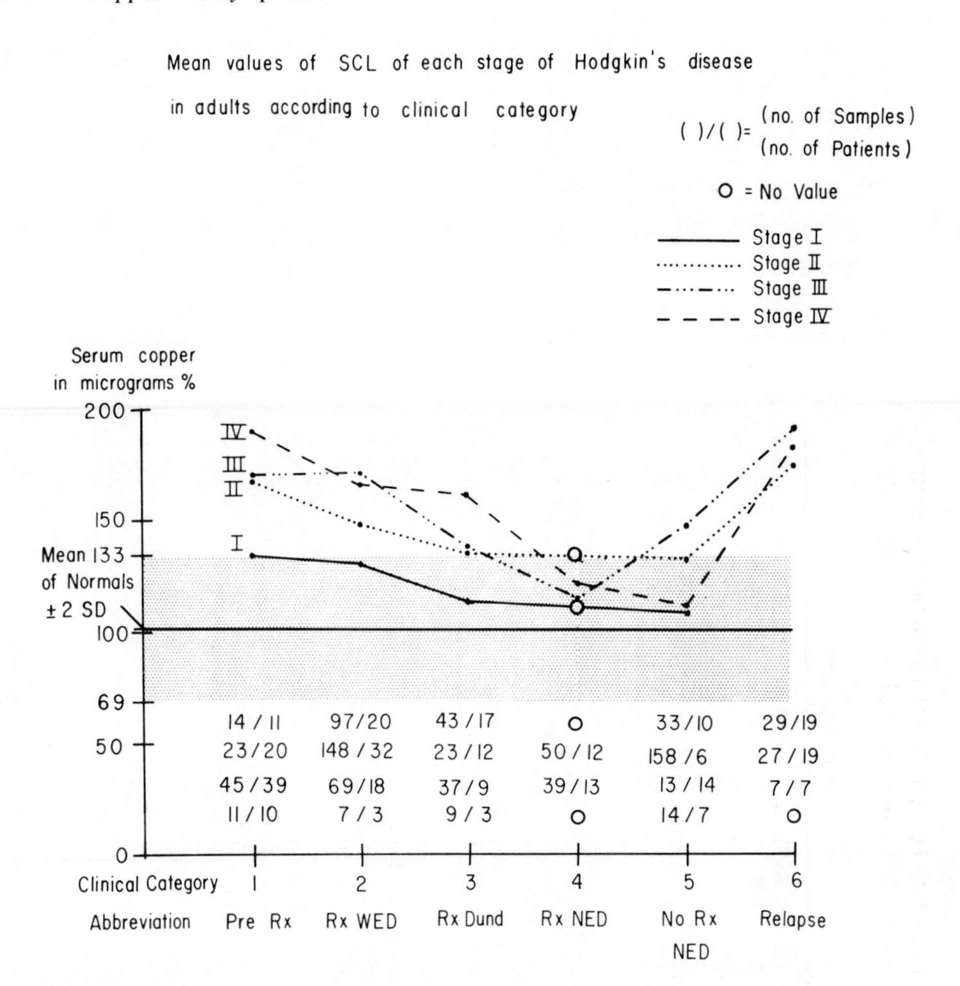

FIGURE 8. The mean value of SCL in each stage (Rye/Ann Arbor classification) show generally similar relationship to the clinical evidence of disease. In addition, there is a relationship between the SCL and extent of disease ("tumor volume"), showing higher serum copper values in stages representing more extensive disease, particularly in the pretreatment period. (From Hrgovcic, M., Tessmer, C. F., Thomas, F. B., Fuller, M. L., Gamble, J. F., and Shullenberger, C. C., *Cancer*, 31, 1337, 1973. With permission.)

cases, however, are in fact hypercupremic in a prerelapse state, all six cases proceeding to relapse. These cases effectively illustrate the predictive nature of an elevated serum copper level under such circumstances.

A plot of sequential serum copper levels from two remaining patients with Hodgkin's disease (mixed cellularity, stage IIB), marked by initials, presents a clinical course of inactive Hodgkin's disease complicated by various chronic inflammations, infections, and a second primary cancer. Plot F.P., which is also depicted in Figure 19, presents a patient in a long-term remission (still free of disease) who has been on thyroid medication due to hypothyroidism and has had erythematous nodular skin lesions on the extremities which were characterized by periodic exacerbations. Each episode was accompanied by increased serum copper levels. Plot V.S. presents a patient in whom the course of disease was complicated by frequent pyelonephritic infections which ceased after a right nephrectomy was performed, with chronic nonspecific hepatitis proven by a liver biopsy and arteriosclerotic occlusive disease which was controlled by an aortofemoral bypass. The patient continues to experience symptoms and signs of either a chronic inflammation (complicated by radiation vasculitis) or possible active occult Hodgkin's disease, the latter proven to be a second primary ma-

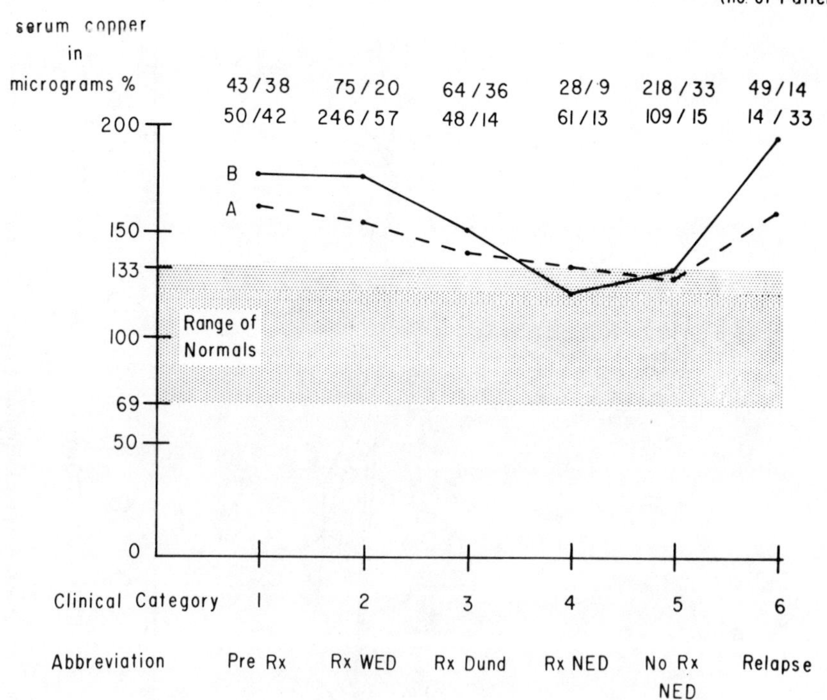

FIGURE 9. A comparison of mean values of SCL of all cases in all stages according to the clinical category and presence (B) or absence (A) of systemic symptoms attributable to Hodgkin's disease shows (in addition to reflection of disease activity) higher SCL in patients with systemic symptoms compared to those without.

lignancy (carcinoma of the pancreas). No evidence of Hodgkin's disease was found at autopsy.

Control data consisting of 45 cases of carcinoma of the prostate undergoing radiotherapy are shown in Table 13 and Figure 13.

Serum copper levels were studied during radiotherapy of patients in two categories of Hodgkin's disease, those showing complete remission and those showing partial remission.[9] Increased serum copper levels in untreated Hodgkin's disease patients returned to the normal range in complete remission during irradiation. On the other hand, in partial remission serum copper level was reduced, but did not reach normal range. The category of complete remission was represented by six patients, one of whom had combined radiotherapy and chemotherapy (Figure 14). Five patients were untreated before this radiotherapy and one was in first relapse.

As indicated in Figure 14 this is a relatively uniform trend, but with individual responses reaching the normal range of serum copper levels as early as the 10th day of radiotherapy and as late as several days following the completion of therapy. The serum copper level of all patients in this group reached the normal range. It is believed that even greater usefulness

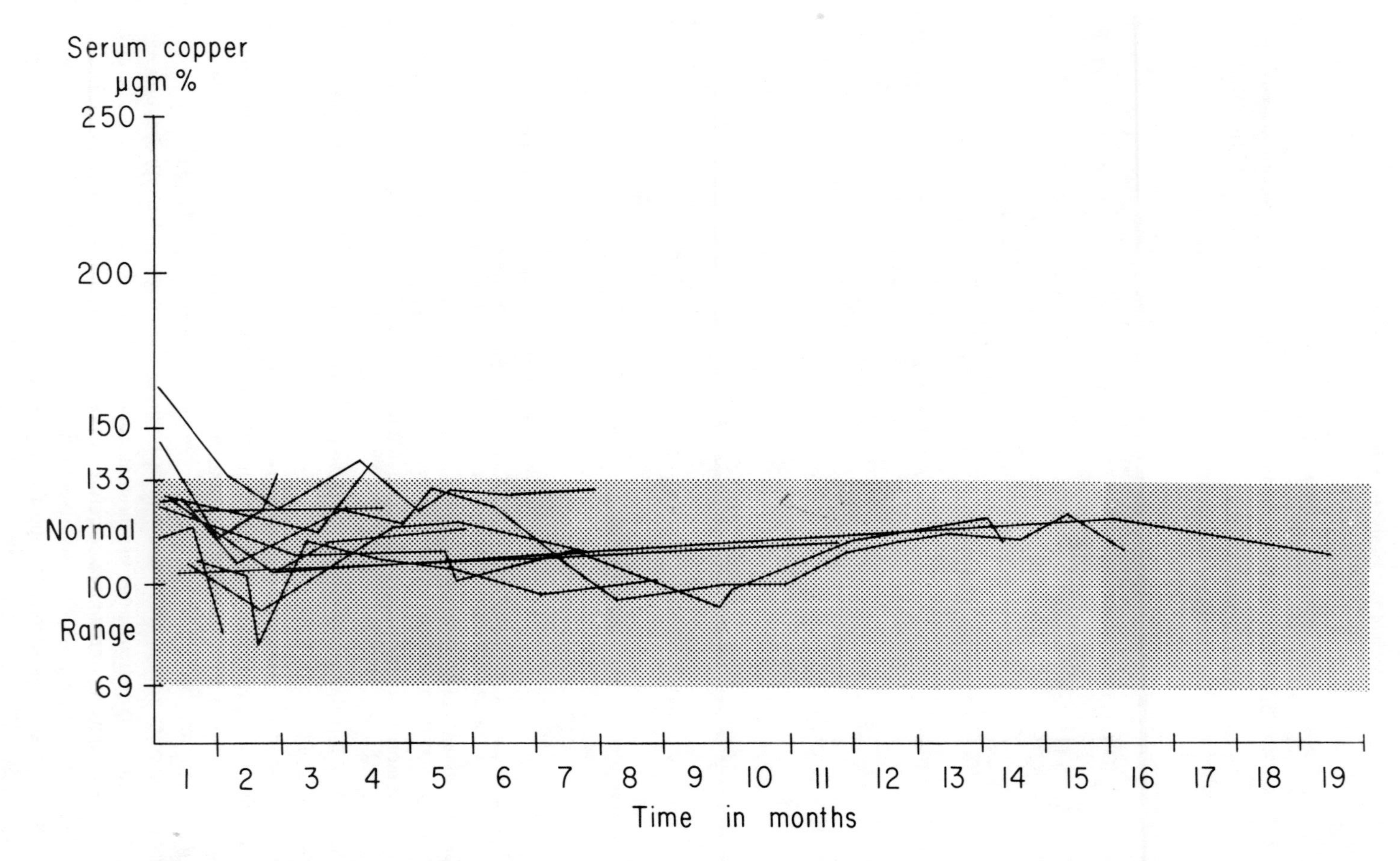

FIGURE 10. Sequential serum copper values in 12 patients with two or more serum copper levels during a period of 2 months or longer while remaining in remission with maintenance therapy. No significant variations with time or maintenance therapy in serum copper levels occur in the same individuals. (Reproduced with permission from Hrgovcic, M., Tessmer, C. F., Mumford, D. M., Ong, P. S., Gamble, J. F., and Shullenberger, C. C., *Texas Medicine*, 71, 53, 1975.)

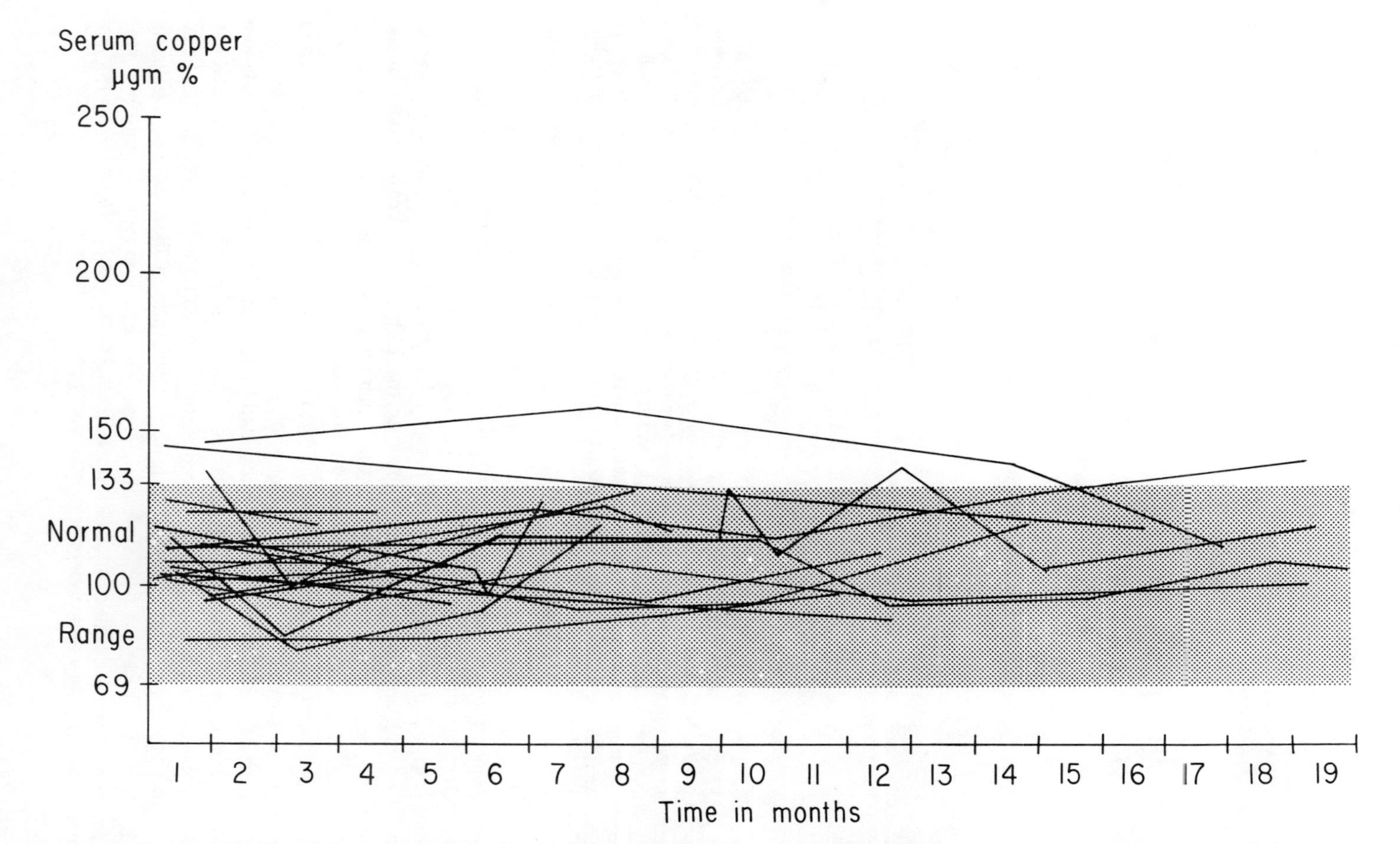

FIGURE 11. Sequential serum copper levels in 20 patients with two or more serum copper levels during a period of 2 months or longer while remaining in remission without maintenance therapy. No significant variations in serum copper levels occur with time in the same individual. (Reproduced with permission from Hrgovcic, M., Tessmer, C. F., Mumford, D. M., Ong, P. S., Gamble, J. F., and Shullenberger, C. C., *Texas Medicine,* 71, 53, 1975.)

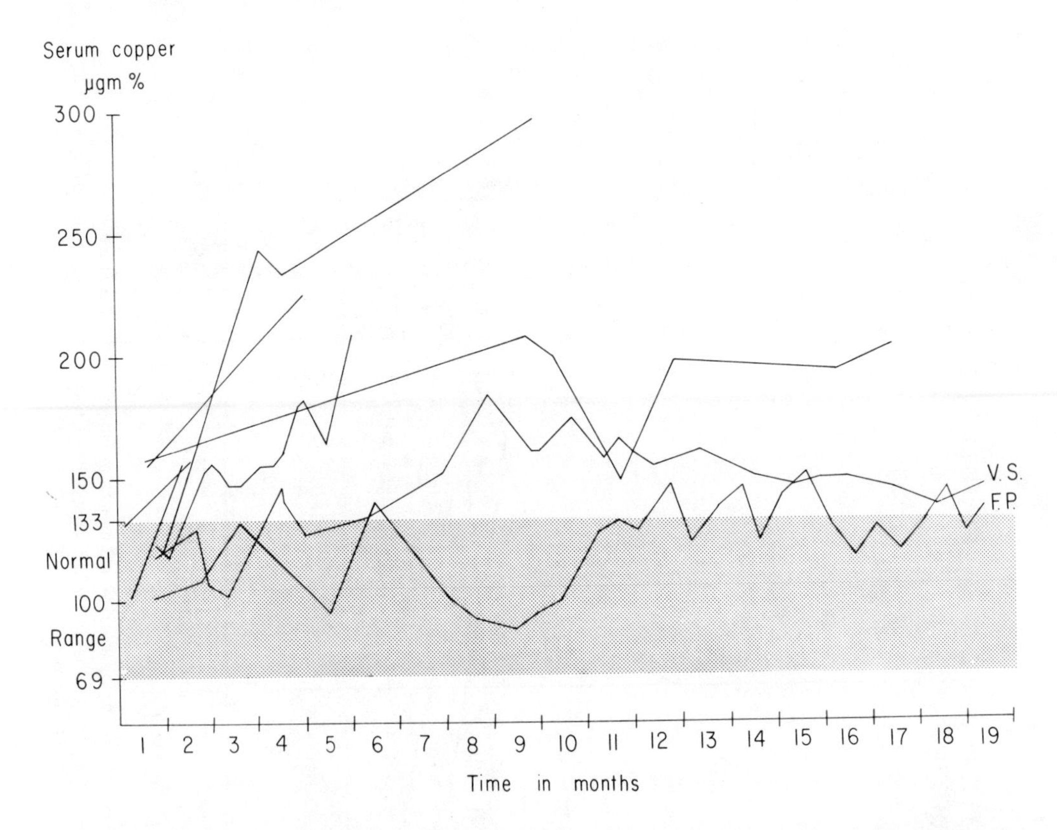

FIGURE 12. Sequential serum copper values in eight adult patients with Hodgkin's disease and hypercupremia in category 5 (no Rx, NED) show these cases with two or more SCL over a period of 2 months or longer while remaining in a single category. Six of these patients are in fact hypercupremic in a prerelapse state, all six cases preceding relapse. The two remaining patients marked by initials present the clinical course of inactive Hodgkin's disease complicated by chronic inflammations and infections (see text). (From Hrgovcic, M., in *Zinc and Copper in Medicine,* Karcioglu, Z. A. and Sarper, R. M., Eds., 1980, 481. Courtesy of Charles C Thomas, Publisher, Springfield, Ill.)

can be achieved with more frequent determinations, thereby increasing the "resolution" of the serum copper response and perhaps showing relationships to shorter-term clinical events.

In the complete remission group, patient MJ, who underwent combined therapy, had a somewhat less definitive response to radiotherapy. The pattern of the serum copper level suggests, that with an apparently rising copper on the 14th day and complications of the primary disease becoming manifest (pericarditis), that the subsequent clinical response and remission was, in fact, a result of combined therapy and not due to radiotherapy alone. This does not in any way appear to be an exception to the group pattern in which the serum copper level has reflected the clinical response to radiotherapy.

The category of partial response was represented by five patients (Figure 15) in whom a downward trend of serum copper level was noted, without reaching the normal level at any time. The patients in this group were judged at the completion of radiotherapy to be in partial remission. This is further indicated in the graphic plot of the means of this group which is shown in comparison with those of the complete remission group (Figure 16). This representation of the mean values of serum copper levels during therapy provides an effective and early index of tumor response to radiotherapy.

2. Illustrative Case Reports

Figure 17 shows the disease course, with pertinent data, of a 20-year-old white male with

Table 13
SERUM COPPER LEVELS IN 33 PATIENTS WITH PROSTATIC CARCINOMA DURING 8 WEEKS OF RADIOTHERAPY

	PreRx	1st week	2nd week	3rd week	4th week	5th week	6th week	7th week	8th week	Total
No. of samples	4	59	65	63	64	64	56	43	9	426
Mean	113.5	115.6	112.8	114.7	113.7	110.9	109.8	112.2	107.6	112.8
SD	11.30	22.24	19.98	19.83	19.08	18.87	15.02	15.44	16.14	13.7
No. of patients	3	33	33	33	33	33	30	26	7	33

From Hrgovcic, M., Tessmer, C. F., Thomas, F. B., Fuller, M. L., Gamble, J. F., and Shullenberger, C. C., *Cancer*, 31, 1337, 1973. With permission.

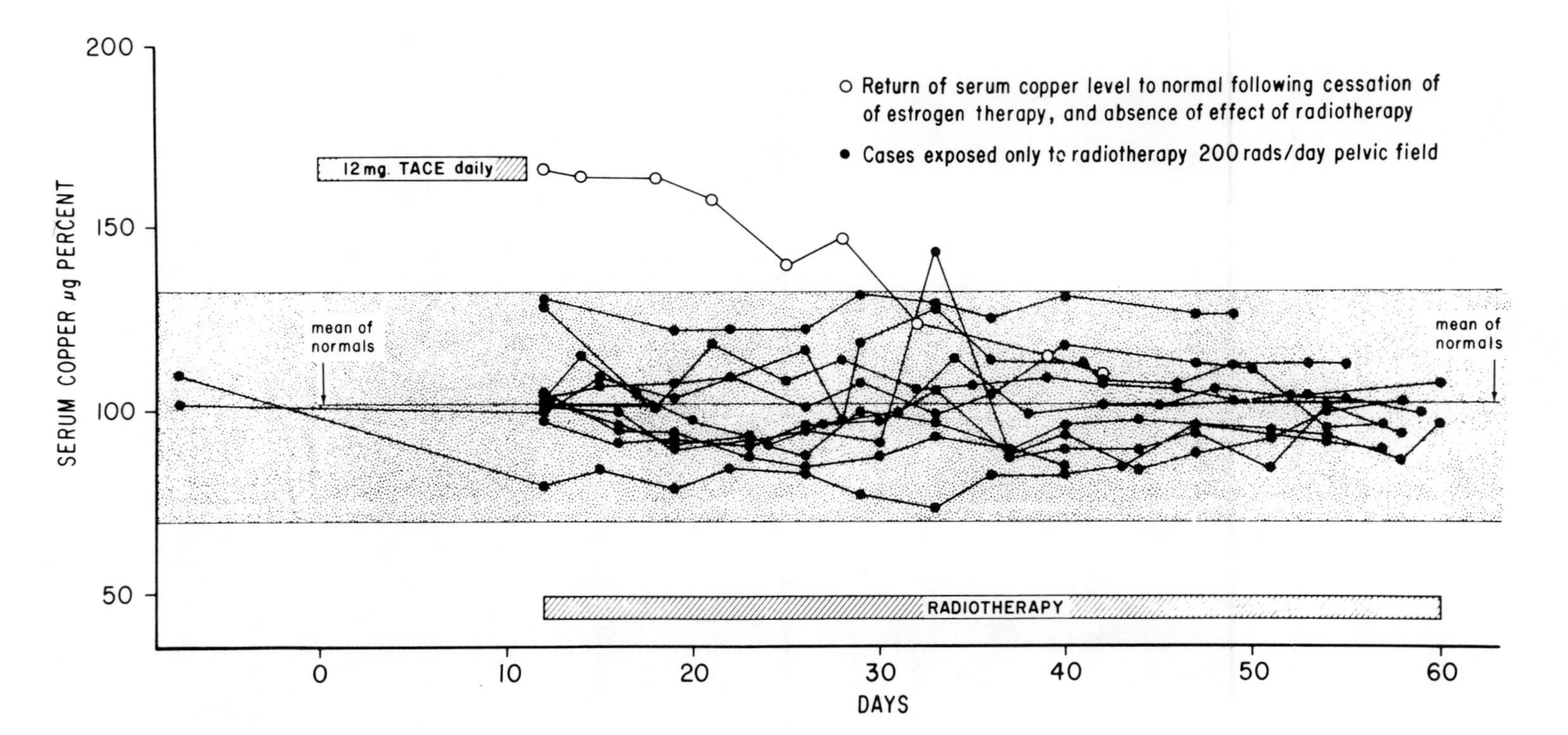

FIGURE 13. Serum copper levels in ten cases of prostatic carcinoma during the period of radiotherapy (total 7000 rad). Values were essentially in the normal range with a mean of all values of 100 mg/100 mℓ and showed no relation to radiotherapy. One additional case was included to show the return to normal serum copper of a patient in whom the administration of TACE (chlorotrianisene) was terminated at the onset of radiotherapy. (From Tessmer, C. F., Hrgovcic, M., Thomas, F. B., Fuller, M. L., and Castro, J. R., *Radiology,* 106, 635, 1973. With permission.)

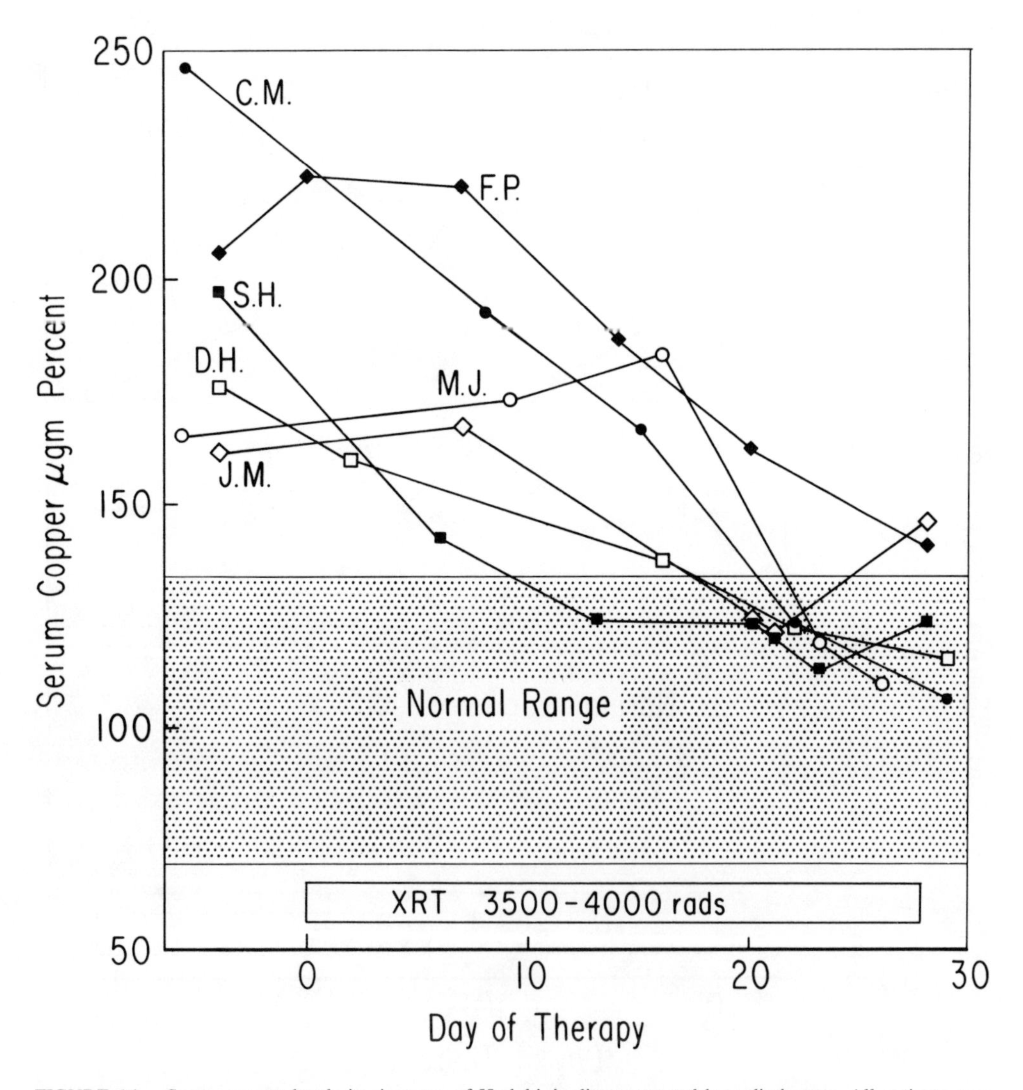

FIGURE 14. Serum copper levels in six cases of Hodgkin's disease treated by radiotherapy. All patients were considered to respond to therapy by complete remission of Hodgkin's disease. One patient (M.J.) received chemotherapy in addition and is discussed in the text. The general trend may be seen to be a return of serum copper to normal levels. Although data presentation has been limited to the period of radiotherapy, all cases reached the normal range. Two patients (F.P. and J.M.) returned to normal in 2 and 6 days, respectively, following the cessation of radiotherapy. (From Tessmer, C. F., Hrgovcic, M., Thomas, F. B., Fuller, M. L., and Castro, J. R., *Radiology*, 106, 635, 1973. With permission.)

Hodgkin's disease, nodular sclerosing type. The right axillary mass measuring 13 × 7 cm in diameter was the only abnormal physical finding.

The patient was symptom-free and underwent staging procedures which failed to reveal any abnormality, thus, establishing this case as a stage IA. The only abnormal biochemical test performed was an elevated serum copper level of 205 µg/100 mℓ. A clinical remission, obtained by irradiation, was of short duration. Six weeks after the completion of radiotherapy the patient returned with systemic symptoms of liver and bone marrow involvement with Hodgkin's disease. A reinduction remission was obtained by administering six courses of MOPP (nitrogen mustard, Vincristine®, procarbazine and prednisone) and this was maintained with Actinomycin-D®. This case provides a typical example of serum copper level determinations serving as an index of the status of disease activity which changed four times

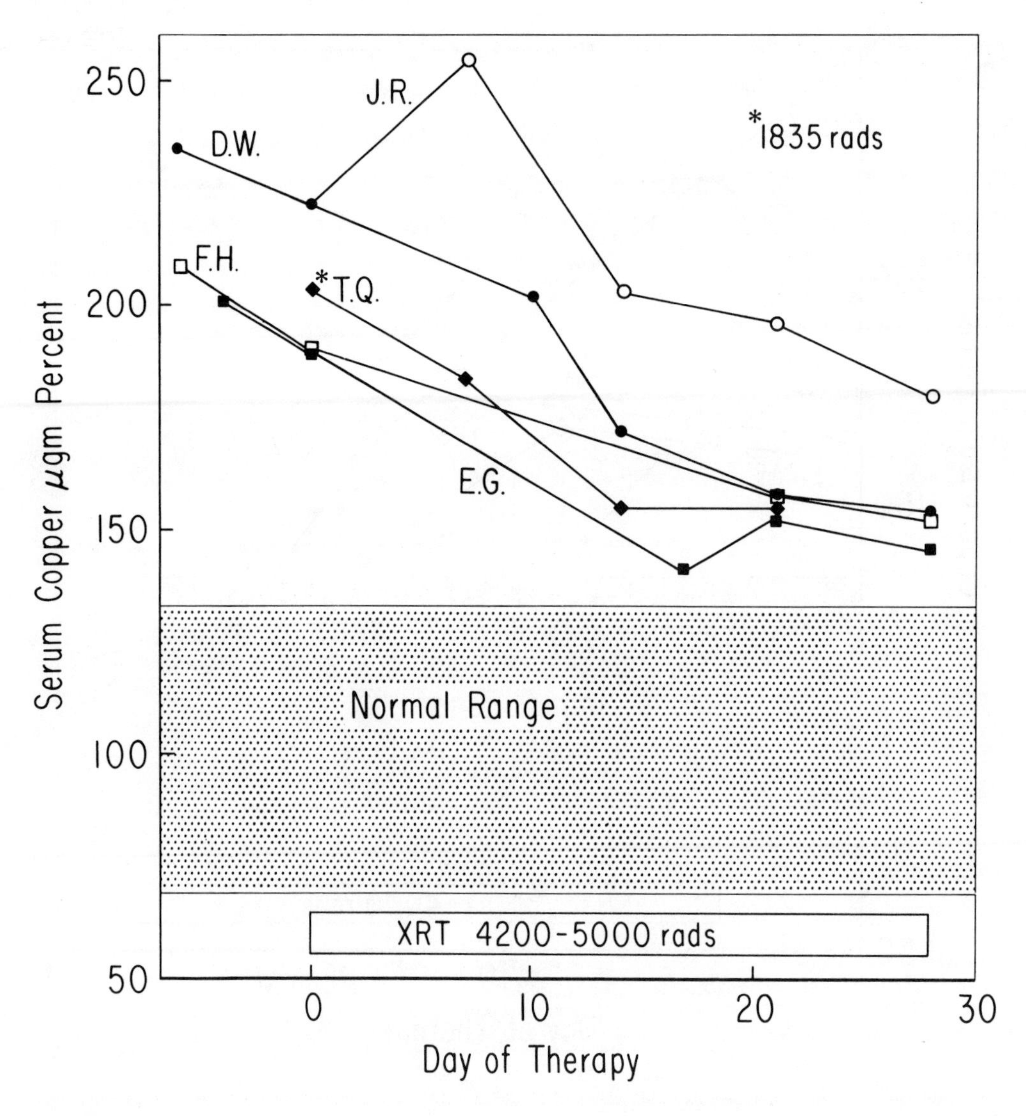

FIGURE 15. Serum copper levels in five cases of Hodgkin's disease treated by radiotherapy showing partial remission. While the trend is a reduction in serum copper levels, the values did not reach the normal range. One patient (T.Q.) received an incomplete course of radiotherapy (1825 rad) due to development of thrombocytopenia. (From Tessmer, C. F., Hrgovcic, M., Thomas, F. B., Fuller, M. L., and Castro, J. R., *Radiology*, 106, 635, 1973. With permission.)

during a 6-month period. It will be noted that the values in this case fall into all clinical categories. Each phase of disease activity was accompanied by a significant change in the serum copper level.

Figure 18 represents the clinical course of a 44-year-old male with generalized Hodgkin's disease, mixed cellularity type with pertinent clinical findings, who responded well to MOPP chemotherapy. As can be seen, the active phase of disease was accompanied by an elevated serum copper level. A decrease in the serum copper levels to normal limits preceded the clinical parameters of response to treatment. A complete remission was obtained and thc serum copper levels were within normal range.

Figure 19 illustrates the course of Hodgkin's disease (mixed cellular type), stage IIB, in a 32-year-old white woman who responded well to radiation therapy. The patient had low-grade fever after completion of X-ray therapy, and a small nodular density in the posterior

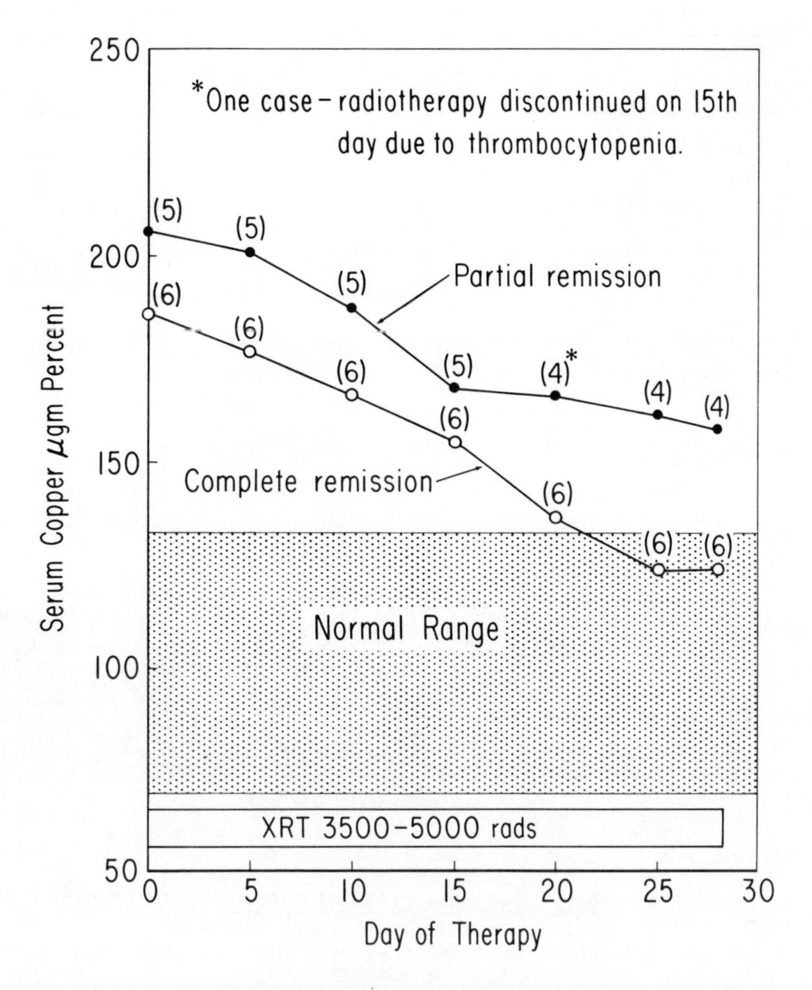

FIGURE 16. Mean serum copper levels of each group represented in Figures 14 and 15 complete and partial remission in 11 cases of Hodgkin's disease. The return to normal levels during radiotherapy in the complete remission group is contrasted to the mean values of the patients responding by partial remission who had a downward trend but did not reach the normal range. Values for the one patient in the partial remission group whose radiotherapy course was interrupted are included. (From Tessmer, C. F., Hrgovcic, M., Thomas, F. B., Fuller, M. L., and Castro, J. R., *Radiology,* 106, 635, 1973. With permission.)

lung field was seen in February 1971. The serum copper levels, however, remained normal during this period, indicating an inactive disease status. Five years later, hypothyroidism with exacerbation of keratotic skin lesions were found with a slight serum copper level elevation. No evidence of Hodgkin's disease reactivation was noted and further follow-up indicated that this patient continues to be in a remission with normal serum copper levels: the last one determined in May 1982.

Favorable response to the applied treatment, regardless of the form of therapy, is accompanied by normalization of the serum copper levels.

Figure 20 represents a 22-year-old man with mixed cellularity type Hodgkin's disease. His serum copper levels were followed from the time of diagnosis until death. While a decrease in the serum copper levels accompanied the clinical improvement during the initial chemotherapy, they did not reach normal levels at any time. Even when the patient was apparently symptom-free and with negative clinical findings (in July and August 1971), the

SYMPTOMS		
Absent	Fever Chills	Absent
CLINICAL FINDINGS		
Right axill mass 15x10 cm. Chest x-ray IVP Lymphangiogram Bone marrow biopsy (ALL NEGATIVE)	Liver 8cm. bcm Spleen 1cm. Liver and Bone marrow biopsy positive for Hodgkin's disease	Negative Bone marrow rebiopsy negative for Hodgkin's disease

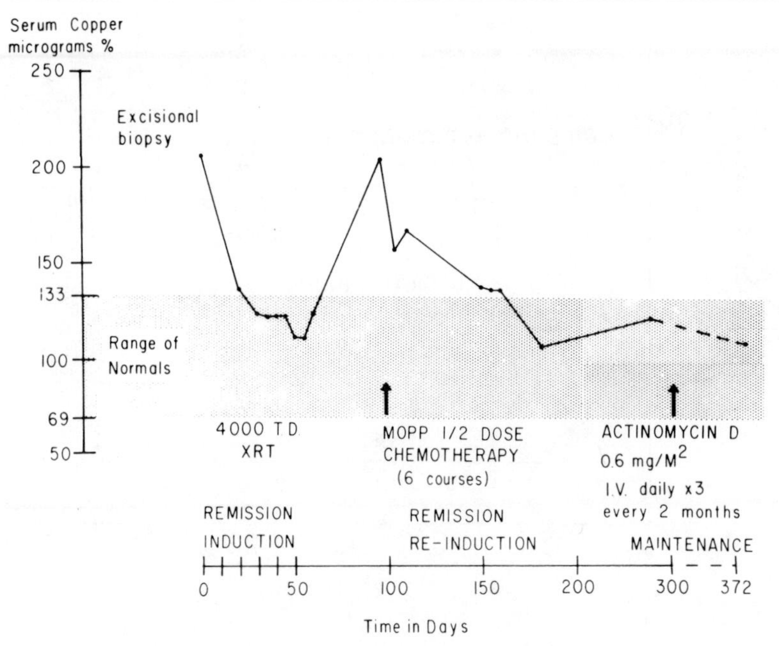

FIGURE 17. Sequential SCL in a patient with Hodgkin's disease, nodular sclerosing type, state IA with pertinent data. This case provides a typical example of SCL determinations serving as an index of the status of disease activity which was changed four times during a 6-month period. Each phase of disease activity was accompanied by significant change in the SCL. The possible significance of a marked hypercupremia in localized Hodgkin's disease is alluded to in the discussion. (From Hrgovcic, M., Tessmer, C. F., Thomas, F. B., Fuller, M. L., Gamble, J. F., and Shullenberger, C. C., *Cancer,* 31, 1337, 1973. With permission.)

serum copper level did not dip to normal. Indeed, the serum copper levels increased during maintenance MOPP treatment and fell to a normal range only on the day of death.

Figure 21 illustrates a 66-year-old male with generalized Hodgkin's disease who did not respond to chemotherapy. A persistently elevated serum copper level was a reflection of the continued presence of disease activity and an adverse disease course. This was an indication of the inefficacy of the chemotherapy with a fatal prognosis.

Correlation between the serum copper level as the indication of an adverse clinical course in a 30-year-old white female with stage IIA lymphocytic depletion Hodgkin's disease is presented in Figure 22. The normal initial serum copper level apparently reflects a good clinical response to treatment. However, a slowly increasing serum copper level preceded definite signs of clinical relapse by a very significant period. The latter period of the clinical course included estrogen therapy which is known to induce hypercupremia in man and may mask disease-induced serum copper variations. Nonetheless, in this case the sequential

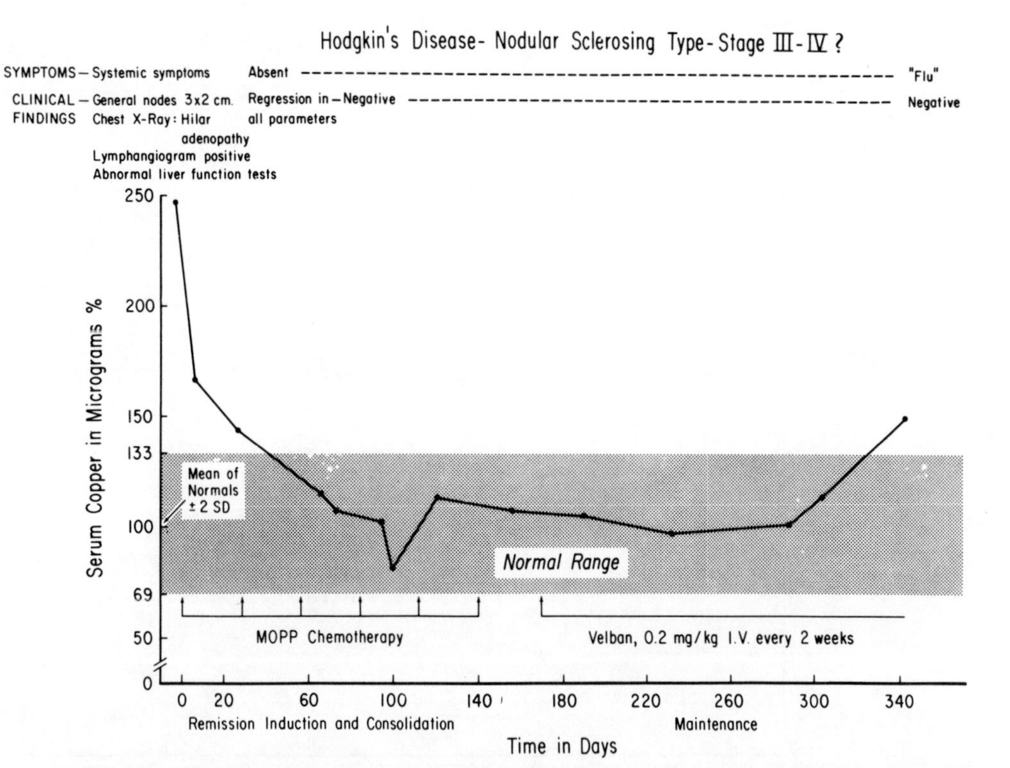

FIGURE 18. The SCL pattern reflecting the status of disease activity and the effectiveness of chemotherapy in a patient with generalized Hodgkin's disease. (From Hrgovcic, M., in *Zinc and Copper in Medicine,* Karcioglu, Z. A. and Sarper, R. M., Eds., 1980, 481. Courtesy of Charles C Thomas, Publisher, Springfield, Ill.)

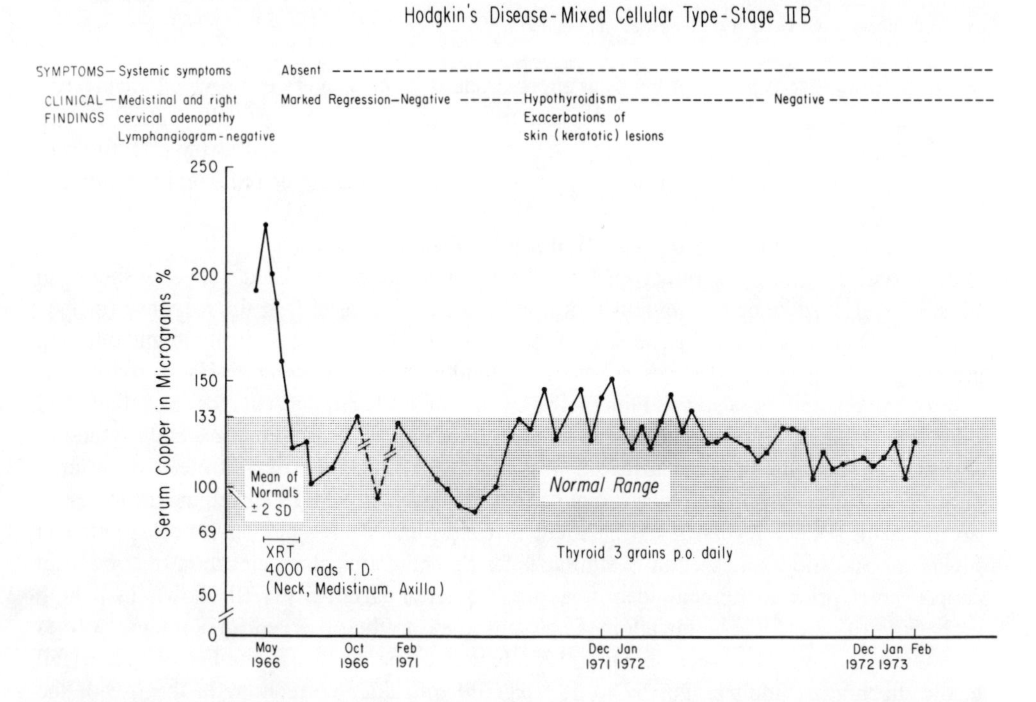

FIGURE 19. Sequential SCLs in long-term follow-up of a Hodgkin's disease patient in complete remission. (From Hrgovcic, M., in *Zinc and Copper in Medicine,* Karcioglu, Z. A. and Sarper, R. M., Eds., 1980, 481. Courtesy of Charles C Thomas, Publisher, Springfield, Ill.)

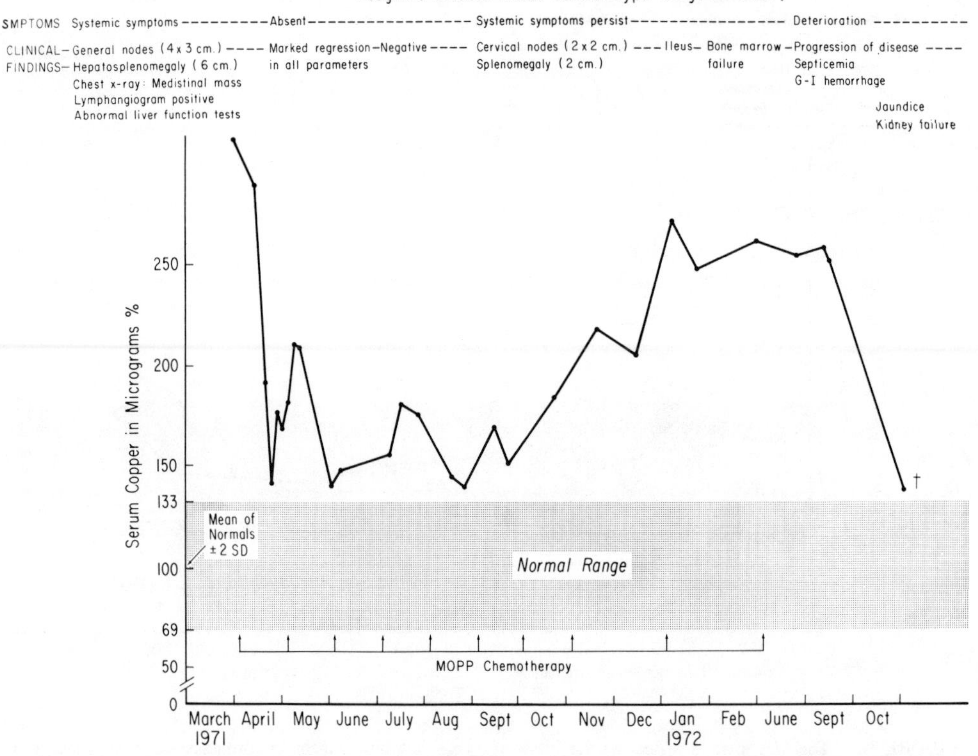

FIGURE 20. The decrease in SCL during therapy, without reaching the normal range at any time, suggests residual active disease. A fall of serum copper level as a terminal event has been observed in some patients. (Reproduced with permission from Hrgovcic, M., Tessmer, C. F., Mumford, D. M., Ong, P. S., Gamble, J. F., and Shullenberger, C. C., *Texas Medicine,* 71, 53, 1975.)

determination of serum copper levels at approximately weekly intervals provided the clearest index of the steady deterioration of the patient's condition. These cases illustrate the close correlation between the serum copper level and disease activity as well as the effectiveness of therapy and are examples of an early indication of a favorable or unfavorable course.

B. Serum Copper Observations in Hodgkin's Disease in Children

The basic relation of serum copper to the clinical categories listed above is shown in Figure 23. The sequence is obviously designed to show categories in the response (or lack of response) to therapy and the course of disease. Stage of disease, classification, and duration are not distinguished. Sequences in several individual cases changing disease activity (category) are marked by arrows. Plot of the means of each clinical category substantiates, however, the basic overall relationship of the serum copper level to disease activity and/or response to therapy. In order to avoid possible distortion of the overall mean by a large number of values contributed by a single patient in a given clinical category, the mean values of all serum copper levels of each patient in each clinical category have been plotted in Figure 24 and show an essentially similar pattern. The anticipatory elevation of the serum copper level prior to clinical identification of relapse is equally well shown in a high proportion of cases. The application of age correction to the observed (uncorrected) values of Figures 23 and 24 is seen in Figure 25. It will be noted that a number of values fall below the normal adult range (69 to 133 μg/100 mℓ) after correction. In the use of the normal adult range in Figure 25, the relationship of serum copper to disease activity is well established. However, the presentation serves to point out cases which seem inconsistent

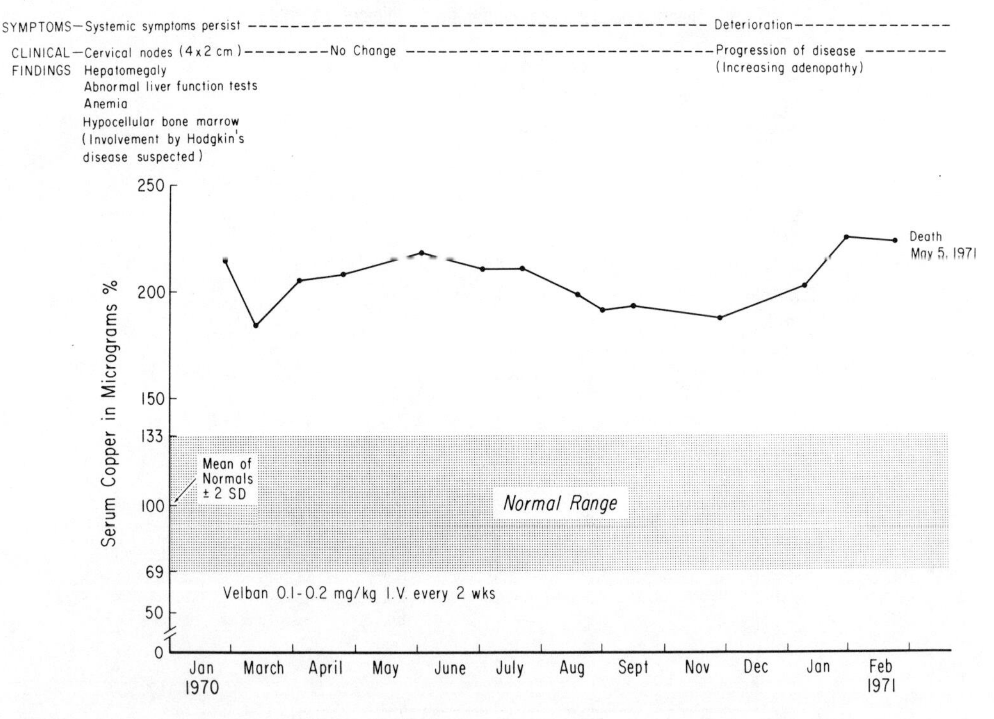

FIGURE 21. A persistently elevated serum copper level in this patient indicated the continued presence of disease activity and inefficacy of therapy used. (From Hrgovcic, M., in *Zinc and Copper in Medicine,* Karcioglu, Z. A. and Sarper, R. M., Eds., 1980, 481. Courtesy of Charles C Thomas, Publisher, Springfield, Ill.)

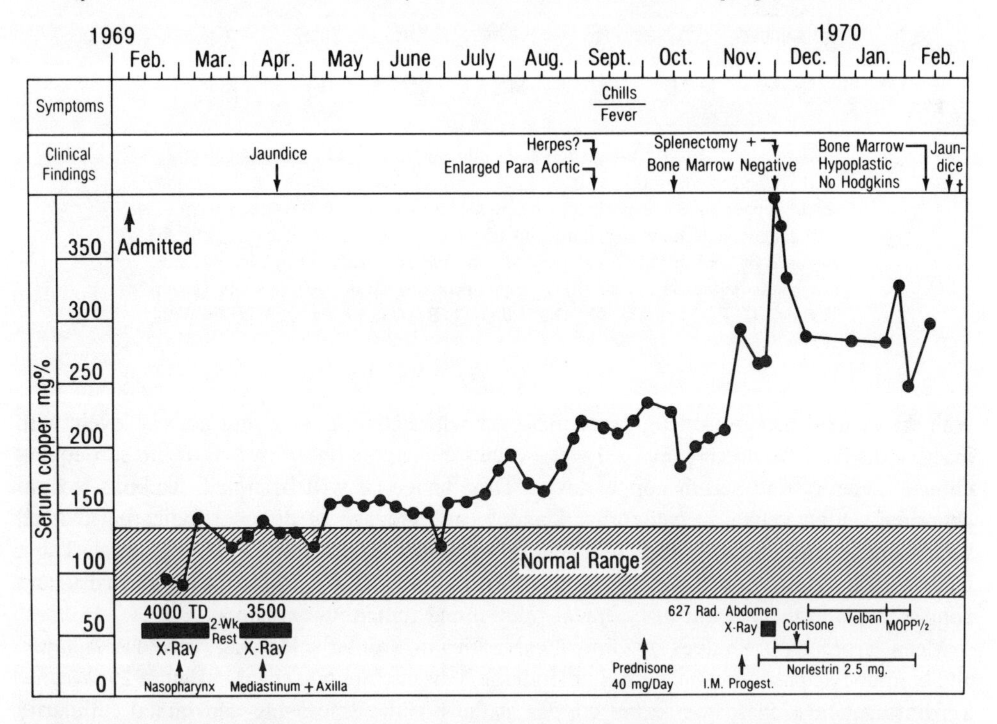

FIGURE 22. Sequential serum copper levels with pertinent clinical data illustrating serum copper determinations as a predictive index of adverse course of disease (see text; Illustrative Cases). (From Hrgovcic, M., Tessmer, C. F., Brown, W. B., Wilbur, J. R., Mumford, D. M., Thomas, F. B., Shullenberger, C. C., and Taylor, G. H., in *Progress in Clinical Cancer,* Ariel, I. M., Ed., Grune & Stratton, New York, 1973, 121-153. With permission.)

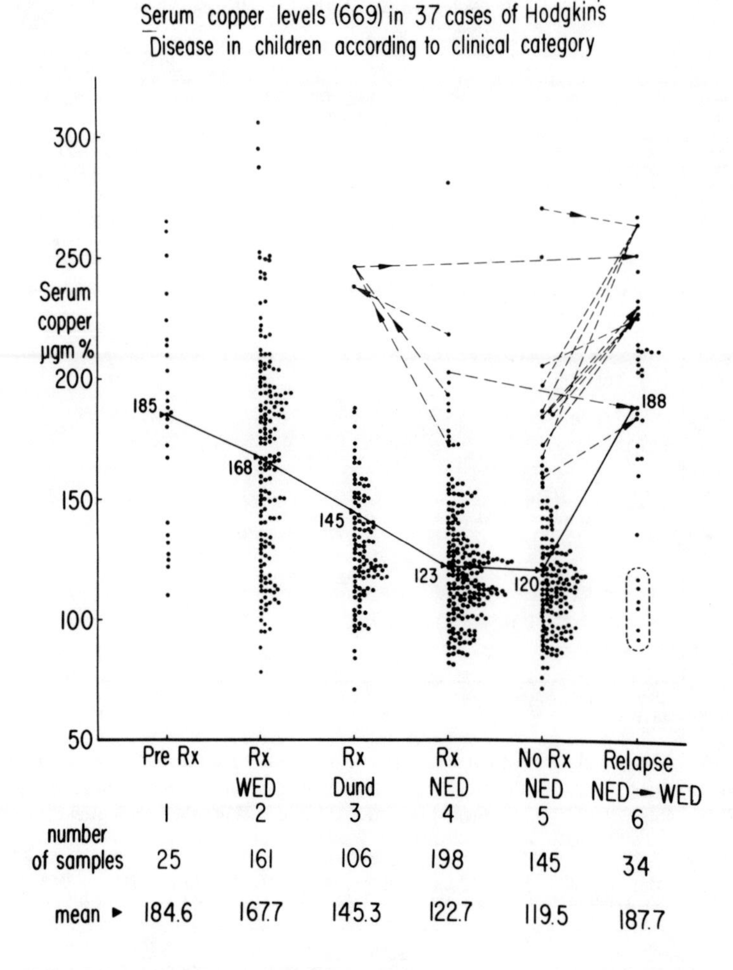

FIGURE 23. All SCL are plotted in the clinical category to which assigned (see text). Represented by the mean SCL of each category, it may be seen that the serum copper shows response to therapy by lower levels, and a rise in relapse. The dotted lines show that almost all of the high SCL in those cases with no evidence of disease (NED, category 5) and thus essentially unexplained are, in fact, predicting relapse as the subsequent clinical history and SCL indicate. (From Tessmer, C. F., Hrgovcic, M., and Wilbur, J. R., *Cancer,* 31, 303, 1973. With permission.)

with the general premise of high serum copper with active disease and normal levels with inactive disease. As noted previously, the values connected by arrows show the subsequent clinical category and serum copper level. This device, it will be noted, accounts for all abnormally high values in categories 4 and 5 (no evidence of disease) indicating that all such values were, in fact, predicting relapse or return of clinically evident disease. Three inconsistent serum copper patterns were noted in category 6 (relapse) with normal serum copper levels in the presence of clinical relapse and remain unexplained.

Mean serum copper values by clinical categories in nodular sclerosing, mixed cellularity, and lymphocyte predominance types of Hodgkin's disease are presented in Figure 26 showing a significant relation to the serum copper in the nodular sclerosing and mixed cellularity types, whereas no relation exists in a limited number of patients with lymphocytic predominance type.

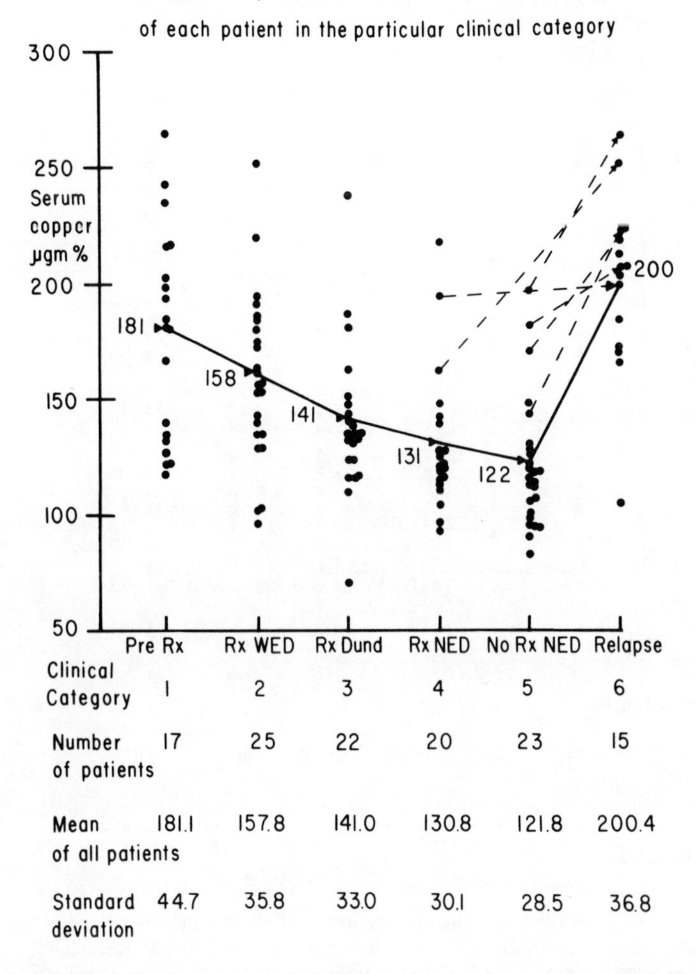

FIGURE 24. All SCL of a given patient in a clinical category are represented by a mean value to prevent undue effect by a large number of values from one case. This procedure clarifies some aspects of the plot of the entire number (669) of SCL in Figure 23. These data are not age-corrected, and are not strictly comparable, nor subject to comparison with mean and range of normals. (From Tessmer, C. F., Hrgovcic, M., and Wilbur, J. R., *Cancer,* 31, 303, 1973. With permission.)

The relationship between serum copper and stage of disease is similar to that seen in adult populations with Hodgkin's disease, with the clearest relation in stage IV as shown in Figure 27. Patients with systemic symptoms had higher serum copper levels in the pretreatment period, whereas in relapse, no clear distinction between categories A and B is noted.

An illustrative case report with sequential serum copper levels is shown in Figure 28 as an example of relationship during the entire course of disease. Figure 28 represents data obtained over a 4-year period from a 14-year-old black boy with Hodgkin's disease (mixed cellularity type), stage IVB. Treatment with cyclophosphamide (Cytoxan®) for 1 month was ineffective, and the serum copper level remained increased. After changing treatment to Velban® (vinblastine sulfate), a decrease in the serum copper level to normal limits resulted within 2 weeks, followed by the disappearance of all positive symptomatology and clinical

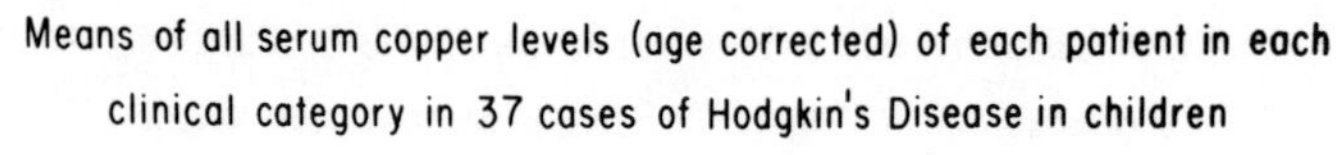

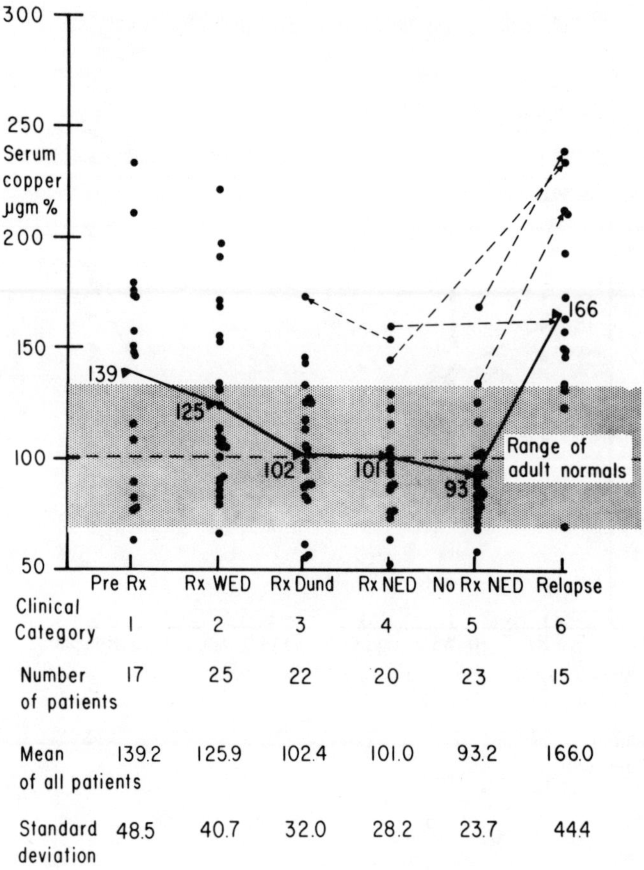

FIGURE 25. This plot represents the means of all SCL of a given patient within the particular clinical category, thus each dot, as in Figure 24, represents one patient — in this instance age-corrected (see text). Such data presumably permit the application of mean and range of normal adult values, as indicated in this figure. The results are helpful in analysis and point to the areas of seeming inconsistency. All high values in categories 4 and 5 (no Rx NED) are indications of impending relapse. (Values in the normal range in categories 1 and 6 (Rx WED, Pre Rx, or relapse) have additional basis for evaluation (see text). (From Tessmer, C. F., Hrgovcic, M., and Wilbur, J. R., *Cancer,* 31, 303, 1973. With permission.)

findings a few weeks later. The serum copper level increased slowly over a period of a year, reaching abnormal values several months before the first signs of active Hodgkin's disease appeared. Wide fluctuation of the elevated serum copper level suggest that the patient's response to subsequent therapy regimes was less than satisfactory. A highly significant elevation of the serum copper level is evident in the terminal phase of the disease. The 18-year-old patient died soon after the last serum copper determination. This case provides an excellent example of the usefulness of sequential serum copper determinations as: (1) an index of the efficacy of the various chemotherapy agents employed in obtaining the first treatment response, and (2) an early indicator of exacerbation of disease, in this case preceding clinical relapse by 6 months. More detailed data analysis of serum copper in children can be seen in the original publication.[8]

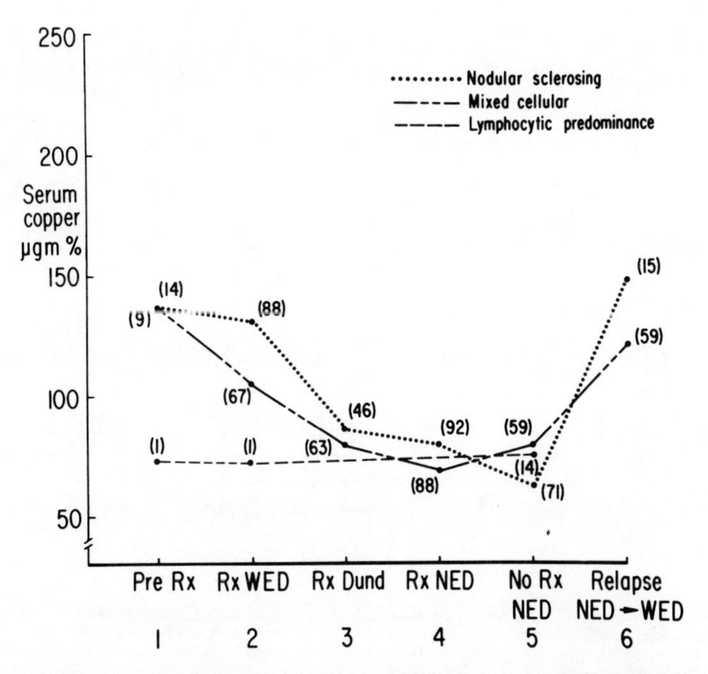

FIGURE 26. The plot of mean values of each of the three major histologic classifications of Hodgkin's disease according to assigned clinical category reveals the relation of SCL to untreated and treated disease which is seen in other analyses. The major observation, unfortunately based on very limited numbers, is the relatively unchanged SCL in the lymphocytic predominance category. This can only be considered a generalization in view of the lack of precision in both histologic characterization of the disease and the clinical categorization. (From Tessmer, C. F., Hrgovcic, M., and Wilbur, J. R., *Cancer,* 31, 303, 1973. With permission.)

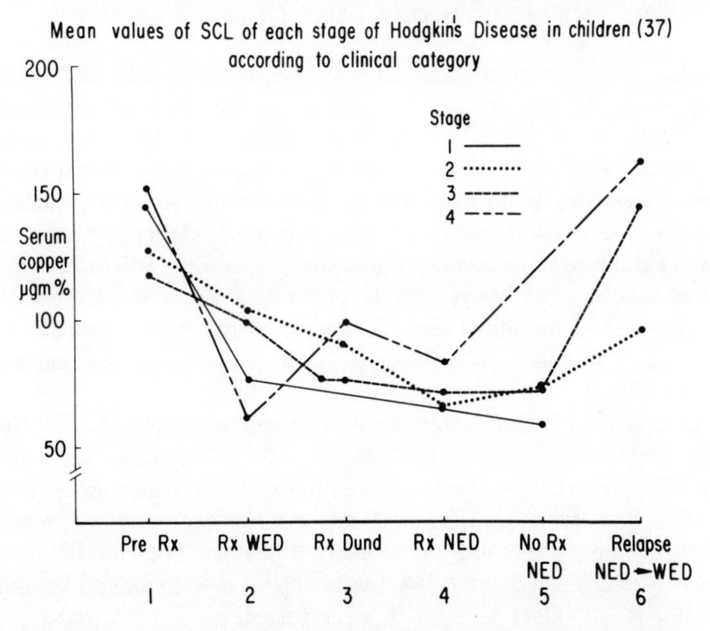

FIGURE 27. The mean values of SCL in each stage (Rye/Ann Arbor classification) show generally similar relation to the clinical evidence of disease, falling with response to treatment and rising with relapse. Clearest relation appears in stage IV, which may also suggest in the pretreatment group, a relation to the presumably greater volume of disease tissue. (From Tessmer, C. F., Hrgovcic, M., and Wilbur, J. R., *Cancer,* 31, 303, 1973. With permission.)

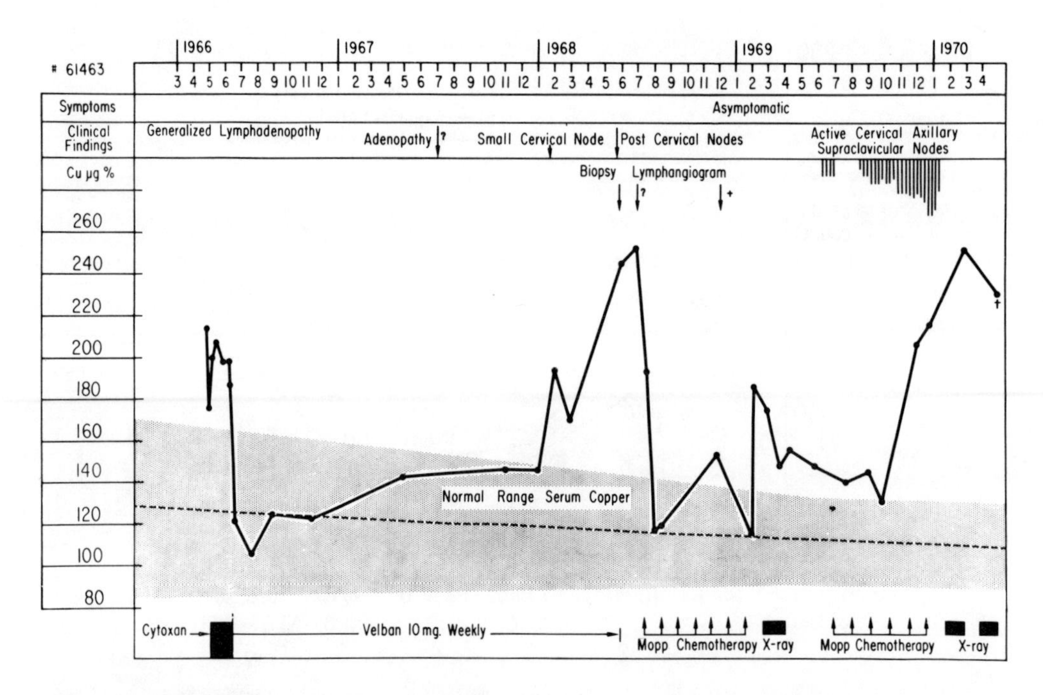

FIGURE 28. Sequential SCLs in a case of Hodgkin's disease (A. H.) illustrate sustained levels in the failure to respond to initial form of treatment, and the subsequent reduction to normal values with satisfactory treatment response. The mean of age corrected values ± 2 SD is indicated. The graph suggests that comparative values in the same individual are equally important for maximum use of SCLs in the evaluation of Hodgkin's disease. (From Hrgovcic, M., Tessmer, C. F., Brown, W. B., Wilbur, J. R., Mumford, D. M., Thomas, F. B., Shullenberger, C. C., and Taylor, G. H., in *Progress in Clinical Cancer*, Ariel, I. M., Ed., Grune & Stratton, New York, 1973, 121-153. With permission.)

III. NON-HODGKIN'S LYMPHOMA PATIENTS

In a pilot study, in addition to Hodgkin's disease and the acute leukemia patients, 14 cases with non-Hodgkin's lymphoma were studied. Six of the nine patients with reticulum cell sarcoma did not show altered serum copper levels. Therefore, and based on limited number of patients with this cell type, the significance of serum copper levels in this diagnostic group was not clear.[2,42] In subsequent studies of 236 adult patients (96 females and 140 males) with a tissue diagnosis of malignant non-Hodgkin's lymphoma, of various types, we analyzed the relation between the serum copper level and disease activity.[14] The youngest patient was 15 years of age, and the oldest 85 years of age; the average age being 54 years. At the time of the initial serum copper level determinations, 88 of the patients were untreated and 148 had been previously treated. Other pertinent clinical information can be seen in Table 14.

Clinical and laboratory investigations were performed as indicated. The same methods, normal values, and clinical categories of disease activity were used as in the Hodgkin's disease group. The serum copper level was determined in this patient group at varying times during their clinical course (regardless of disease activity, its extent, or previous therapy), at intervals ranging from every 3rd day to every 6 months. The number of serum copper levels for each patient ranged from 168 single copper determinations to as many as 14 observations on one individual during a 10-week period.

A. Observations (Non-Hodgkin's Lymphoma Patients)

In Figure 29 the serum copper levels in all of the patients with non-Hodgkin's lymphoma are plotted according to the clinical category, including the number of samples, the number

Table 14
STUDY GROUP WITH SERUM COPPER LEVEL IN RELATION TO HISTOLOGIC GROUPS OF MALIGNANT LYMPHOMA, NON-HODGKIN'S TYPE

	No. of patients	No. of SCL		Clinical category					
				PreRx	Rx WED	Rx Dund	Rx NED	No Rx NED	Relapse
Nodular									
Lymphocytic									
Poorly differentiated	39	55	#SCLs	24	15	4	7	4	1
			Mean	159.0	154.7	141.9	109.9	113.4	118.7
			SD	40.7	29.6	40.7	19.9	21.2	—
Mixed type	30	41	#SCLs	8	18	4	3	6	2
			Mean	144.3	160.0	153.0	116.1	128.0	182.5
			SD	20.8	45.0	32.9	6.8	28.4	52.4
Histiocytic RCS[a]	21	35	#SCLs	9	11	3	6	3	3
			Mean	146.5	150.9	134.5	104.0	117.7	159.8
			SD	56.8	17.7	13.0	7.5	29.4	13.2
Nodular (follicular) lymphoma (unclassified)	5	7	#SCLs		2	1		—4	
			Mean		128.0	93.0		120.7	
			SD		1.4			17.4	
Subtotal	95	138	#SCLs	41	46	12	16	17	6
			Mean	153.4	154.7	139.7	108.8	121.0	160.5
			SD	41.5	34.0	39.5	14.4	23.2	34.1
Diffuse									
Lymphocytic									
Well differentiated	45	65	#SCLs	16	42	1	3	2	1
			Mean	165.6	157.9	194.1	118.2	113.8	118.7
			SD	52.3	31.0		7.1	10.3	
Poorly differentiated	18	23	#SLCs	8	8	1	3	2	1
			Mean	131.2	141.6	111.1	109.3	101.9	207.9
			SD	26.6	56.8		2.1	2.2	
Histiocytic RCS[a]	64	103	#SCLs	26	49	1	10	15	2
			Mean	164.3	143.0	161.1	117.6	122.1	154.7
			SD	39.1	27.7		18.0	27.6	16.6

Table 14 (continued)
STUDY GROUP WITH SERUM COPPER LEVEL IN RELATION TO HISTOLOGIC GROUPS OF MALIGNANT LYMPHOMA, NON-HODGKIN'S TYPE

				Clinical category					
	No. of patients	No. of SCL		PreRx	Rx WED	Rx Dund	Rx NED	No Rx NED	Relapse
Undifferentiated RCS[a]	14	24	#SCLs	6	9		2	4	3
			Mean	176.0	141.7		131.3	130.0	136.3
			SD	46.0	59.5		15.2	21.0	10.9
Subtotal	141	215	#SCLs	56	108	3	18	23	7
			Mean	161.2	148.6	155.4	117.8	121.0	149.3
			SD	43.5	35.3	41.8	15.1	24.5	30.1
Total	236	353							

[a] Reticulum cell sarcoma.

From Hrgovcic, M., Tessmer, C. F., Thomas, F. B., Ong, P. S., Gamble, J. F., and Shullenberger, C. C., *Cancer*, 32, 1512, 1973. With permission.

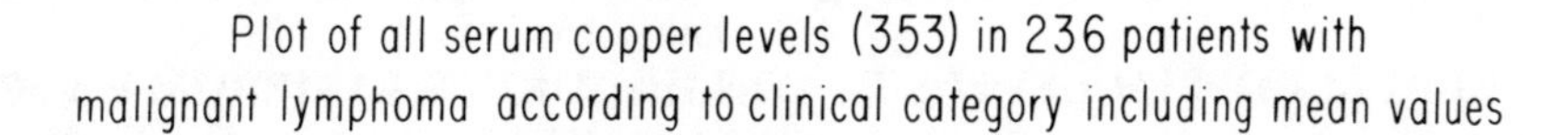

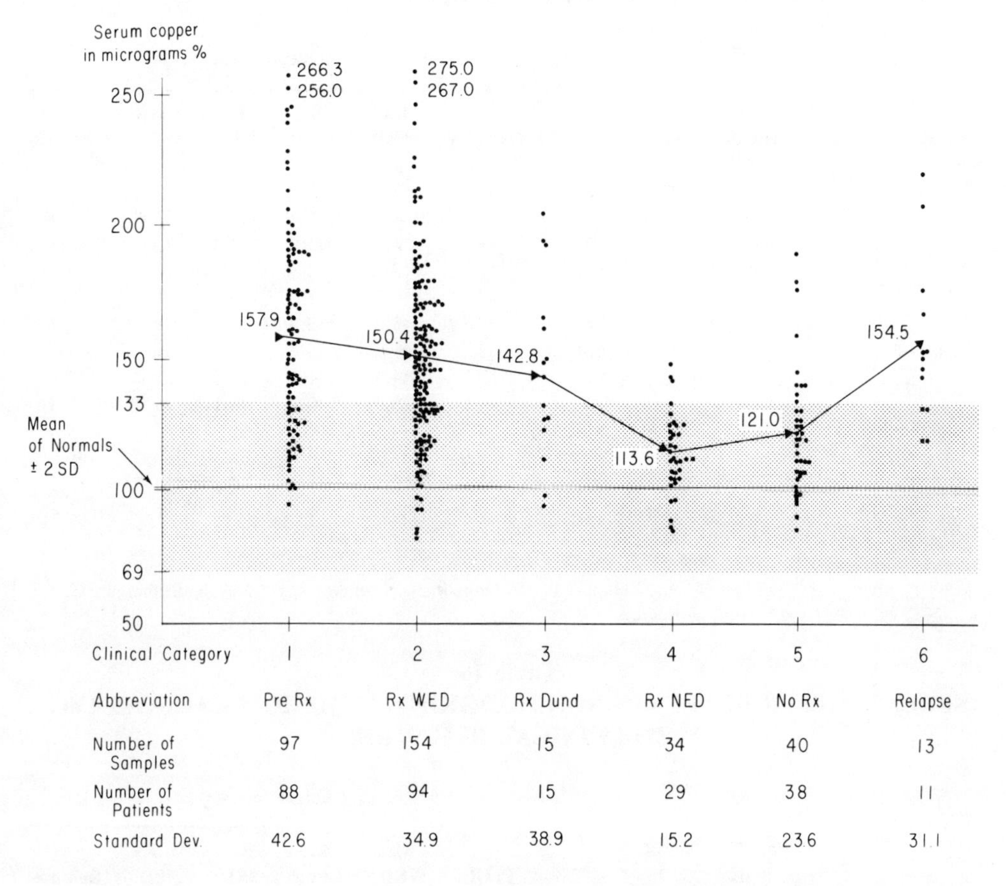

FIGURE 29. Serum copper values in patients with non-Hodgkin's lymphoma. Increased SCL in the active phase of disease represented by the mean SCL of each category decreases with response to treatment and rises with relapse. (From Hrgovcic, M., Tessmer, C. F., Thomas, F. B., Ong, P. S., Gamble, J. F., and Shullenberger, C. C., *Cancer,* 32, 1512, 1973. With permission.)

of patients, and the mean as well as the standard deviations. The basic overall relationship of the serum copper level to the disease activity is represented by the mean of the serum copper level to each category. Again, it may be seen that active phases of disease are accompanied by elevated serum copper levels. The serum copper level is normal in the absence of disease manifestations (remission) and elevated in relapse. The same sequence of the basic relationship of the serum copper level to the disease activity in each of the initial histologic classifications of malignant lymphoma (all groups and types) may be seen in Table 14.

A comparison of the serum copper levels according to the nodular or diffuse pattern (Table 14), cell type (Table 15), and the clinical behavior (Table 16) of malignant non-Hodgkin's lymphoma shows essentially similar values, with slightly higher levels in diffuse lymphoma, undifferentiated reticulum cell sarcoma, and rapidly progressive types of malignant lymphoma.

An examination of the relation between the serum copper level and extent of disease (stage), in addition to the reflection of disease activity noted in the previous figures and tables, shows that patients with generalized disease (stages III and IV) have a consistently higher serum copper level than those with localized disease (stages I and II) (Figure 30).

Table 15
SERUM COPPER LEVELS IN NON-HODGKIN'S LYMPHOMA ACCORDING TO CELL TYPE

	No. of patients	No. of SCL		Clinical category					
				PreRx	Rx WED	Rx Dund	Rx NED	No Rx NED	Relapse
Lymphocytic	107	150	#SCLs	48	67	7	13	12	3
			Mean	156.6	154.3	137.8	111.7	114.0	148.4
			SD	43.9	34.2	45.6	14.8	16.1	51.5
Mixed	30	41	#SCLs	8	18	4	3	6	2
			Mean	144.3	160.0	153.0	116.1	128.0	182.5
			SD	20.8	45.0	32.9	8.6	28.4	52.4
Histiocytic	85	138	#SCLs	35	60	4	16	18	5
			Mean	152.7	144.5	141.1	112.5	121.4	157.7
			SD	44.0	26.2	40.3	16.1	27.0	12.8
Undifferentiated	14	24	#SCLs	6	9		2	4	3
			Mean	176.0	141.7		131.3	130.0	136.3
			SD	46.0	59.5		15.2	21.0	10.9
Total	236	353							

From Hrgovcic, M., Tessmer, C. F., Thomas, F. P., Ong, P. S., Gamble, J. F., and Shullenberger, C. C., *Cancer*, 32, 1512, 1973. With permission.

Table 16
SERUM COPPER LEVELS IN NON-HODGKIN'S LYMPHOMA ACCORDING TO CLINICAL BEHAVIOR

	No. of patients	No. of SCL		Clinical category					
				PreRx	Rx WED	Rx Dund	Rx NED	No Rx NED	Relapse
		Slowly or Moderately Progressive Disease							
Nodular lymphomas (all types) and diffuse lymphocytic well- and poorly differentiated lymphoma	158	226	#SCLs	65	96	14	22	21	8
			Mean	153.6	155.0	141.5	110.2	118.5	161.2
			SD	43.8	34.9	40.0	12.8	21.7	37.4
		Rapidly Progressive Disease							
Diffuse histiocytic and undifferentiated lymphoma	78	127	#SCLs	32	58	1	12	19	5
			Mean	166.5	142.8	161.1	119.9	123.8	143.7
			SD	31.9	33.8		17.7	26.0	15.1
Total	236	353							

From Hrgovcic, M., Tessmer, C. F., Thomas, F. B., Ong, P. S., Gamble, J. F., and Shullenberger, C. C., *Cancer*, 32, 1512, 1973. With permission.

A group of 18 patients with sequential serum copper levels was studied in order to evaluate the relation of the serum copper level to the response to therapy. The patients in this group were clinically evaluated before and after the last serum copper level determination. The

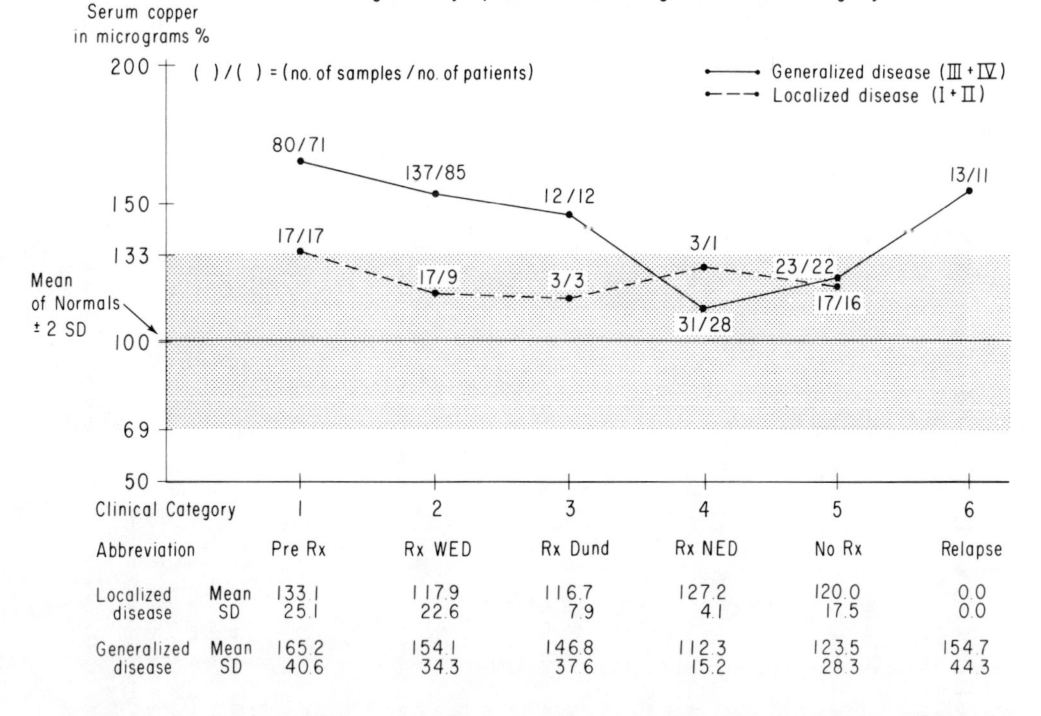

FIGURE 30. The mean values of SCL according to stage show relationship to the clinical evidence of disease. In addition, relationship between SCL and extent of disease (tumor volume) shows a definitely higher SC value in generalized vs. localized disease. (From Hrgovcic, M., Tessmer, C. F., Thomas, F. B., Ong, P. S., Gamble, J. F., and Shullenberger, C. C., *Cancer,* 32, 1512, 1973. With permission.)

clinical findings by which the patients were placed in the ''responding'' and ''nonresponding'' groups are summarized in Table 17. The sequential serum copper level in these two groups of ''responders'' and ''nonresponders'' are presented in Figure 31. This analysis confirms, in general, the close correlation between the changes in the serum copper level and the disease activity noted in the previous figures and tables. The group of 10 patients responding to therapy (left portion in Figure 31, patients 1 to 10 in Table 17) was accompanied by a rapid or gradual decrease in the serum copper level which frequently preceded clinical improvement and indicated the effectiveness of the applied treatment, regardless of the form of therapy. As indicated in Figure 31, this is a relatively uniform trend with individual responses reaching normal ranges of serum copper levels as early as the 2nd day of therapy and as late as 13 days following the initiation of treatment.

The highest serum copper levels were observed in the patients with undifferentiated non-Hodgkin's lymphomatous disease involving the various organs (patients 1 to 4). The response to radiotherapy in patient 5 (McG. N.) was accompanied by normalization of the serum copper level and relapse (not shown in Figure 31) which occurred 18 months later with an elevation of the serum copper level (159 μg%). In patient 8 (S.R.), a serum copper level of 129 ng%, probably relates to the prior therapy (3000 rad to the mediastinum) and a good clinical response. The remaining two patients in the responding group, 9 and 10 (L.M. and J.M.) with localized active disease and a good clinical response to irradiation, failed to show a clear correlation between the serum copper level and disease activity. The details will be presented in the discussion.

The most interesting and perhaps instructive are the serum copper levels in patient 6

Table 17
SUMMARY OF CLINICAL FINDINGS IN NON-HODGKIN'S LYMPHOMA PATIENTS

Patient	Age	Sex	Lymphoma disease	Prior Rx and results	Pre-observation clinical findings	Stage	Therapy	Post-observation clinical findings	Response to Rx	SCL (μg%)	
										Before Rx	After Rx
1.	40	F	Undifferentiated (RCS)	None	Generalized adenopathy (8 × 8); left anterior chest wall mass (15 × 15) Mediastinal mass and left pleural effusion Hepatomegaly (4)	IV	COP (2 courses)	Marked regression in all parameters	Good	218	122
2.	18	M	Undifferentiated (RCS)	XRT to mediastinum 1900 r; upper right abdomine quadrant 1250 r, PR COP: 4 courses PR	Cervical and axillary nodes (3 × 2) Mediastinal widening Hepatomegaly (4) Petechiae-epistaxis Bone marrow involved (leukemia pattern)	IV	Prednisone VCR L-Asparaginase (1 course)	Negative	Good	245	97
3.	27	M	Undifferentiated (RCS)	None	Severe abdominal cramps Melena Stomach, ileum, cecum ascending colon and liver involvement at laparatomy	IV	CTS + VCR (1 course)	Asymptomatic malignant gastric ulcer smaller	Good	191	134
4.	64	M	Undifferentiated (RCS)	None	Large retroperitoneal tumor mass with involvement and partial obstruction of the third portion of duodenum Hepatomegaly (5)	IV	COP (1 course)	Marked regression	Good	184	124
5.	35	F	Nodular histiocytic (RCS)	None	Left cervica node (4 × 4) Paraaortic and coeliac nodes	III	XRT 4000 r — left neck 3000 r — upper abdomen 1000 r — boost to paraaortic nodes	Negative	Good	175	111
6.	65	M	Histiocytic (RCS) (left testis)	None	Left testis absent (orchiectomy) Left paraaortic nodes	II	CTX + VCR (2 courses) XRT 3000 r to the upper abdomen 1400 r boost to the paraaortic nodes 3500 r to the pelvis	Nodes reduced	Good	155	148
7.	54	M	Histiocytic (RCS)	None	Right lower abdominal mass (11 × 8)	II	CTX + VCR (1 course)	Right lower abdominal mass (6 × 4)	Good	143	111

					Marked pelvis and paraaortic nodes enlargement						
8.	65	F	Nodular poorly differentiated lymphocytic	XRT — in progress	Mediastinal pelvic and paraaortic nodes	III	XRT 3000 r to the mediastinum	Mediastinal regression Abdominal nodes not treated	Good	128	93
9.	47	M	Histiocytic (RCS)	None	Right cervical nodes (6 × 4)	II	XRT 4000 r	Negative	Good	121	118
10.	83	M	Undifferentiated (RCS)	None	Left cervical nodes (6 × 7)	II	XRT 4000 r	Negative	Good	118	118
11.	69	M	Histiocytic (RCS)	None	Large oropharyngeal mass	IV	XRT 1800 r	Progression of disease	No response	189	269
					Bone marrow involvement		COP (1 course)	Relief of symptoms	Improvement	269	181
12.	67	F	Nodular lymphoma mixed type	COP-PR	Mediastinal mass Pleural effusion Hepatomegaly (8)	IV	COP	Progression of disease	No response	188	194
13.	50	F	Lymphocytic lymphoma well-differentiated	XRT-CR Leukeran-CR Cytoxan®-PR	General nodes (4 × 3) (leukemia pattern)	IV	Leukeran prednisone	Increasing lymphadeno–pathy Splenomegaly (4)	No response	185	190
14.	32	F	Nodular	Prednisone Nitrogen mustard Cytoxan® } PR	General nodes (6 × 6) Abdominal mass (12 × 12) Hepatomegaly (6) Splenomegaly (6)	IV	COP (3 courses)	No change	No response	179	171
15.	68	M	Histiocytic (RCS)	None	Esophagus, stomach and duodenum involvement Hepatomegaly (8)	IV	XRT 4000 r to the upper abdomen	No change	Slight palliation without objective improvement	175	186
16.	78	M	Nodular histiocytic (RCS)	COP-PR	Pleural effusion Abdominal mass (10 × 8)	IV	COP	Progression of disease	No response	170	171
17.	59	F	Nodular histiocytic (RCS)	MOPP-PR CTX-PR	General nodes (3 × 3)	III IV	CTX	Progression	No response	165	162
18.	50	M	Nodular histiocytic (RCS)	XRT-CR COP-PR	Cervical nodes (3 × 3) Axillary nodes (5 × 5) Splenomegaly (6) Lung involvement	IV	5-Azacytidine (3 courses)	Transient reduction in nodes and spleen size Unchanged lung lesions	Mixed response	227	185

Note: Patients 1-10 responded favorably to therapy (left portion, Figure 31); patients 11-18 failed to respond to therapy (right portion, Figure 31). Nodal and organ size in centimeters. PR, partial response; CR, complete response; CTX, Cytoxan® (cyclophosphamide); VCR, vincristine; MOPP, combination treatment consisting of: mechlorethamine hydrochloride (Mustargen®), vincristine sulfate (Oncovin®), procarbazine hydrochloride (Matulane®), and prednisone; COP, combination of three drugs: Cytoxan®, Oncovin®, and prednisone; XRT, stands for X-ray therapy; and r stands for rad, which refers to the dose of ionizing radiation absorbed in tissue.

From Hrgovcic, M., Tessmer, C. F., Thomas, F. B., Ong, P. S., Gamble, J. F., and Shullenberger, C. C., *Cancer,* 32, 1512, 1973. With permission.

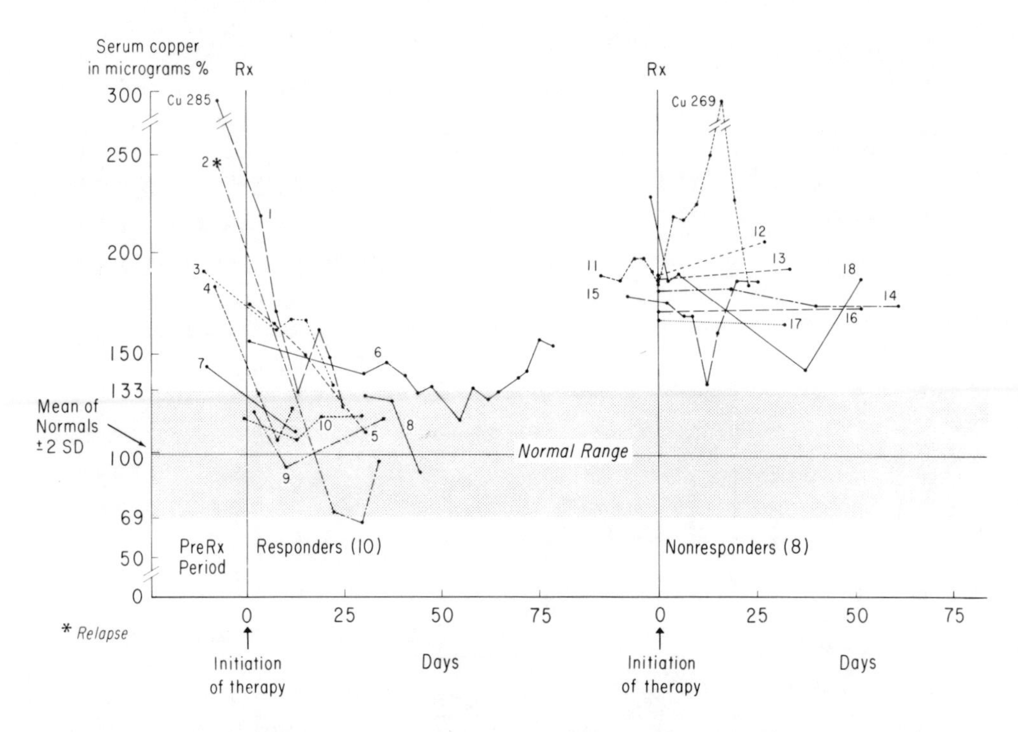

FIGURE 31. Sequential SC values of each group with malignant lymphoma, representing responders and nonresponders, are contrasted. In general, a relatively uniform downward trend of SCL during therapy may be seen in all patients considered to respond to treatment. Exceptional are the last two cases, 9 and 10, with localized disease, which show variation within normal range. An increased of the SCL in a clinically responding patient (6) points to the presence of occult disease and an adverse course. In contrast to the responding group, those nonresponding were associated with persistent active disease and elevated SCL. Rise and fall of SCL in three nonresponding patients (11, 15, and 18) were accompanied by transient increase and decrease of disease activity, respectively (see text, Section III.A. and Table 17). (From Hrgovcic, M., Tessmer, C. F., Thomas, F. B., Ong, P. S., Gamble, J. F., and Shullenberger, C. C., *Cancer,* 32, 1512, 1973. With permission.)

(Y.A.) on whom serial sequential serum copper level determinations (N = 14) were made while he was receiving combined chemo- and radiotherapy during a 10-week period. The normalization of the serum copper levels accompanied a clinical improvement and by clinical evaluation he was properly included in the "responder" group. While the patient was receiving radiotherapy, the serum copper level was increased and reached initial abnormal levels at the completion of irradiation. Three months later this patient relapsed with liver and multiple lung involvement, but the serum copper levels were not available at that time.

In striking contrast to the responding group, in which the serum copper level generally preceded or accompanied clinical improvement, stands the nonresponding group of 8 patients (right portion of Figure 31, patients 11 to 18 in Table 17) characterized by persistent disease activity and by persistently high serum copper levels, indicating the ineffectiveness of therapy. Patient 11 (B.C.), in the nonresponding group, had 5 serum copper level determinations performed during the pretreatment period within 2 weeks. All of these were markedly elevated. Clinically, the patient was thought to have localized disease in the form of a large mass involving the nasopharyngeal region (the only positive physical findings) and, therefore, underwent radiotherapy. During irradiation, the patient's condition became worse and the serum copper levels increased throughout the radiotherapy period. A continued workup revealed bone marrow invasion by neoplastic cells from two sites of aspiration, the sternum and the right iliac crest. The treatment was changed to COP chemotherapy (Cytoxan®, Oncovin®, and prednisone) resulting in a marked clinical improvement. A decrease in the serum copper level of this patient was noted to accompany the favorable response to chemotherapy.

Patient 14 (S.B.) had sequential serum copper level determinations during a 2-month period at 3-week intervals while receiving COP therapy. The patient's condition, characterized by disseminated active disease, did not improve and the serum copper level remained high without significant fluctuations.

Patient 15 (H.M.), from the nonresponding group, whose serum copper level dropped following the initiation of radiotherapy and within 3 days rose to an abnormally high level during the continued irradiation, experienced a slight palliation without objective clinical improvement; 6 weeks after the completion of irradiation this patient returned with the same pretreatment symptomatology and clinical findings. The serum copper level was not determined at that time.

Patient 18 (A.O.) showed some fluctuation in the size of the lymph nodes and spleen during chemotherapy with 5-azocytidine. The lung lesions remained unchanged. A transient reduction in the size of the spleen and nodes was associated with a temporary decrease in the serum copper level. Since the clinical evaluation before and after the observation period did not reveal any change in measurable parameters, the patient was included in the ''nonresponder'' group. These and other details will be presented in the discussion.

B. Illustrative Case Report

The disease course of a 7-year-old Caucasian girl with localized lymphocytic lymphoma diagnosed by right cervical node biopsy examination on April 25, 1966, but terminating in acute lymphocytic leukemia, is given in Figure 32. Bone marrow analysis on admission was within normal limits. Elevated serum copper dropped to normal limits during radiotherapy. The patient was in clinical remission for 2 months. On July 13, 1966 the patient was asymptomatic with negative physical findings and normal bone marrow. A month later the bone marrow showed 91% blast cells and the serum copper levels were increased to 217 μg/100 mℓ. During chemotherapy the copper levels again dropped to normal limits and complete clinical and hematological remission was achieved. In this case the increasing serum copper levels during remission appeared to be an early indicator of relapse, preceding by 1 month the leukemic transformation of bone marrow.

IV. ACUTE LEUKEMIA GROUP

A portion of our 1968 pilot study examined the serum copper relationship in 28 patients with acute leukemia.[2] The majority of these were in the diagnostic category of acute lymphocytic leukemia. The first group of data shown in Figure 33 represents serum copper levels from 12 pediatric patients who had received previous treatment and who were in complete bone marrow remission when first seen. The serum copper levels of these patients were all within normal limits.

The second group (Figure 33) represents paired serum copper levels and bone marrow blast cell percentages from eight patients with acute lymphocytic leukemia before and after conventional chemotherapy. Increased percentages of blast cells in the bone marrow were regularly accompanied by high serum copper levels. Within 4 weeks following effective chemotherapy, all serum copper levels dropped to normal limits, and bone marrow examination indicated that the patients were in remission.

The third group comprised three patients designated as indeterminate because treatment had been initiated 1 or 2 weeks before the serum copper levels were determined.

Figure 34 shows more detailed findings in the nine patients with acute lymphocytic leukemia. Weekly serum copper determinations and percentage of blast cells in bone marrow are recorded before, during, and after the induction of remission. Six patients had untreated

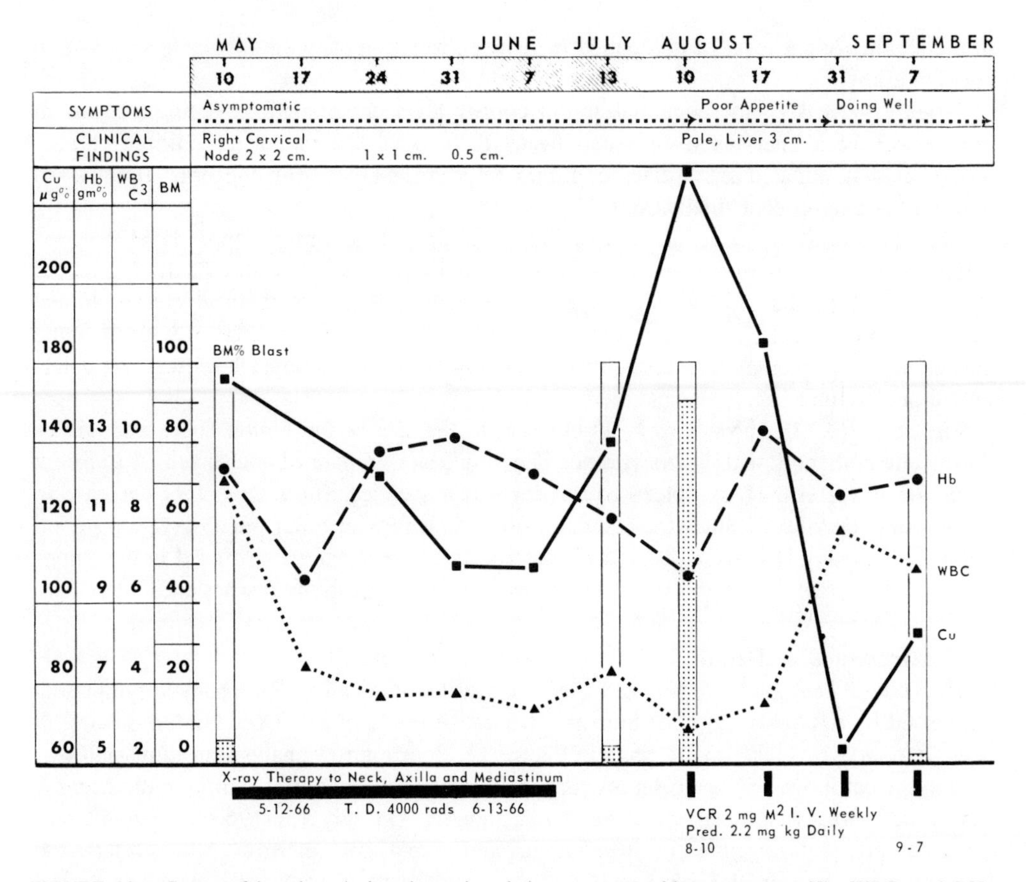

FIGURE 32. Course of lymphocytic lymphoma in relation to marrow blast percentage Hb, WBC, and SCL terminating in acute leukemia. (From Hrgovcic, M., Tessmer, C. F., Minckler, M. T., Mosier, B., and Taylor, H. G., *Cancer,* 21, 743, 1968. With permission.)

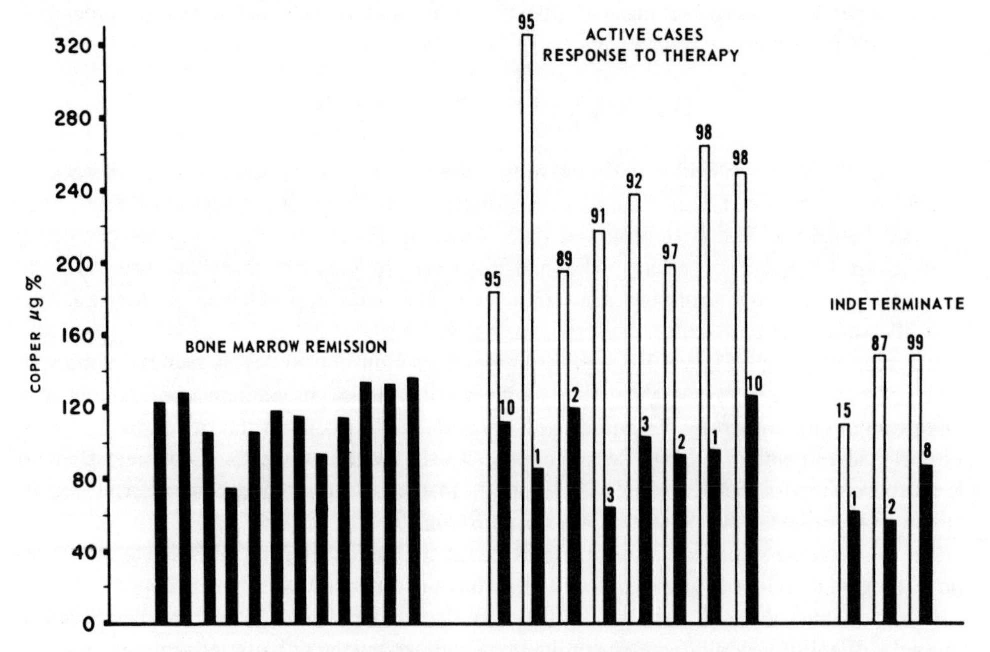

FIGURE 33. Graph of 23 pediatric patients with acute lymphatic leukemia studied (comparing bone marrow blast cells percentage to SCL). (From Hrgovcic, M., Tessmer, C. F., Minckler, M. T., Mosier, B., and Taylor, H. G., *Cancer,* 21, 743, 1968. With permission.)

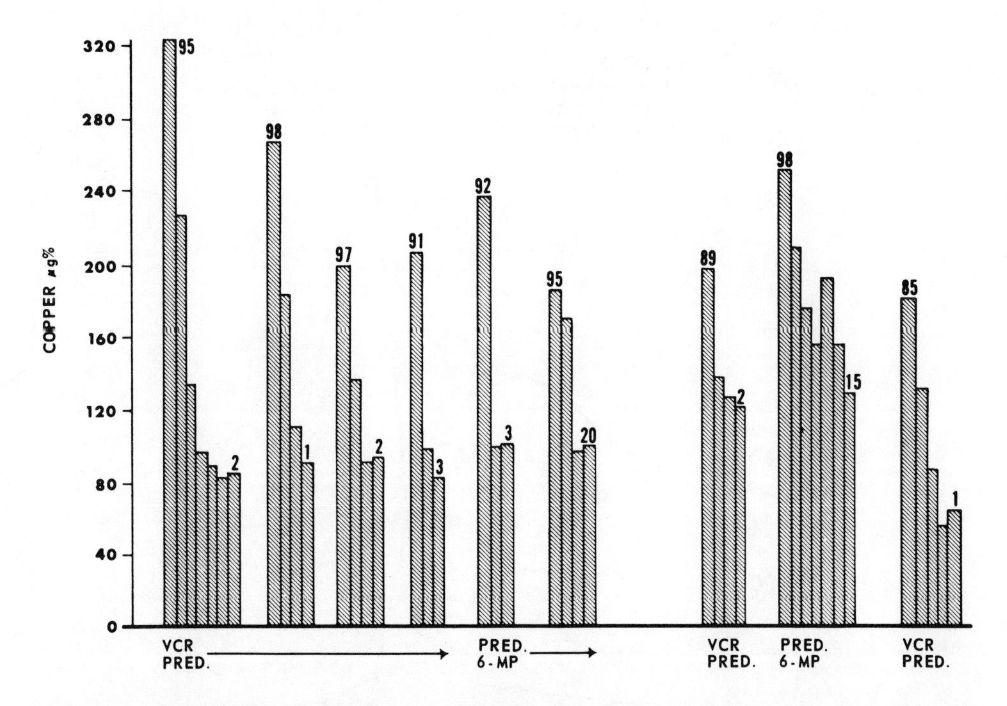

FIGURE 34. Weekly serum copper levels and percentage of blast cells in bone marrow of nine patients with acute leukemia treated with chemotherapy. (From Hrgovcic, M., Tessmer, C. F., Minckler, M. T., Mosier, B., and Taylor, H. G., *Cancer*, 21, 743, 1968. With permission.)

Table 18
PEARSON CORRELATION COEFFICIENTS FOR SCL WITH BONE MARROW PERCENTAGES IN ACUTE LEUKEMIA

	% Granulocytic	% Lymphocytes	% Erythrocytic	% Blast cells
Pediatrics				
Serum copper	−0.2073	−0.1231	−0.5393	+0.5776
N = 265	($p < 0.001$)	($p < 0.023$)	($p < 0.001$)	($p < 0.001$)
Adult serum				
Copper	−0.4066	−0.1033	−0.1996	+0.5110
N = 82	($p < 0.001$)	N.S.[a]	($p < 0.038$)	($p < 0.001$)

[a] N.S. = not significant.

From Tessmer, C. F., Hrgovcic, M., Brown, B., Wilbur, J., and Thomas, F. B., *Cancer*, 29, 173, 1972. With permission.

acute lymphocytic leukemia and three were in relapse. Serum copper was reduced to normal levels following effective chemotherapy; frequently the reductions preceded the bone marrow remissions obtained by 4 to 6 weeks.

In an effort to examine certain quantitative aspects of the serum copper relationship with the marrow blast percentage, a statistical analysis was made using 354 observations on 116 patients (64 pediatric and 52 adult) with acute leukemia.[6]

Pearson correlation coefficients were calculated for the 265 pediatric and 89 adult paired serum copper level/bone marrow blast cell determinations and are presented in Table 18. Correlation coefficients significantly different from zero at the 0.001 level were obtained in the analysis of both pediatric and adult serum copper levels with the percentage of blast cells in the bone marrow. There was also significant correlation of serum copper levels with

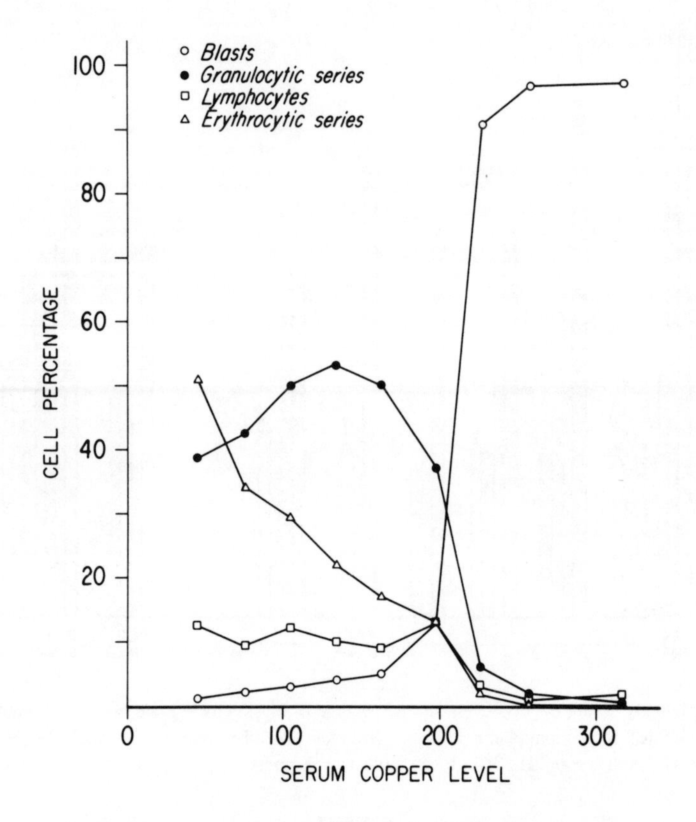

FIGURE 35. Bone marrow cell components (median %) in relation to serum copper in pediatric group (N = 265) with acute leukemia. (From Tessmer, C. F., Hrgovcic, M., Brown, B., Wilbur, J., and Thomas, F. B., *Cancer,* 29, 173, 1972. With permission.)

both pediatric and adult granulocyte percentages and with the pediatric erythrocyte percentage. These relationships are shown in Figures 35 and 36 as median values of the cell differential at each time of concomitant serum copper values. Further analysis of this comparison is of interest. The serum copper vs. blast cell data in the pediatric patient group with acute leukemia and a plot of the mean, range, standard deviation, and the number of samples in each of ten equal intervals are included in Figure 37. This study strongly confirmed the relationship of the serum copper level to the bone marrow blast percentage. The correlation between the serum copper level and the percentage of blast cells in both the pediatric and adult patients is high. The spread of the individual values (Figures 38 and 39) suggests that the relationship is complex.

The relationships which have been presented here may be stated in terms of probability in several clinically useful categories based on the following assumptions:

0—5% Blasts	Acute leukemia, BM remission
5—25% Blasts	Acute leukemia, partial BM remission, or partial BM relapse
25—100% Blasts	Acute leukemia, untreated BM disease, or relapse

The use of the serum copper level in this study in predicting the blast cell population in these categories is indicated in Tables 19 and 20. In the adult and pediatric patients, the proportion of readings in the remission range generally decreases as the copper reading increases. Similarly, the proportion of readings in the relapse range steadily increases with the serum copper level.

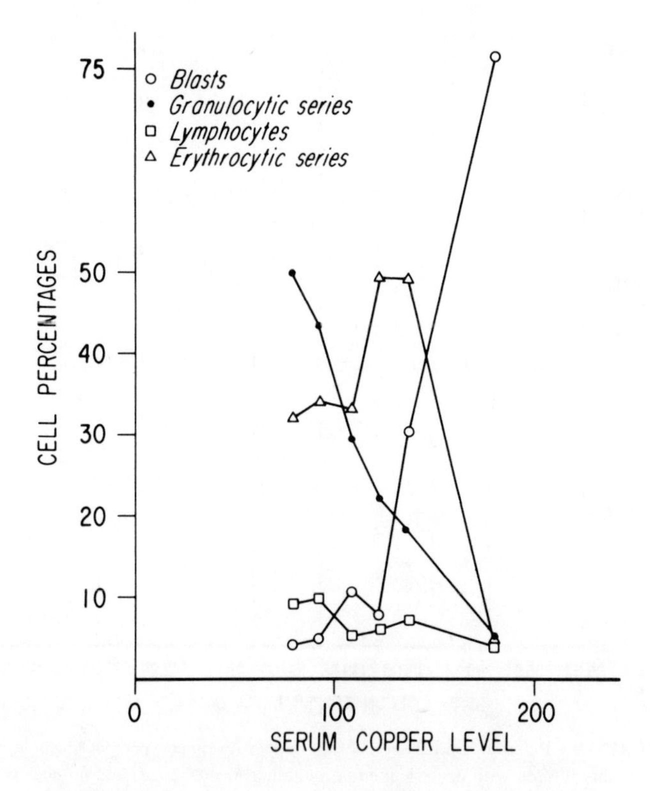

FIGURE 36. Bone marrow cell components (median %) in relation to serum copper in adult group (N = 89) with acute leukemia. (From Tessmer, C. F., Hrgovcic, M., Brown, B., Wilbur, J., and Thomas, F. B., *Cancer,* 29, 173, 1972. With permission.)

As can be seen in Figure 40 there may be some difference in the location of the curves of the adult (> 15 years) and pediatric (≤ 15 years) data. Since a previous study[2] had suggested a relationship of serum copper level to age (Figure 1), a correction factor was applied, thus normalizing all serum copper level data to age 20. This resulted in close agreement between the age-corrected and adult curves.

Data from the 25 pediatric patients with multiple paired observations were analyzed on the basis of our categories of general agreement or disagreement (Table 21). Correlation implies that the serum copper and the blast percentages are both within or both outside their defined clinically normal ranges. For copper, we have used the previously defined[2] 95% confidence interval for a normal adult population of 69 to 133 μg%. The term "correlation indeterminate" was applied to situations in which either the serum copper level or the blast percentage fell just outside the normal range. For copper, the limit for this deviation was set as determined by the 97.5% confidence interval for a normal adult population — 62 to 140 μg%. For the blast percentage, the interval was set between 6 and 10%. Explained noncorrelation describes situations which are not included in the criteria of the two previous categories but are explainable from information on a major clinical event, believed to relate directly to an increased serum copper level, derived from the patient's medical record. Unexplained noncorrelation was used in cases which could not be explained from entries in the medical record.

Table 21 and Figure 41, representing all cases, shows the degree of correlation noted between serum copper level and bone marrow blasts throughout the entire analytic period (those periods of clinical observation in which adequate paired serum copper level and bone marrow values existed). In 61% of the observations there was complete correlation, that is,

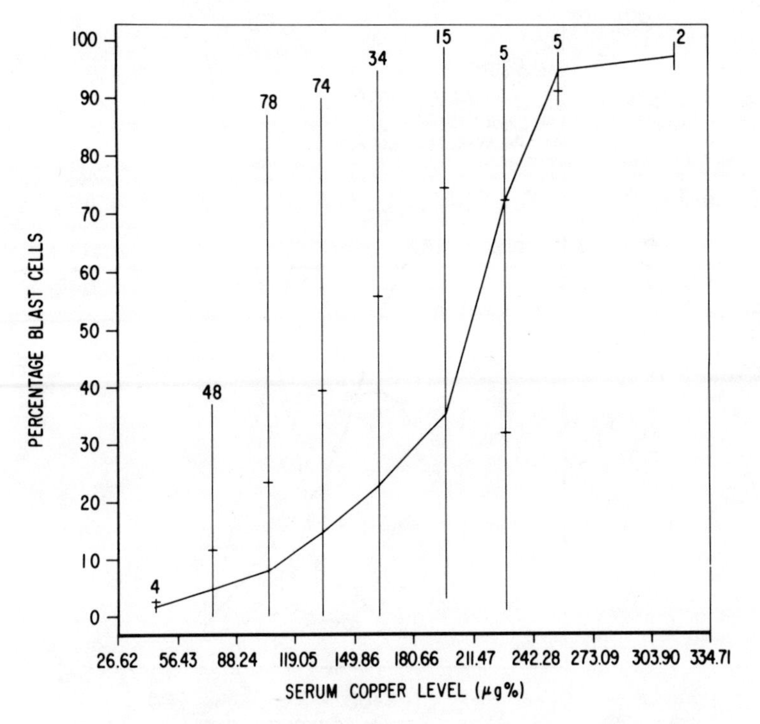

FIGURE 37. Relationship between serum copper levels and percent blasts in bone marrow of children with acute leukemia including mean, range, and number within group. (From Tessmer, C. F., Hrgovcic, M., Brown, B., Wilbur, J., and Thomas, F. B., *Cancer,* 29, 173, 1972. With permission.)

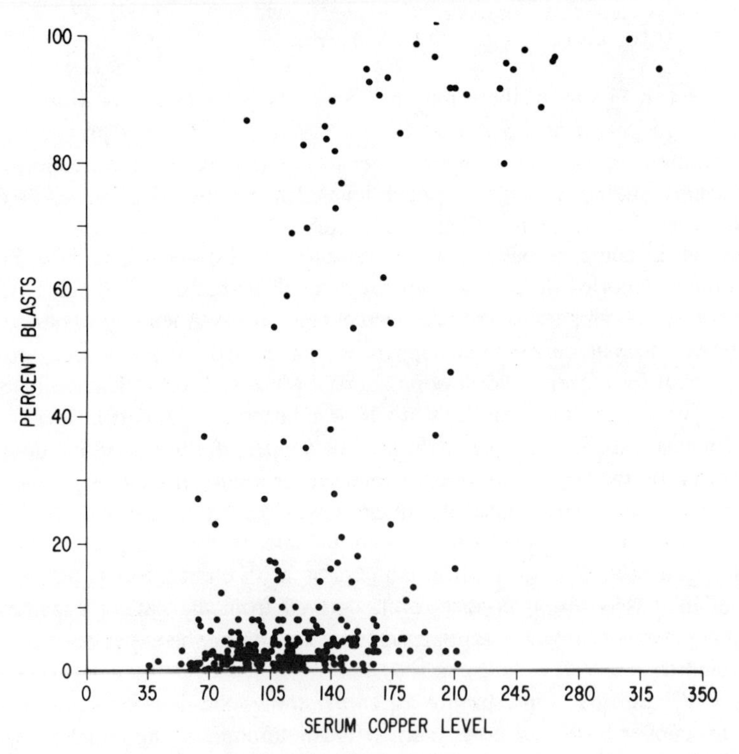

FIGURE 38. Scatter plot of bone marrow blast cells vs. serum copper level in the pediatric group with acute leukemia. (From Tessmer, C. F., Hrgovcic, M., Brown, B., Wilbur, J., and Thomas, F. B., *Cancer,* 29, 173, 1972. With permission.)

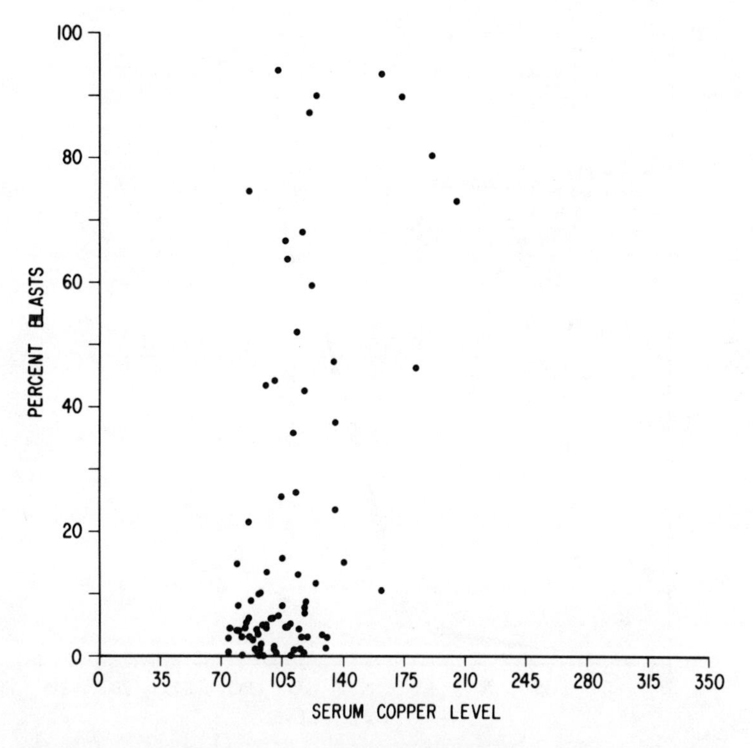

FIGURE 39. Scatter plot of bone marrow blast cells vs. serum copper in the adult group with acute leukemia. (From Tessmer, C. F., Hrgovcic, M., Brown, B., Wilbur, J., and Thomas, F. B., *Cancer*, 29, 173, 1972. With permission.)

Table 19
PROPORTION OF CLINICAL BLAST CATEGORIES (IN %) BY COPPER VALUES (PEDIATRIC ACUTE LYMPHATIC LEUKEMIA, AGE CORRECTED)

		Serum copper values (μg%)			
		<69	**69—101**	**101—133**	**133—∞**
Bone marrow	0—5	77.3	79.1	64.2	41.4
blast cell (%)	5—25	13.6	18.6	25.4	22.4
grouping	25—100	9.1	2.3	10.4	36.2
	N =	22	43	67	58

From Tessmer, C. F., Hrgovcic, M., Brown, B., Wilbur, J., and Thomas, F. B., *Cancer,* 29, 173, 1972. With permission.

Table 20
PROPORTION OF CLINICAL BLAST CATEGORIES (IN %) BY COPPER VALUES (ADULTS)

		Serum copper values (μg%)			
		<69	**69—101**	**101—133**	**133—∞**
Bone marrow	0—5	—	63.9	40.0	0
blast cell (%)	5—25	—	30.6	22.9	30.0
grouping	25—100	—	5.6	37.1	70.0
	N =	0	36	35	10

From Tessmer, C. F., Hrgovcic, M., Brown, B., Wilbur, J., and Thomas, F. B., *Cancer,* 29, 173, 1972. With permission.

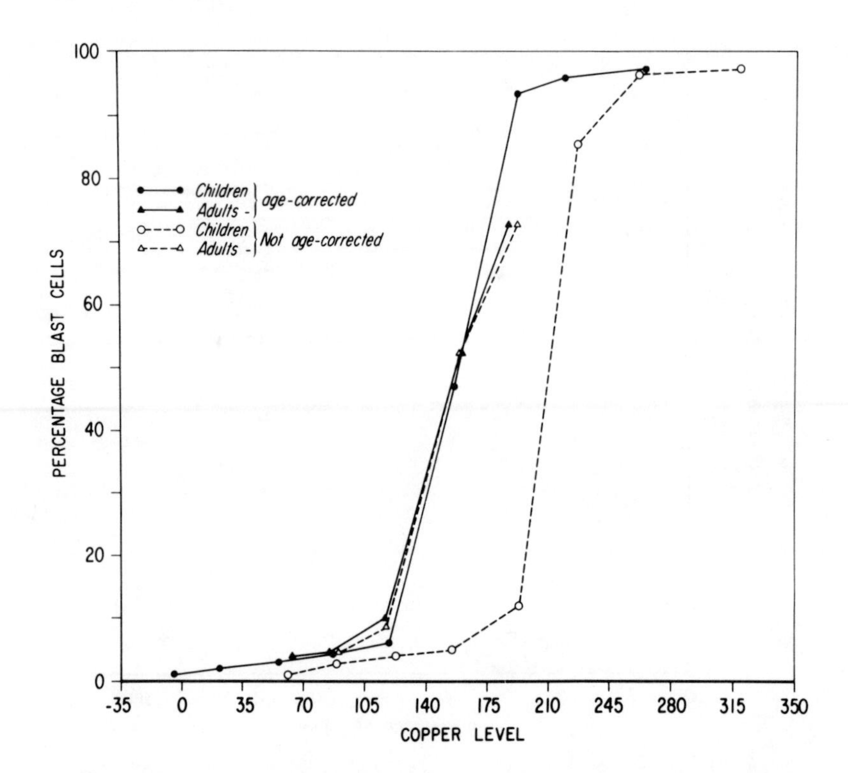

FIGURE 40. Comparison of adult and pediatric marrow blast cells vs. serum copper in acute leukemia with age-corrected serum copper values. (From Tessmer, C. F., Hrgovcic, J., Thomas, F. B., Wilbur, J. R., and Brown, B., *Cancer,* 29, 173, 1972. With permission.)

serum copper levels coincided with normal or abnormal or both levels of bone marrow blasts, and would, in effect, predict normal or abnormal bone marrow blast cells. In 17.6% of the determinations, no adequate explanation for values diverging from the concept of correlation between serum copper levels and bone marrow blasts was found. A portion of the divergence, although not explainable, is found particularly in the terminal period. The category of correlation indeterminate (11.7%) is one in which data neither preclude correlation nor are adequate for a satisfactory conclusion. The group of explained noncorrelation is that one in which a major clinical event was noted at the time of the serum copper elevation, the nature of which was believed to relate directly to the serum copper elevation.

One of the categories of great clinical importance is the response to therapy which relates particularly to the interval immediately following the institution of therapy (Table 22). Adequate data were available only for the period following intitial therapy, believed to be the most important period. These data were further divided into two periods, one consisting of the first 30 days after initial treatment was started and the second, the total period after initial therapy for which adequate paired values were available. There is complete correlation (100%) between serum copper and bone marrow blast percentages in the initial period. With more extended intervals, the correlation decreases. This complete correlation in the initial phase of the first therapy course is of great value as a means of measuring treatment response.

Another clinically significant category which may be readily identified is that of remission, which has been defined as a bone marrow blast percentage of 0 to 5%. In the paired values of serum copper level and bone marrow blast percentages in Table 23, 116 of 151 (76.8%) are correlated, 4 (2.7%) are correlation indeterminate, and 16 (11.6%) are noncorrelated explained.

The evaluation of potential relapse is important, and the examination of these conditions

Table 21
PAIRED VALUES OF SERUM COPPER AND BONE MARROW BLAST CELLS IN 25 CASES OF ACUTE LEUKEMIA IN CHILDREN

Case	Type of leukemia	No. of paired SC and bone marrow determinations	Correlated	Correlation indeterminate	Noncorrelated explained	Noncorrelated unexplained
1 JV	ALL	18	11	3	1	3
2 GC	AGL	6	5	0	1	0
3 CB	ALL	15	12	2	0	1
4 KP	ALL	8	5	3	0	0
5 RJ	AUL	14	3	1	3	7
6 SS	ALL	8	2	0	3	3
7 PS	ALL[b]	11	7	1	1	2
8 DP	ALL	10	8	0	2	0
9 DH	ALL	5	4	0	1	0
10 SC	AGL	5	3	0	2	0
11 ST	AUL	13	7	3	1	2
12 BM[a]	ALL	1[a]	1	0	0	0
13 LZ	ALL	6	3	2	0	1
14 MP	ALL	13	7	2	1	3
15 TJ	ALL	10	5	3	0	2
16 SO	ALL	14	8	1	3	2
17 SP	ALL	13	13	0	0	0
18 AW	AMonL	10	10	0	0	0
19 PR	ALL	4	4	0	0	0
20 DB	ALL	6	1	3	0	2
21 PW	ALL[c]	9	6	1	0	2
22 FP	ALL	9	3	0	0	6
23 CL	ALL	9	5	0	3	1
24 SJ	ALL[b]	17	11	2	1	3
25 RC	ALL	4	2	1	0	1
Totals						
25		238	146 (61.3%)	28 (11.7%)	23 (9.7%)	41 (17.3%)

[a] The exception to criteria for inclusion appears justified in view of the total number of serum copper values (8) and bone marrow examinations (12) which together provide particularly valuable clinical evaluation of a degree not seen in other unpaired data.

[b] Malignant lymphoma (lymphocytic) terminating in acute lymphatic leukemia (ALL).

[c] Malignant lymphoma (reticulum cell sarcoma) terminating in acute lymphatic leukemia (ALL).

From Tessmer, C. F., Hrgovcic, M., Thomas, F. B., Wilbur, J. R., and Mumford, D. M., *Cancer*, 30, 358, 1972. With permission.

believed to be related to high serum copper level is of interest. Table 24 shows the clinical conditions in each noncorrelated explained case at the time paired values were determined. One patient may present several conditions. For example, one patient (case 5) showed two noncorrelated paired values at the time of central nervous system involvement, and one at the time of varicella. The clinical conditions with elevated serum copper and normal bone marrow blast percentages may be grouped as neoplastic extramedullary proliferation,[43] viral infections, and miscellaneous infections (respiratory, otitis, pericarditis, and unspecified).[7]

V. SERUM COPPER LEVEL INTERPRETATIONS

Represented by the means on Figures 5, 6, 24, 25, 29, and 31 it is obvious that the active

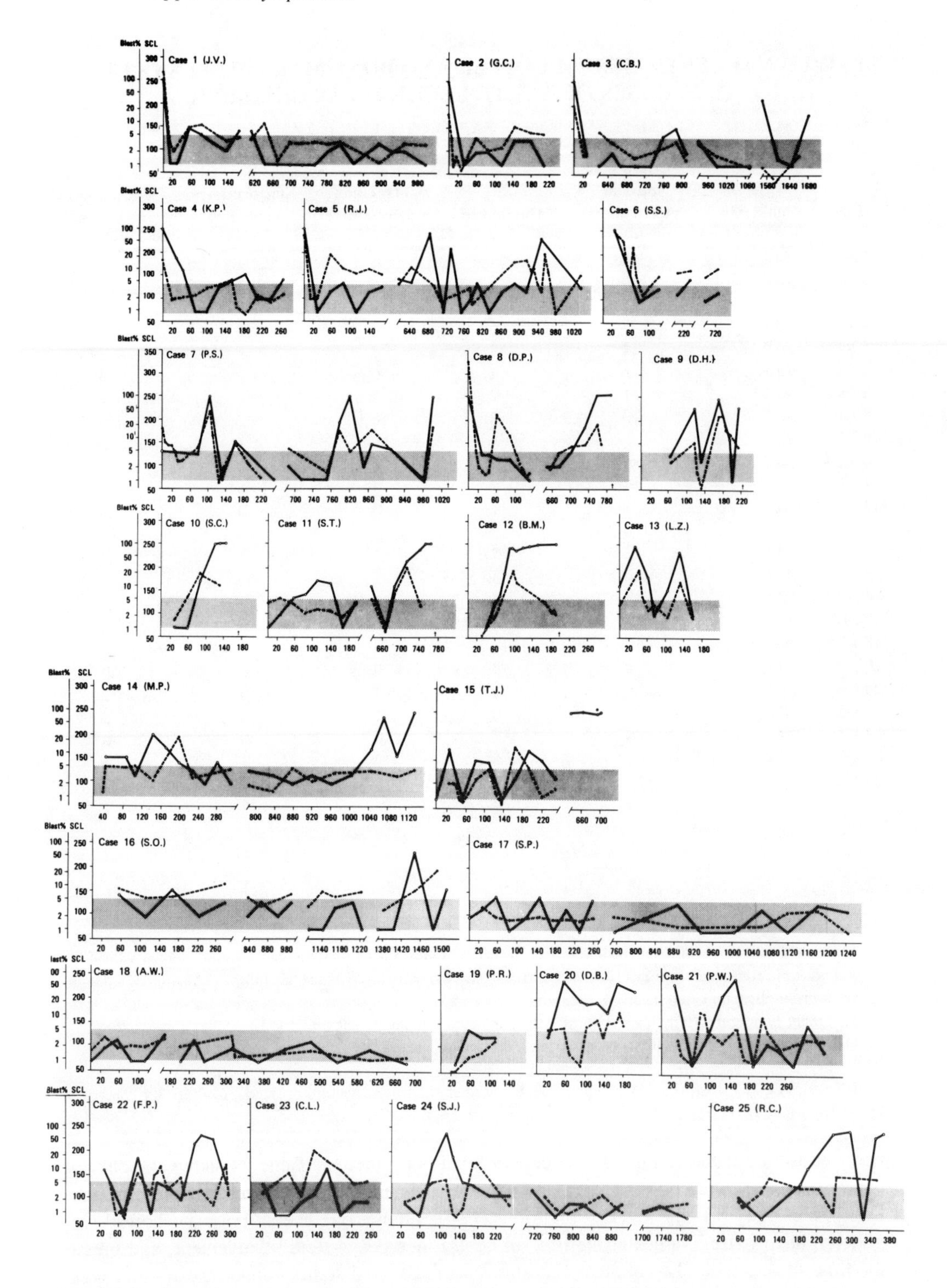

FIGURE 41. Graphic representation of the serum copper levels (broken line) and bone marrow blast percentages (solid line) on ordinate scales. Approximate normal range for both is indicated by shaded area. Abscissa in days following initial data for periods of approximate paired data (see text). For example, in case 18 (A.W.), satisfactory paired data are available only in two periods of bone marrow remission. Terminal cases are indicated by + on the abscissa (cases 7 to 12). (From Tessmer, C. F., Hrgovcic, J., Thomas, F. B., Wilbur, J. R., and Mumford, G. M., *Cancer,* 30, 358, 1972. With permission.)

Table 22
PAIRED VALUES OF SERUM COPPER AND BONE MARROW BLAST CELLS IN 8 CASES OF ACUTE LEUKEMIA IN CHILDREN FOLLOWING INSTITUTION OF INITIAL THERAPY

Case	No. of paired SC and bone marrow determinations	Interval of analysis (days)	Correlated	Correlation indeterminate	Noncorrelated explained	Noncorrelated unexplained
1	2	30	2	0	0	0
2	1	30	1	0	0	0
3	2	27	2	0	0	0
4	1	30	1	0	0	0
5	2	30	2	0	0	0
6	1	30	1	0	0	0
7	2	30	2	0	0	0
8	2	30	2	0	0	0
Total	13		13 (100%)			
1	6	169	3	3	0	0
2	7	208	4	3	0	0
3	2	27	2	0	0	0
4	8	266	5	1	0	2
5	7	169	2	0	0	5
6	3	124	2	1	0	0
7	3	56	2	1	0	0
8	6	132	4	0	2	0
Total	42		24 (57.1%)	9	2	7

From Tessmer, C. F., Hrgovcic, J., Thomas, F. B., Wilbur, J. R., and Mumford, D. M., *Cancer*, 30, 358, 1972. With permission.

phase of disease is characterized by increased serum copper levels, and the inactive phase with serum copper levels within normal range. Furthermore, the means of the serum copper levels in each category show the progressive reduction with response to therapy indicating reduced disease activity and an elevation in relapse, with recurrence of disease.

The widespread of the individual values of the serum copper levels in each category, as indicated by the scatter plots in Figures 6, 24, and 29 is probably a reflection of the multiple factors involved. These, in our opinion, include the degree of disease activity and its extent, as well as the volume of diseased tissue, clinical behavior and disease progression, histiologic type, presence or absence of systemic symptoms, the effect of medication and interfering substances, response to therapy, the individual pattern of the serum copper level and the differences between individuals, infections, and impending relapse without manifest clinical findings.

It is of value to examine in detail serum copper variations seemingly inconsistent with the hypothesis that serum copper level mirror the activity of disease process. Approximately 20% of the cases in this series were found to be exceptions to the general pattern. Considering the serum copper level in untreated Hodgkin's disease cases (category 1), 17 of 80 patients were within the normal range (Table 11). Of these 17 cases, 8 were considered to have localized disease stage I, and 6 of these 8 patients were of the lymphocytic predominance type. Since there is substantial evidence of a difference in serum copper levels in localized

Table 23
PAIRED VALUES OF SERUM COPPER AND BONE MARROW BLAST CELLS IN THE REMISSION PHASE OF 25 CASES OF ACUTE LEUKEMIA IN CHILDREN

Case	No. of paired SC and bone marrow determinations	Correlated	Correlation indeterminate	Noncorrelated explained	Noncorrelated unexplained
1	8	7	0	1	0
2	5	3	1	1	0
3	10	9	1	0	0
4	4	4	0	0	0
5	11	2	0	3	6
6	6	1	0	3	2
7	8	5	2	1	0
8	7	5	0	2	0
9	2	2	0	0	0
10	1	1	0	0	0
11	6	6	0	0	0
12[a]	1	1	0	0	0
13	2	2	0	0	0
14	7	7	0	0	0
15	4	3	0	0	1
16	11	7	0	2	2
17	11	11	0	0	0
18	10	10	0	0	0
19	3	3	0	0	0
20	1	1	0	0	0
21	5	5	0	0	0
22	6	3	0	0	3
23	8	5	0	2	1
24	12	11	0	1	0
25	2	2	0	0	0
			Totals		
25	151	116 (76.8%)	4 (2.7%)	16 (10.6%)	15 (9.9%)

[a] The exception to criteria for inclusion appears justified in view of the total number of serum copper values (8) and bone marrow examinations (12) which together provide particularly valuable clinical evaluation of a degree not seen in other unpaired data (Figure 41).

From Tessmer, C. F., Hrgovcic, J., Thomas, F. B., Wilbur, J. R., and Mumford, D. M., *Cancer,* 30, 358, 1972. With permission.

(borderline serum copper level) vs. generalized disease (high serum copper level), we feel that in 8 of 17 cases the apparently inconsistent values in the pretreatment period represent low volume of diseased tissue. The remainder of this group (nine cases) are without explanation for what we consider an inconsistent pattern of serum copper level.

In considering other apparently inconsistent values, in this instance normal serum copper level in relapse, 4 of 44 patients were identified. Two of them were in the terminal phase of disease, and one had serum copper levels of 133 and 129 ng%, which dropped to 80 ng% accompanying remission reinduction. The reason for the rapid fall of serum copper level as a terminal event observed in some cases is not known and may be a reflection of many factors, particularly liver failure and compromised immune mechanisms. The occur-

Table 24
CONDITIONS COINCIDING WITH ELEVATED SERUM COPPER VALUES NOT CORRELATED WITH ELEVATED BONE MARROW BLASTS IN 25 PATIENTS WITH ACUTE LEUKEMIA IN REMISSION

	Paired SC values and bone marrow blast determinations
Neoplastic extramedullary proliferation	6
Infection (viral)	
Varicella	1
Rubella	1
Infection (miscellaneous)	
Respiratory	2
Otitis	2
Pericarditis	1
Unspecified	2
Lymphoma	1
Total	16

From Tessmer, C. F., Hrgovcic, J., Thomas, F. B., Wilbur, J. R., and Mumford, D. M., *Cancer*, 30, 358, 1972. With permission.

rence of a "relative" change, even within normal ranges, will be alluded to subsequently. One case is without apparent explanation.

Clinical category 2 consists of active disease undergoing treatment, including those entering this category from 1 (untreated) and 6 (relapse). Analysis of patients in category 2 disclosed three groups of cases: persistent activity of disease without response to therapy and with persistently high serum copper level values; partial response to therapy with a gradual decrease in the serum copper level accompanying clinical improvement, and a complete response to therapy characterized by a rapid fall in the serum copper level preceding a clinical remission.

There were 25 cases in category 3 (Rx, disease activity undetermined) who presented hypercupremia (Table 10). This is admittedly a difficult group to analyze, inasmuch as clinical findings themselves, as far as disease activity is concerned, are inconclusive. Of the 25 in this group, 11 show adequate explanation, 7 prerelapse, and 4 manifest infection, with the remaining 14 (56%) without adequate explanation. Incomplete control of disease and occult infection or interfering substances are possibilities, particularly in this category.

There were 15 patients with Hodgkin's disease in complete clinical remission (category 4 and 5) (Table 10) who presented high serum copper levels; 7 of these 15 cases were in prerelapse, as determined by subsequent clinical course. An additional five cases showed infection. The remaining three cases of this group (20%) were without explanation of the hypercupremia in either history or clinical findings, (unless the smoking habit, which was not taken into consideration, caused serum copper level elevations).

Relatively high (and unexplained) serum copper levels in the absence of clinical evidence of active Hodgkin's disease in children in category 4 (treatment with no evidence of disease) and in category 5 (no treatment and no evidence of disease), are in fact an indication of relapse, and precede by a very significant period of time the high levels in clinically recognized relapse. As noted previously, the values connected by arrows show the subsequent clinical category and serum copper levels.

The group of pretreatment serum copper values in children show, perhaps, the largest inconsistent group by the age-corrected values (Figure 25), 7 of 17 values falling within the normal range. Six of these seven were without systemic symptoms. Four of the seven have never shown an abnormal serum copper level in their period of observation. Two showed significant response to therapy (122 to 49 and 64 to 36 μg/100 mℓ), and one showed a value above normal shortly after the initiation of therapy. While not proposing that all inconsistencies can be explained, there are considerations which suggest that each case be evaluated by both general criteria and by individual characteristics. It is also our experience that greater "resolution" of certain clinical patterns is attainable by more frequent serum copper determinations, although not always possible in a clinical setting. Serum copper shows a significant relation to the disease activity and therapy as indicated by the six clinical categories in the nodular sclerosing, mixed cellularity, and lymphocytic depletion types of Hodgkin's disease (Figure 7). Similar values occur in the pretreatment period, while somewhat different in relapse, reflecting the differences in clinical behavior of the histologic types (lowest values in nodular sclerosing, highest in lymphocytic depletion). This observation has been also confirmed by subsequent studies.[44,45]

The lymphocytic predominance group seems to show no significant variations, with serum copper level in the normal range. Due to a limited number of patients in this group and since most of these cases had localized Hodgkin's disease (stage IA), it is not possible at the present time to interpret the lack of correlation in this group, which is probably related to small tumor volume (localized disease) rather than histologic type itself (lymphocytic predominance). Perhaps biopsy removed most of the tumor.

Considering the relation of serum copper level to stage of Hodgkin's disease, as shown in Figure 8, in addition to the general relationship of serum copper level to disease activity, there is an obvious relationship of the serum copper level in the pretreatment period to stage. In stage I, eight of ten untreated patients had normal serum copper levels (average, 115 ng%) and have been in clinical remission since completion of radiotherapy. Two remaining patients with localized disease and marked hypercupremia relapsed in a short period of time after completion of radiotherapy (one of these cases is depicted in Figure 17). This observation suggests that marked hypercupremia in localized disease may represent either occult foci of disseminated Hodgkin's disease or subsequent adverse course of disease. If this holds true, serum copper level could be helpful in distinguishing between true localized disease and other stages, or a basis for prognosis of the course of the disease.

The full significance of the systemic symptoms (fever, night sweats, weight loss, and perhaps generalized pruritis) is not established as yet, except as they indicate more extensive disease both below and above the diaphragm.[19,20] Furthermore, they are of utmost importance in the early detection of relapse prior to the advent of clinical signs, especially in abdominal Hodgkin's disease. In this series, values of the serum copper levels are higher in patients with systemic symptoms than without (Figure 9).

The pretreatment category in patients with non-Hodgkin's lymphoma shows perhaps the largest number of inconsistent values, 29 of 88 being within the normal range (Figure 29). Of these 29, 12 were considered to have localized disease. Since there is substantial evidence of a difference in serum copper levels in localized (borderline serum copper level) vs. generalized (high serum copper level) disease (Figure 30), we believe that in 12 of the 29 patients the apparently inconsistent values in the pretreatment period represent a low volume of diseased tissue. The remainder of this group (17 of 29 patients) is without explanation for what we consider an inconsistent serum copper level pattern.

Malignant non-Hodgkin's lymphomas represent a heterogeneous group of disorders of the lymphoreticular system.[21,46-51] Analysis of histologic groups (nodular or diffuse proliferation and various cell types) as well as clinical behavior failed to reveal a significant difference, although patients with undifferentiated lymphoma have somewhat higher serum copper level compared to those with other histologic types (Table 16). The general relationship of serum copper level to disease activity remains.

A review of the relationship of serum copper level (mean value) to clinical category by stage of disease (Figure 30) shows, in addition to the reflection of disease activity in the serum copper level, a clear correlation with the extent of disease. The lowest serum copper levels in pretreatment period were found in stage I, mean value 130.4 μg% (four patients). In stage II, mean value was 133.9 μg% (13 patients), in stage III mean was 155.6 μg% (30 patients), and in stage IV mean was 171.1 μg% (41 patients).

Due to a limited number of patients with stage I disease in the pretreatment period (four), and difficulties encountered in this essentially retrospective study in staging and distinguishing cases with generalized disease (stages III and IV), patients with stages I and II are combined and compared as a group of localized disease vs. generalized disease (stages III and IV) (Figure 30). A significant difference in serum copper levels between localized and generalized disease was noted, being consistently higher in stages representing more extensive disease. While no data are available in this study to determine the relationship of serum copper level to tumor volume, stages I through IV may be considered to be increasing volumes of tumor or proliferating tissue, particularly in the pretreatment period.

With the exception of two cases with localized disease (Figure 31, patients 9 and 10), the most informative group was 18 patients with sequential serum copper level determinations who showed a high correlation between the serum copper level and disease activity as well as effectiveness of therapy.

The usefulness of serum copper level is best presented in case 6 (T.A.) (Figure 31), who was followed with serial serum copper level determinations during 10 weeks while undergoing combined chemotherapy and radiotherapy. One interpretation of the serum copper level in this patient is that sustained elevation of the serum copper level at the completion of irradiation indicates that active disease was still present although the treatment was effective for clinical remission. An increase in the serum copper level during treatment seems to represent an unfavorable sign indicating already disseminated foci or adverse course of disease, as shown by subsequent clinical course.

Observation of the serum copper level in a non-responding patient 11 (B.C.) suggests that marked hypercupremia in clinically localized disease is unlikely, and probably represents disseminated disease, as shown in this and other individual cases (Figure 17) and by comparison of collective presentations of serum copper levels (Figure 31). In addition, this case provides an example of serum copper determinations as an index of the efficacy of the various forms of therapy.

Patient 14 (S.B.), who had four serum copper values at regular 3-week intervals while receiving COP chemotherapy, shows that high serum copper level may be maintained when the clinical disease status does not change.

A fall in serum copper levels which is not followed by persistence of normal serum copper levels indicates continued active disease and need for further therapy, as shown by subsequent clinical course in patient 14 (H.M.) and 18 (A.O.). Analyses of the individual cases of Hodgkin's disease and non-Hodgkin's lymphoma in this clinical material (Figures 17, 22, 28, 31, 32, and 46) confirms the close correlation between the changes in the serum copper level and the disease activity process noted in the collective presentations by clinical categories (Figures 5, 6, 23 to 25, and 29).

A comparison of mean serum copper level in 191 adult patients with Hodgkin's disease and mean serum copper level of 256 patients with non-Hodgkin's lymphoma indicates a similar (almost identical) serum copper level pattern, however, being consistently higher in all clinical categories in patients with Hodgkin's disease (Figure 42). The striking correlation between serum copper level and the percentage of blast cells in the bone marrow noted in the pilot study[2] was extended and confirmed in the subsequent investigation.[6]

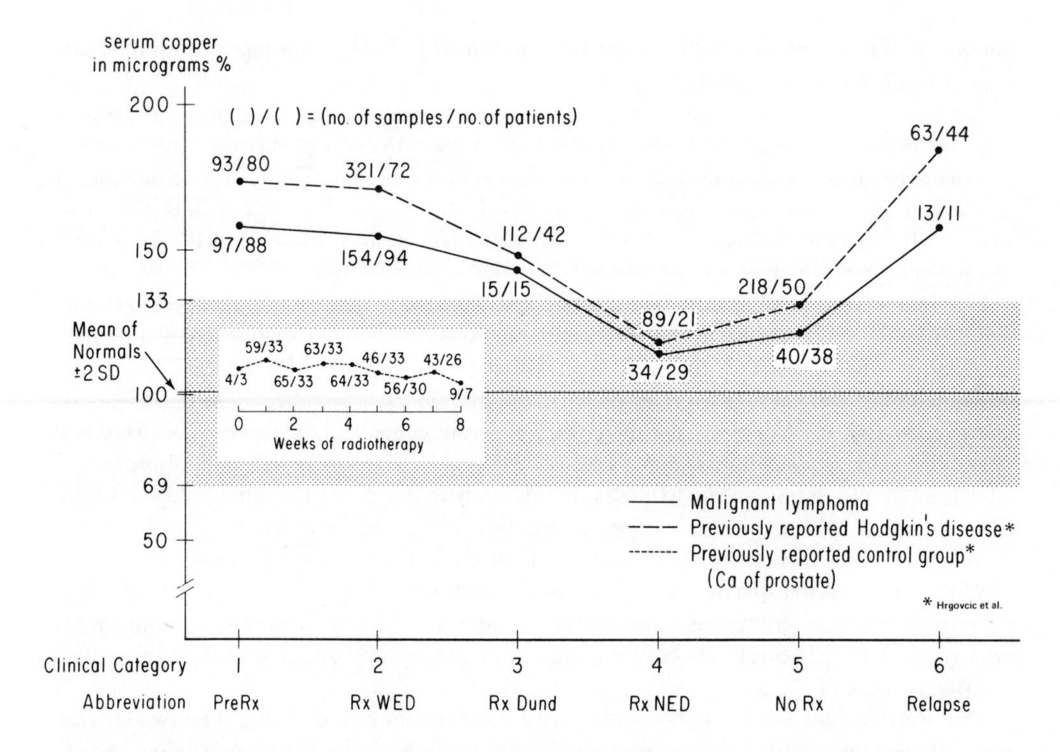

FIGURE 42. Mean SCLs in patients with malignant lymphoma (353) in contrast to the mean values of the patients with Hodgkin's disease (896) and prostatic carcinoma (426). A similar basic overall relationship and the same sequence of SCL with disease activity is noted in both groups (malignant lymphoma and Hodgkin's disease), except that patients with Hodgkin's disease show SCLs consistently higher than patients with malignant lymphoma. Unlike Hodgkin's disease and malignant lymphomatous disease, patients with carcinoma of the prostate showed no relationship of SCL to disease or therapy. The mean SCL for each of the 8 weeks of therapy varied from 107.6 to 115.6 mg% and did not differ significantly from pretreated values. (From Hrgovcic, M., Tessmer, C. F., Thomas, F. B., Ong, P. S., Gamble, J. F., and Shullenberger, C. C., *Cancer,* 32, 1512, 1973. With permission.)

Based on these initial observations,[2,6] it was hoped that the serum copper level might be of clinical value in the evaluation of leukemic disease activity. This wishful expectation was not proven in long-term studies.[7] Impressive initial correlation between the serum copper level and percentage of blast cells in the bone marrow in untreated acute leukemia patients seems to be reduced in the later course of leukemic disease and generally lost in the terminal phases.

Therefore, at the present time, serum copper level determinations in follow-up of leukemic disease activity have no useful clinical applications in daily oncological practice. Clinical and bone marrow evaluation in the management of patients with acute leukemia remains unchallenged.

In contrast to the lympho- and myeloproliferative malignant diseases, in which the serum copper level reflects disease activity, stands a group of patients with prostatic carcinoma who showed no relationship of serum copper level to disease or radiotherapy employed throughout a period of 8 weeks (Table 13, Figures 13 and 44). The contrast between hematological neoplasms and carcinoma of the prostate gland raises the matter of background and significance of serum copper level changes with the degree of disease activity observed.

There are certain general observations to be made regarding the relationship of the serum copper level to disease activity based on the cumulative experience and observations in this area, in many instances illustrated in detail in this and other reports published previously.[1,2,4,5,8,9,12-17]

A rapid fall of the serum copper from the abnormal high levels to the normal range, followed by the persistence of a normal level over many months, indicates that the response has been good (Figures 16, 19, and 28).

A slow fall in the serum copper level does not necessarily exclude a complete response, particularly if it is followed by the persistence of a normal serum copper level. In some cases, a temporary increase in the serum copper level was noted soon after therapy was initiated, then falling with response to treatment. The decrease in the serum copper level during therapy, without reaching the normal range at any time, suggests residual active disease (Figures 15 and 20). The only exception to the general relationship of serum copper levels and disease activity which seems to be clearly identified is the fall of the serum copper level, which occurs at times as a terminal event in active lymphomatous disease (Figure 20).

A persistently elevated or steadily rising serum copper level is an indication of the continued presence of disease activity or impending relapse and is related to an adverse course of disease, in the absence of a known alternative cause (Figures 21 and 22).

An increase of the serum copper level in patients apparently in remission indicates the presence of occult or recurrent disease and the need for further investigations (if otherwise unexplained) (Figures 22 and 46).

The highest serum copper levels of 474 and 526 μg% were seen in two female patients (one with Hodgkin's and another with non-Hodgkin's lymphoma) with extensive active disease associated with pregnancy or estrogen medication. A significant fall in the serum copper level of over 100 μg was observed as early as the 3rd day following effective combination chemotherapy. In the majority of patients who responded well to radiotherapy or chemotherapy and achieved complete remission, serum copper level normalized within 4 to 6 weeks. In some cases, a slow fall of the serum copper level over a couple of months was observed. Therefore, a satisfactory control of disease is indicated not by the rapidity of serum copper normalization, but by the downward direction, and the speed of fall is unrelated to the final outcome.

It is important to determine, where possible, the baseline serum copper levels related to the individual. We have observed Hodgkin's disease patients, stages I and II, with small tumor volume to have serum copper levels which were borderline or within normal range in pretreatment period which dropped to below normal limits following effective therapy. A relative increase or decrease in serum copper levels during a patient's clinical course is more important in the evaluation of disease activity than the absolute height of the levels. Serum copper variations in normal subjects from week to week and month to month under ordinary circumstances are insignificant.[52-54] Each individual, thus, possesses his own "normal" range — a value that may be more important than the overall normal range. For example, within the range of two standard deviations of "normality" commonly used as a laboratory guide, a Hodgkin's disease patient's serum copper level changing from a low normal to a level in the upper portion of the normal range may be a "significant" change, or vice versa. Among aspects which have been associated with quantitative changes in serum copper levels in Hodgkin's disease are the degree and extent of disease activity, particularly the volume of the diseased tissue, the clinical behavior and the degree of disease progression, and the histologic type and presence, or absence, of systemic symptoms.[8,13,14,25,34,41,42] These variables are obviously somewhat interdependent.

Certain stipulations must be stressed. Single determinations of serum copper levels are of little value. Only serial changes in serum copper levels are revealing. Review of individual patient records demonstrates that absolute and relative levels of serum copper are both important.

An appreciation of the dynamics of both the manner and the direction of serum copper level changes enhances its discriminatory usage. Moreover, it must be remembered that serum copper levels are a nonspecific correlate and can be altered by a number of factors

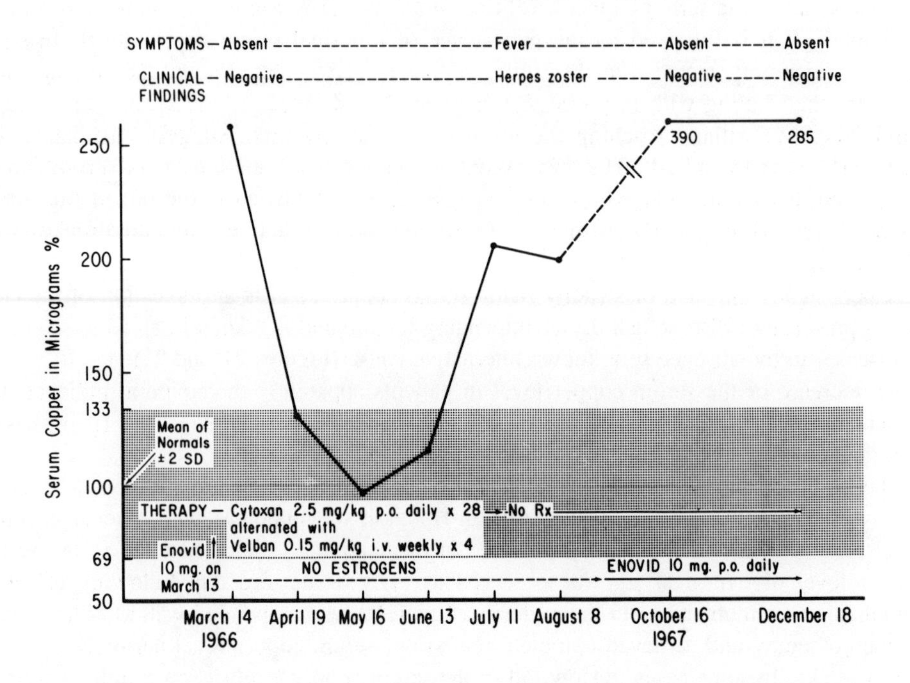

FIGURE 43. The "interference" of estrogen therapy and herpes zoster infection with serum copper measurements. Increased serum copper level returns to normal in 4 weeks after cessation of estrogen therapy (Enovid®). Herpes zoster infection also elevated serum copper levels. (Reproduced with permission from Hrgovcic, M., Tessmer, C. F., Mumford, D. M., Ong, P. S., Gamble, J. F., and Shullenberger, C. C., *Texas Medicine,* 71, 53, 1975.)

— many of them nononcological events.[55-63] These interfering factors must be considered before changes in serum copper levels can be specifically related to lymphoreticular disease activity.

Figures 10 and 11 exemplify serum copper alterations in Hodgkin's disease remission with and without maintenance therapy. These two figures document the reliability of normal serum copper levels as a reflection of the absence of disease activity when unaffected by interfering substances or conditions. In addition to the time factor, the results show the lack of possible direct effect of cancer chemotherapy maintenance on serum copper levels. The 12 patients in remission and on maintenance therapy (Figure 10) are compared to the 20 patients in remission (Figure 11) not given maintenance chemotherapy. It can be seen that with time, no significant variation in serum copper levels occurred in individual patients regardless of maintenance status. Obviously drug therapy itself did not alter serum copper levels.

A number of nonneoplastic variables are known to influence serum copper levels. Pregnancy, estrogen therapy (including contraceptive drugs), androgens, viral infections, thyroid hormones, acute inflammation (chemical, viral, and bacterial), chronic inflammation, and substances coloring body fluids (colorimetric method) are the most frequent causes of serum copper elevations. These have been summarized in Chapter 6.

Figure 43 presents two common causes of copper value misinterpretation — hormone effect and infections. The disease course of an 18-year-old woman with mixed cellularity Hodgkin's disease in complete remission while on maintenance therapy is schematically portrayed. An initially high serum copper level slowly returned to normal during a 4-week

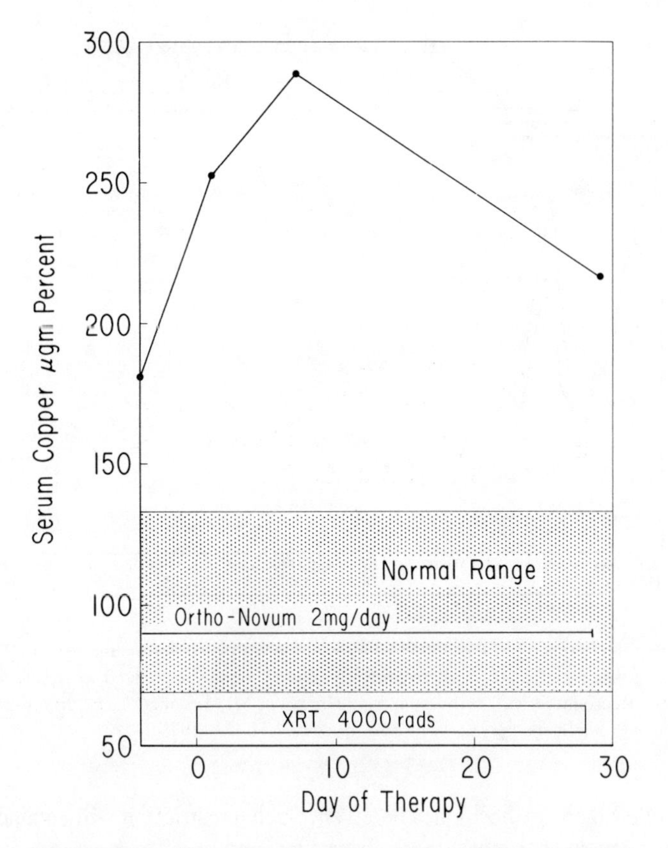

FIGURE 44. The serum copper values during radiotherapy of one patient with Hodgkin's disease responding by complete remission during radiotherapy, in which an estrogenic substance, Ortho-Novum® (norethindrone with mestranol) 2 mg/day was being taken. In this instance, any relation between SCL and radiotherapy response appears to be masked by the known serum copper response to estrogens. (From Tessmer, C. F., Hrgovcic, M., Thomas, F. B., Fuller, M. L., and Castro, J. R., *Radiology,* 106, 635, 1973. With permission.)

period after discontinuation of the estrogen medication (Enovid®-norethynodrel with mestranol). Shortly thereafter another complication ensued. A herpes zoster infection resulted in a definite serum copper elevation. Three weeks later this viral infection began to subside. Unfortunately, repeat serum copper levels were not done at this time. Upon readministration of Enovid®, however, the serum copper levels, instead of continuing downward, again became elevated. There was no corroborative evidence of increased Hodgkin's disease activity in this patient despite the elevated serum copper level caused by Enovid® and the viral infection. Changes in serum copper levels in response to intercurrent infections, however, fluctuate more rapidly than the pattern following estrogen therapy. Thus, the manner as well as the direction of serum copper changes has significance.

Another example of estrogen therapy interference with serum copper level is graphically presented in Figure 44. This patient was taking the estrogen norethindrone with mestranol (Ortho-Novum® 2 mg/day) during irradiation therapy. In this case, there was a prompt response to radiotherapy, establishing clinical remission. While the number of serum copper levels is less than desirable, it does serve to indicate a sustained high serum copper level such that no evaluation of treatment response by this means can be carried out.

At each follow-up visit, female patients should be asked about their contraceptive usage. In female patients it seems reasonable, therefore, to suggest that when possible a form of

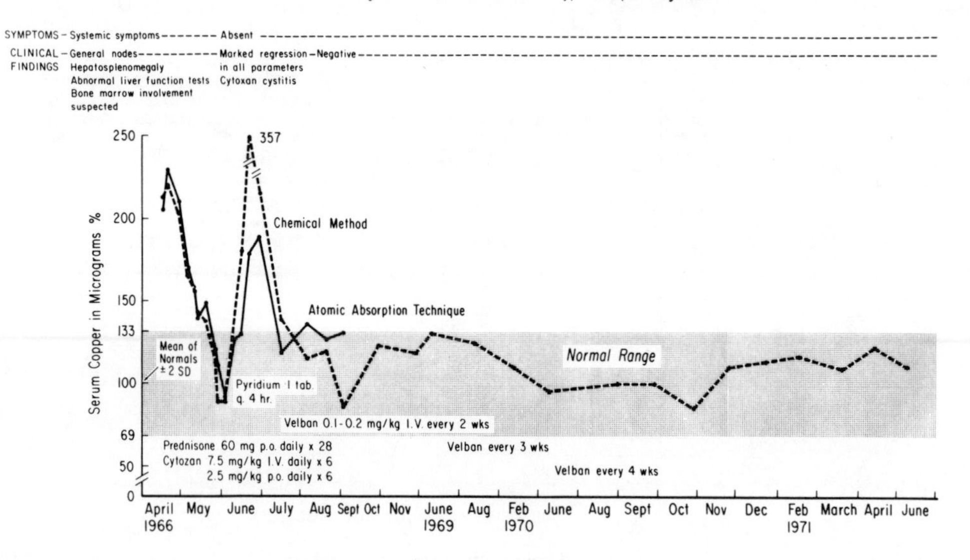

FIGURE 45. The ''interference'' of Pyridium® administration with the colorimetric method. The SCL performed by chemical method was significantly higher during Pyridium® therapy compared to atomic absorption spectrophotometry values. (Reproduced with permission from Hrgovcic, M., Tessmer, C. F., Mumford, D. M., Ong, P. S., Gamble, J. F., and Shullenberger, C. C., *Texas Medicine,* 71, 53, 1975.)

contraceptive other than sex hormones be given. Such a restriction will enhance the accuracy of serum copper levels as a measure of disease activity.

Figure 45 (an 18-year-old man with mixed cellularity Hodgkin's disease) shows two examples of serum copper fluctuations. The first is the type of serum copper level change induced by alteration in disease status — from activity to remission. The second is an example of ''interference'' with serum copper levels in a patient with good response to chemotherapy.

In the disease-related change, the serum copper level drop to normal preceded and accompanied the remission state. Note, however, an instance of initial transient copper elevation. This has been observed at the onset of therapy.

During maintenance Cytoxan® (cyclophosphamide) therapy, the patient suffered severe hemorrhagic cystitis and an ''interference'' copper phenomenon. The resultant (chemical) inflammation elevated the serum copper levels in two assays — chemical (colorimetric) and atomic absorption spectrophotometry. Moreover, treatment with Pyridium® (phenazopyridine HCl), a substance known to color body fluids, interfered with the serum copper determinations performed by the colorimetric method. The chemically determined copper levels were abnormally high when compared to those by atomic absorption.

To confirm this conclusion, chemical and atomic absorption measurements of serum copper in a normal volunteer given Pyridium® were done. With the administration of Pyridium® by mouth, only in the chemical assay did serum copper levels rise precipitously. A rapid return to normal preingestion levels occurred after cessation of the test done.

A final illustrative individual case is that of a 27-year-old white man with a nodular sclerosing-type of Hodgkin's disease on whom five serial serum copper level determinations were performed during a prolonged period when he was judged to be ''clinically'' free of disease. Figure 46 depicts his medical course with pertinent negative and positive clinical findings. The initial normal serum copper levels indicate a good response to irradiation. Subsequently, a gradual rise in the serum copper levels and a fall in the hemoglobin levels

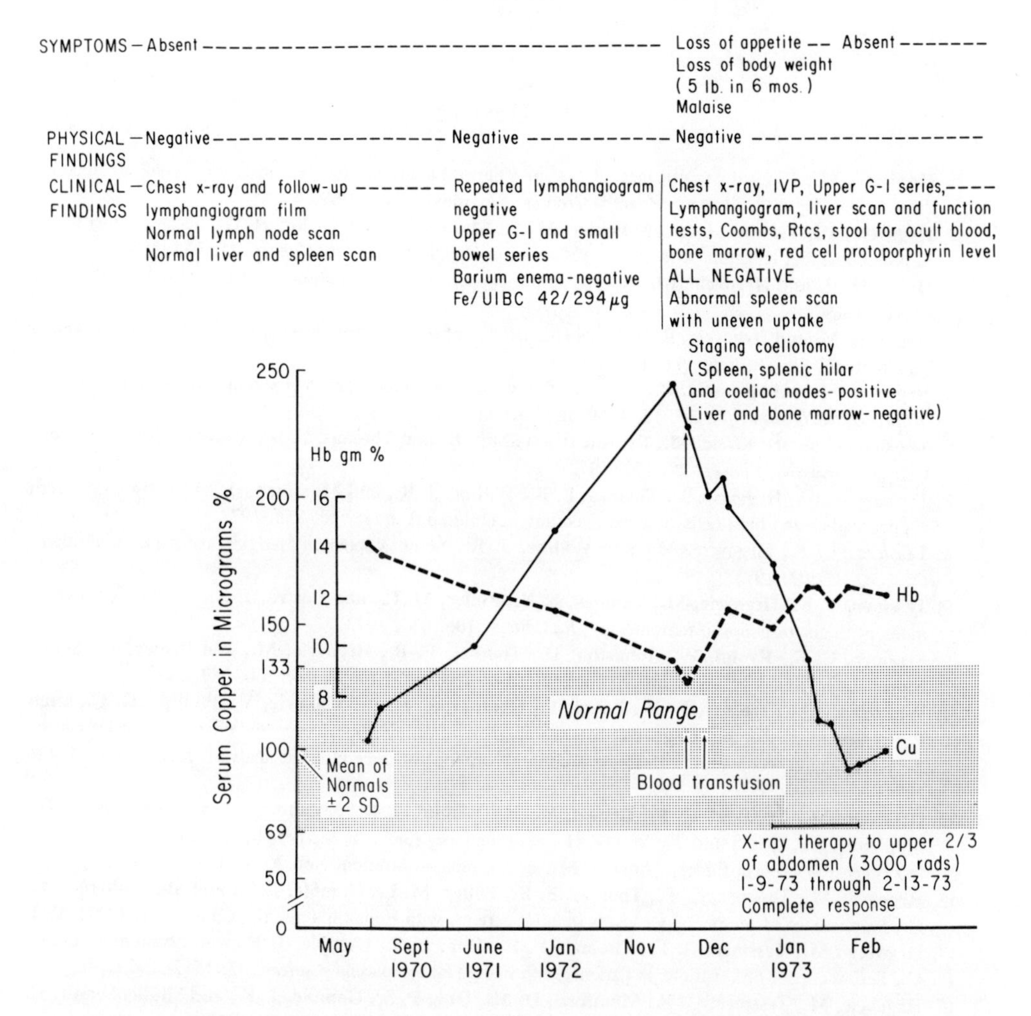

FIGURE 46. Sequential SCLs with pertinent clinical data illustrating serum copper determinations as a predictive index of impending relapse. (Reproduced with permission from Hrgovcic, M., Tessmer, C. F., Mumford, D. M., Ong, P. S., Gamble, J. F., and Shullenberger, C. C., *Texas Medicine*, 71, 53, 1975.)

were noted. This preceded definite signs of a clinical relapse by a significant period of 16 months. In December, 1972 following a spleen scan, a staging celiotomy was performed and the spleen, as well as the splenic hilar and celiac nodes, were found to be diseased. Bone marrow and liver biopsies were negative. Interestingly, the removal of the heavily diseased spleen alone apparently precipitated an appreciable decrease in the high serum copper levels. Subsequent abdominal irradiation succeeded in effecting the return of his serum copper level to normal levels. In retrospect, the increase in the serum copper level and the fall in the hemoglobin level were the only prodromal indexes which anticipated the verified clinical relapse. Few correlations are this dramatic, but serum copper levels are frequently among the first to show a response.

Therefore, in view of the possible influences of interfering substances and conditions and the nonspecificity of the laboratory tests, including serum copper, we emphasize that abnormal values of any laboratory indicator should not be taken as an indication for treatment, but rather as an indication to search for relapsing disease.

REFERENCES

1. **Hrgovcic, M.,** Clinical Significance of Serum Copper Levels in Hemoblastoses (in Croatian) doctoral dissertation, Zagreb University, Zagreb, Croatia, Yugoslavia, July 1968.
2. **Hrgovcic, M., Tessmer, C. F., Minckler, M. T., Mosier, B., and Taylor, H. G.,** Serum copper levels in lymphoma and leukemia with special reference to Hodgkin's disease, *Cancer,* 21, 743, 1968.
3. **Hrgovcic, R. and Hrgovcic, M.,** Normal serum copper levels in childhood (in Croatian), *Yug. Pediatr.,* 11, 83, 1969.
4. **Hrgovcic, M. and Hrgovcic, R.,** Clinical significance of serum copper levels in acute childhood leukemia (in Croatian), *Yug. Pediatr.,* 11, 145, 1969.
5. **Hrgovcic, M. and Hrgovcic, R.,** Clinical significance of serum copper in Hodgkin's disease, *Bull. Transfusiol. (Bilten Transfuzije),* 24, 19, 1969, in Croatian.
6. **Tessmer, C. F., Hrgovcic, M., Brown, B., Wilbur, J., and Thomas, F. B.,** Serum copper correlations with bone marrow, *Cancer,* 29, 173, 1972.
7. **Tessmer, C. F., Hrgovcic, J., Thomas, F. B., Wilbur, J. R., and Mumford, D. M.,** Long-term serum copper studies and blast cells in acute leukemia in children, *Cancer,* 30, 358, 1972.
8. **Tessmer, C. F., Hrgovcic, M., and Wilbur, J. R.,** Serum copper in Hodgkin's disease in children, *Cancer,* 31, 303, 1973.
9. **Tessmer, C. F., Hrgovcic, M., Thomas, F. B., Fuller, M. L., and Castro, J. R.,** Serum copper as an index of tumor response to radiotherapy, *Radiology,* 106, 635, 1973.
10. **Tessmer, C. F., Krohn, W., Johnston, D., Thomas, F. B., Hrgovcic, M., and Brown, B.,** Serum copper in children 6-12 years: an age correction factor, *Am. J. Clin. Pathol.,* 60, 687, 1973.
11. **Wheeler, R. M., Liebert, R. B., Zabel, T., Chaturvedi, R. P., Valkovic, V., Phillips, G. C., Ong, P. S., Cheng, E. L., and Hrgovcic, M.,** Comparison of X-ray fluorescence and charged particle-induced X-ray excitation techniques with flame atomic absorption for trace element analysis, *Med. Phys.,* 1, 68, 1974.
12. **Hrgovcic, M., Tessmer, C. F., Brown, W. B., Wilbur, J. R., Mumford, D. M., Thomas, F. B., Shullenberger, C. C., and Taylor, G. H.,** Serum copper studies in the lymphomas and acute leukemias, in *Progress in Clinical Cancer,* Ariel, I. M., Ed., Grune — Stratton, New York, 1973, 121-153.
13. **Hrgovcic, M., Tessmer, C. F., Thomas, F. B., Fuller, M. L., Gamble, J. F., and Shullenberger, C. C.,** Significance of serum copper levels in adult patients with Hodgkin's disease, *Cancer,* 31, 1337, 1973.
14. **Hrgovcic, M., Tessmer, C. F., Thomas, F. B., Ong, P. S., Gamble, J. F., and Shullenberger, C. C.,** Serum copper observations in patients with malignant lymphoma, *Cancer,* 32, 1512, 1973.
15. **Hrgovcic, M., Tessmer, C. F., Mumford, D. M., Ong, P. S., Gamble, J. F., and Shullenberger, C. C.,** Interpreting serum copper levels in Hodgkin's disease, *Tex. Med.,* 71, 53, 1975.
16. **Hrgovcic, M.,** Clinical experiences with serum copper levels in patients with hematologic malignancies, in *12th Annu. Conf. Trace Substance in Environmental Health,* Hemphill, D. D., Ed., University of Missouri, Columbia, 1978.
17. **Hrgovcic, M.,** Copper in myeloproliferative and lymphoproliferative disorders, in *Zinc and Copper in Medicine,* Karcioglu, Z. A. and Sarper, R. M., Eds., Charles C Thomas, Springfield, Ill., 1980, 481-520.
18. **Lukes, R. J. and Butler, J. J.,** The pathology and nomenclature of Hodgkin's disease, *Cancer Res.,* 26, 1063, 1966.
19. **Butler, J. J.,** Histopathology of malignant lymphomas and Hodgkin's disease, in *Leukemia-Lymphoma,* Year Book Medical Publishers, Chicago, 1970, 123-142.
20. **Kaplan, H. S.,** *Hodgkin's Disease,* 2nd ed., Harvard University Press, Cambridge, Mass., 1980.
21. **DeVita, V. T., Jr. and Hellman, S.,** *Hodgkin's Disease and the Non-Hodgkin's Lymphomas,* DeVita, V. T., Jr., Hellman, S., and Rosenberg, S. A., Eds., Lippincott, New York, 1982, chap. 35.
22. **Rosenberg, S. A.,** Report of the committee on the staging of Hodgkin's disease in symposium: obstacles to the control of Hodgkin's disease, *Cancer Res.,* 26, 1310, 1966.
23. **Carbone, P. P., Kaplan, H. S., Mushoff, K., Smithers, D. V., and Tubiana, M.,** Report of the committee on Hodgkin's disease staging, *Cancer Res.,* 31, 1860, 1971.

24. **Peters, M. V., Hasselbach, R., and Brown, T. C.,** The natural history of the lymphomas related to the clinical classification, in *Proc. Int. Conf. Leukemia-Lymphoma,* Zarafonetis, C. J. D., Ed., Lea — Febiger, Philadelphia, 1968, 357-370.
25. **Jensen, K. B., Thorling, E. G., and Anderson, C. F.,** Serum copper in Hodgkin's disease, *Scand. J. Haematol.,* 1, 62, 1964.
26. **Sprague, S. and Slavin, W.,** Determination of iron, copper and zinc in blood serum by an atomic absorption method requiring only dilution, *At. Absorption Newsl.,* 4, 228, 1965.
27. **Cartwright, G. E.,** Copper metabolism in human subjects, in *Copper Metabolism,* McElroy, W. D. and Glass, B., Eds., Johns Hopkins University Press, Baltimore, 1950, 274.
28. **Rice, E. W.,** Spectrophotometric determination of serum copper with oxyldihydrazide, *J. Lab. Clin. Med.,* 55, 325, 1960.
29. **Herring, W. B., Leavell, B. S., Parkar, L. M., and Voc, J. H.,** Trace metals in human plasma and red blood cells. A study of magnesium, chromium, nickel, copper and zinc. I. Observations of normal subjects, *Am. J. Clin. Nutr.,* 8, 846, 1960.
30. **Koch, J. J., Jr., Smith, E. R., Shimp, N. F., and Connor, J.,** Analysis of trace elements in human tissues. I. Normal tissues, *Cancer,* 9, 499, 1956.
31. **Chen, P. E.,** Abnormalities of copper metabolism in Wilson's disease: a preliminary report, *Chin. Med. J.,* 75, 917, 1957.
32. **Parker, M. M., Humoller, F. L., and Mahler, D. J.,** Determination of copper and zinc in biological material, *Clin. Chem.,* 12, 20, 1967.
33. **Dawson, J. B., Ellis, D. J., and Newton-John, H.,** Direct estimation of copper in serum and urine by atomic absorption spectroscopy, *Clin. Chim. Acta,* 21, 33, 1968.
34. **Ray, R. G., Wolf, H. P., and Kaplan, H. S.,** Value of laboratory indicators in Hodgkin's disease: preliminary results, *Natl. Cancer Inst. Monogr.,* 36, 315, 1973.
35. **Fisher, G. L., Byers, V. S., Shifrine, M., and Levine, A. S.,** Copper and zinc levels in serum from human patients with sarcomas, *Cancer,* 37, 356, 1976.
36. **Shah-Reddy, I., Khilanani, P., and Bishop, C. R.,** Serum copper levels in non-Hodgkin's lymphoma, *Cancer,* 45, 2156, 1980.
37. **Niedermeyer, W. and Grigs, J. H.,** Trace metal composition of synovial fluid and blood serum of patients with rheumatoid arthritis, *J. Chronic Dis.,* 23, 527, 1971.
38. **Meret, S. and Henkin, R. I.,** Simultaneous direct estimation by atomic absorption spectrophotometry of copper and zinc in serum, urine and cerebrospinal fluid, *Clin. Chem.,* 17, 369, 1971.
39. **Heinemann, G.,** Eisen-Kupfer-und Zinkanalysen unter Antvendung der Atom-Absorptiones-Spectral-photometrie, *Z. Klin. Chem. Klin. Biochem.,* 10, 467, 1972.
40. **Thomas, F. B., Goka, J. T., Tessmer, C. F., and Hrgovcic, M.,** Copper-binding protein to normal human plasma: identification and preliminary observation (unpublished).
41. **Topham, R. W. and Frieden, E.,** Identification and purification of a nonceruloplasmin ferroxidase of human serum, *J. Biol. Chem.,* 245, 6698, 1970.
42. **Hrgovcic, M. J. and Shah-Reddy, I.,** Serum copper levels in non-Hodgkin's lymphoma (letters to editor), *Cancer,* 47, 636, 1981.
43. **Sullivan, M. P. and Hrgovcic, M.,** Extramedullary leukemia, in *Clinical Pediatric Oncology,* Sutow, W. W., Vietti, T. J., and Fernbach, D. J., Eds., C. V. Mosby, St. Louis, 1973, chap. 11.
44. **Thorling, E. B. and Thorling, K.,** The clinical usefulness of serum copper determinations in Hodgkin's disease, *Cancer,* 38, 225, 1976.
45. **Pisi, E., Di Feliciantonio, R., Figus, E., and Ferri, S.,** Comportamento e significato prognostico della ceruloplasmina sierica in relazione al quadro istopatologico nella malattia di Hodgkin, *Minerva Med.,* 59, 944, 1968.
46. **Lukes, R. J. and Collins, R. D.,** Immunologic characterization of human malignant lymphomas, *Cancer,* 34, 1488, 1974.
47. **Lukes, R. J. and Collins, R. D.,** New approaches to the classifications of the lymphomata, *Br. J. Cancer,* 31 (Suppl. 2), 1, 1975.
48. **Chiz-Yang, L. and Harrison, E. G.,** Histochemical and immunohistochemical study of diffuse large-cell lymphomas (lymphoma heterogeneous group), *Am. J. Clin. Pathol.,* 70, 721, 1978.
49. **Nathwami, B. N., Kim, H., and Rappaport, H.,** Malignant lymphoma, lymphoblastic, *Cancer,* 38, 964, 1976.
50. **Lukes, R. J. and Tindle, B. H.,** Immunoblastic lymphadenopathy. A hyper-immune entity resembling Hodgkin's disease, *N. Engl. J. Med.,* 292, 1, 1975.
51. **Frizzera, G., Moran, E. M., and Rappaport, H.,** Angioblastic lymphadenopathy: diagnosis and clinical course, *Am. J. Med.,* 59, 803, 1975.
52. **Lahey, M. E., Gubler, C. J., Cartwright, G. E., and Wintrobe, M. M.,** Studies on copper metabolism. VI. Blood copper in normal human subjects, *J. Clin. Invest.,* 32, 322, 1953.

53. **Wintrobe, M. M., Cartwright, G. E., and Gubler, C. J.,** Studies on the function and metabolism of copper, *J. Nutr.*, 50, 395, 1953.
54. **Cartwright, G. E. and Wintrobe, M. M.,** Copper metabolism in normal subjects, *Am. J. Clin. Nutr.*, 14, 224, 1964.
55. **Adelstein, S. J. and Vallee, B. L.,** Copper metabolism in man, *N. Engl. J. Med.*, 265, 892, 1961.
56. **Davidoff, G. N., Votaw, M. L., Coon, W. W., Wexler, S. A.,** Elevations in serum copper, erythrocytic copper and ceruloplasmin concentrations in smokers, *Am. J. Clin. Pathol.*, 70, 790, 1978.
57. **Sass-Kortsak, A.,** Copper metabolism, in *Advances in Clinical Chemistry*, Vol. 8, Sobotka, H. and Stewart, C. P., Eds., Academic Press, New York, 1965, 1-67.
58. **Sinha, S. and Gabieli, E.,** Serum copper and zinc levels in various pathologic conditions, *Am. J. Clin. Pathol.*, 54, 570, 1970.
59. **Beisel, W. R.,** Trace elements in infectious processes, *Med. Clin. North Am.*, 60, 831, 1976.
60. **Lahey, M. E., Gubler, C. J., Cartwright, G. E., and Wintrobe, M. M.,** Studies on copper metabolism. VII. Blood copper in pregnancy and various pathologic states, *J. Clin. Invest.*, 32, 329, 1953.
61. **Markovitz, H., Gubler, C. J., Mahoney, J. P., Cartwright, G. E., and Wintrobe, M. M.,** Studies on copper metabolism. XIV. Copper, ceruloplasmin and oxidase activity in sera of normal human subjects, pregnant women and patients with infection, hepatolenticular degeneration and the nephrotic syndrome, *J. Clin. Invest.*, 34, 1498, 1955.
62. **Russ, E. and Raymun, J.,** Influence of estrogens on total serum copper and ceruloplasmin, *Proc. Exp. Biol. Med.*, 92, 465, 1956.
63. **Lee, G. R.,** The anemia of chronic disease, in *Seminars in Hematology*, Vol. 20, Metal Metabolism in Hematologic Disorders, Miescher, P. A. and Jaffe, E. R., Eds., Grune & Stratton, New York, 1983, 61—80.

Chapter 10

SERUM COPPER LEVELS AND OTHER LABORATORY INDICATORS IN LYMPHOMAS

Malignant lymphomas remain an enigmatic disease with protean manifestations and may, following apparently effective therapy, reappear after many years. The state of ''remission'' may not indicate absolute absence of foci of disease.

Following tissue diagnosis and accurate staging procedures, there still may remain some doubt as to the extent of disease, particularly in non-Hodgkin's lymphoma (known for noncontiguous spread) and difficulties in identifying microscopic foci of disease. It is even more difficult to reassess with certainty the occult disease activity by noninvasive radiographic and imaging procedures in the subsequent course of disease, due to the lack of objective measurable parameters to confirm the subjective clinical impression.

With the growing application of radiotherapy and chemotherapy in lymphomatous disease, it has become increasingly important to find methods to detect recurrent or occult (nodal or extranodal), active disease as early as possible. Since an increasing number of chemotherapeutic agents have become available, it is important to predict, insofar as possible, a satisfactory response to cytotoxic drugs before a prolonged course of treatment has compromised functioning bone marrow and, thus, prevented or delayed treatment with other possibly more effective drugs. There certainly exists a need for a sensitive and specific biological indicator which will accurately reflect disease activity and response to therapy and be of help in deciding whether and when a particular therapy should be continued, stopped, or replaced by an alternative treatment regimen. Unfortunately such an ''ideal biological monitor'' has not been found yet.

Various hematological and biochemical indicators have been used to monitor activity of lymphomatous disease.[1-3] None are disease-specific and their reliability and clinical significance are somewhat controversial. Laboratory parameters reported to indicate disease activity include total white blood count,[4,5] absolute neutrophil count,[6] erythrocyte sedimentation rate,[7-10] serum copper,[11-22] and ceruloplasmin,[23-28] all of which have been reported to be elevated and hematocrit,[29,30] the absolute lymphocyte count and the percentage of lymphocytes,[31-34] serum zinc,[35,36] iron,[37-40] and bradykinogen[41] which are said to be depressed during disease activity. The leukocyte[42-46] and serum alkaline phosphatase[47-40] levels have also been reported to be increased, particularly in advanced stages of disease. Bromsulfthalein retention is noted in the presence and the absence of liver disease.[50-52] Abnormalities which occur less often include elevation of serum haptoglobin,[53,54] and alpha-2-globulins,[55-59] hydroxyproline-containing protein,[60] C-reactive protein,[61,62] serum hexosamine,[63] serum protein-bound hexose,[64] serum transcobalamin,[65] acid glycoprotein,[2] and β_2-microglobulin.

The difficulty with these tests is that they are essentially nonspecific in character and similar abnormalities are encountered in a wide range of conditions and disease states, particularly inflammatory and infectious conditions.

Normal blood counts are commonly seen in untreated lymphoma patients, particularly in early stages of disease. On the other hand, abnormal hematologic findings are frequently observed in advanced disease. Kaplan[1] in his series of 100 previously untreated randomly selected Hodgkins's disease patients noted 12% of them to be anemic, with packed cell volume less than 35 vol%. An increased white blood cell (WBC) count over 10,000/mm^3 was found in 27% and leukopenia (WBC less than 5000/mm^3) in only 5% of these patients. In untreated Hodgkin's disease patients, absolute lymphocyte counts less than 1000/mm^3 were observed in approximately 20% of cases, suggesting advanced and/or relatively aggressive (lymphocyte depletion type) disease. Aisenberg[31] reported severe lymphocytopenia of less than 500 lymphocytes in 75% of his Hodgkin's lymphoma patients entering a near-terminal phase of the disease, whereas low normal or only slightly depressed lymphocyte count was noted at the onset of disease. Brown et al.[32] found in 19 out of 50 untreated patients (38%) lymphocyte counts below 1500/mm^3 and similar findings were reported by Young et al.[66] Increased numbers of mononuclear cells (monocytes or large lymphoid cells) have been observed in the differential WBC count, especially on smears of leukocyte concentrates, thus resembling Reed-Sternberg or lymphocarcoma cells. Examination of the peripheral smear in patients with non-Hodgkin's lymphoma may yield evidence of malignant cells in approximately 15% of patents, primarily those with poorly differentiated lymphocytic lymphomas (nodular or diffuse).[67] In later phases of disease, particularly in advanced stages of disease undergoing aggressive treatment (chemotherapy alone or in combination with irradiation), the abnormalities in blood count (anemia, leukopenia, and thrombocytopenia) have been frequently observed as the result of myleosuppression or bone marrow invasion by basic disease, hypersplenism, or autoimmune phenomena. The correlation between peripheral blood counts and marrow involvement by lymphoma is poor. Some abnormality in blood counts have been found in only 37% of patients with bone marrow infiltration by lymphoma.[68] Approximately one half of patients with abnormal blood counts will not have bone marrow involvement on biopsy. Chronic blood loss from the gastrointestinal tract can be a contributing factor in the development of anemia, in addition to disrupted iron metabolism in patients with malignant conditions.[8]

Hematologic findings, therefore, have limited value in following the activity of malignant lymphomatous disease.

Hypoferremia with normal levels of unsaturated iron-binding capacity has been reported to be one of the first signs of activity of Hodgkin's disease.[6,39] This has been attributed to a defective reutilization of iron and indicative of a "systemic syndrome".[6] Jaffe and Bishop[30] found hypoferremia to always occur in relapse but also in "remission" if the disease state persisted over a long period. Normal serum iron levels in remission in advanced stages of Hodgkin's disease could possibly be attributed to such factors as liver disease, recent blood transfusion, iron medication, hemolytic disease, or the administration of certain cancer chemotherapy drugs, e.g., nitrogen mustard, causing a temporary inhibition or the iron incorporation into red blood cells and a depression of the erythrocyte uptake of iron.[37,38,69,70] Low serum iron levels, usually with normal or decreased unsaturated iron-binding capacity, may be seen in patients with chronic infection. Ray et al.[18] reported serum iron to correlate with disease activity in 57% of 112 patients with Hodgkin's disease. Reduced serum iron or iron transferrin concentrations in active disease, and normal values in remission, have been also reported.[40,71] Hughes[72] found no correlation between serum transferrin and ceruloplasmin concentrations. On the other hand, Foster et al.[28] observed good correlation between these two metaloenzymes, although ceruloplasmin elevation has proved to be a more reliable early sign of disease reactivity.

Since serum iron and transferring levels show considerable variation in normal controls with the time of day, state of nutrition, associated diseases, therapy, etc. these observations tend to limit the clinical usefulness of these tests in following lymphomatous disease activity.

Decreased serum zinc levels have been reported in patients with Hodgkin's disease,[35,36] acute leukemia,[73] and some other malignancies.[74-76] Approximately 40% of all the lymphomas studied by Koch et al.[77] had a plasma-zinc level above the normal range. Cappelaere et al.[19] found no significant correlation between serum zinc level and disease activity in his series of 92 Hodgkin's disease patients.

Bucher and Jones[36] stated that the copper:zinc ratio correlated well with disease activity in a series of patients with Hodgkin's disease. Previously untreated and relapsing patients had mean copper:zinc ratio values of 2.1 and 2.2, respectively, whereas those with partial and complete remissions had mean ratios of 1.7 and 1.5, respectively. In view of the wide variation of zinc level caused by nondisease-related factors such as contamination of sera, and hemolysis, serum zinc determination alone does not appear to be useful in monitoring lymphomatous disease activity.

Elevation of leukocyte alkaline phosphatase has been reported to reflect activity of Hodgkin's disease.[42-46] Jaffe and Bishop[30] reported increased leukocyte alkaline phosphatase in 45% of pediatric patients in relapse and, in remission, 55% in boys and 17% in girls. Flury and Wegman[43] reported a mean leukocyte alkaline phosphatase value of 149 in 51 patients with active Hodgkin's disease and 95 in remission, compared to 45 in normal controls. In a series of 16 patients treated with intensive irradiation the mean pretreatment leukocyte alkaline phosphatase value was 192, and after treatment 88. On the contrary, 14 other patients treated with palliative irradiation and 7 with chemotherapy, showed a slight increase or no change from the pretreatment leukocyte alkaline phosphatase level. Ray et al.[18] reported leukoycte alkaline phosphatase to correlate with disease activity in 24% of 112 consecutive untreated patients with Hodgkin's disease. Increased leukocyte alkaline phosphatase is also found in patients with infections and with polycythemia vera. Thus, it is considered to have limited applications and usefulness in monitoring disease activity in lymphoma.

Glatstein et al.,[51] in their review of 65 patients with Hodgkin's disease, laparotomy staged, concluded that liver involvement with basic disease was not well correlated with any one liver function test. Aisenberg et al.[48] investigated the values of serum alkaline phosphatase in 111 patients at the onset of Hodgkin's disease. Increased alkaline phosphatase was found in 14% of cases with stage I and II, 65% in stage III, and 81% of cases with stage IV disease. The source of the elevated alkaline phosphatase was found to be liver in 23, bone in 3, both in 2, and in 3 cases it was not possible to state which isoenzyme was elevated. Levine[52] found abnormal liver function tests (alkaline phosphatase, transaminases, or bromsulphathalein retention) in 9 of 26 patients with Hodgkin's disease and only one case of true stage IV disease was documented. Elevation of alkaline phosphatase has also been found in conditions such as late adolescent bone growth, febrile states, and nonspecific liver or bone disorders.

Serum protein abnormalities are not uncommon in patients with malignant lymphomas. Serum albumin levels may be reduced, particularly in advanced stages whreas the alpha$_1$-, alpha$_2$-, and betaglobulin fractions are often increased at some time during the course of disease.[55-59] Hypergammaglobulinemia,[8] hypogammaglobulinemia,[78] macroglobulinemia,[79] and rare cases of agammaglobulinemia[54] have been reported in patients with malignant lymphoma. Moore et al.[79] reported the data on serum protein electrophoresis performed in 63% of 1069 consecutive patients with lymphoma referred to the University of Texas M.D. Anderson Hospital and Tumor Institute from January 1965 to July, 1969. They found no monoclonal peaks of IgM specificity in sera from 345 patients with nodular lymphoma or Hodgkin's disease. Of 333 patients with diffuse lymphoma, 1.5% had an IgG peak, suggesting a coincidental relationship of IgG peaks to lymphoma. IgM peaks occurred in 3.6% of patients with diffuse lymphoma, an incidence about 60 times more frequent than that in normal subjects. Such peaks were more frequent in older patients. There was a close correlation between lymphoma mass and the level of the IgM peak in individual patients. Irunberry and Colonna[80] reported increased levels of IgG, IgA, and IgM in 32, 19, and 14%, respectively, of samples determined during relapse and the smaller percentage of samples during remission. In two cases of myeloma, already discussed, the majority of significantly increased serum copper was found to be associated with IgG myeloma protein.[81,82]

Alterations of acute phase reactant proteins (haptoglobin, ceruloplasmin, C-reactive protein, protein-bound hexose, alpha$_1$ antitrypsin and alpha$_1$-antichymotrypsin, alpha$_1$-acid glycoprotein, serum hexosamine) have been reported to have potential value in the management of lymphomatous disease.[2,23-28,53-64] Malpas and Fairley[58] studied the total concentration of alpha$_2$-globulin and two of its components, haptoglobin and ceruloplasmin, in the sera from 78 patients with primary malignant lymphoreticular diseases. The total alpha$_2$-globulin fraction was significantly raised only in Hodgkin's disease. The serum haptoglobin was also significantly increased in Hodgkin's disease and some other reticuloses, including acute leukemia, lymphosarcoma, and reticulum cell sarcoma. The serum level of ceruloplasmin was raised in all reticuloses. There was possibly some correlation between the rise in haptoglobin and ceruloplasmin concentrations and the total alpha$_2$-globulin in Hodgkin's disease, but this was not statistically significant. Correlation between alpha$_2$-globulin concentrations and haptoglobin levels revealed a correlation coefficient of 0.29 and a somewhat higher correlation with ceruloplasmin of 0.44.

Snyder and Ashwell[54] reported quantitative studies on the serum carbohydrate and glycoprotein levels of cancer patients and compared them with the values obtained on normal volunteers and chronically ill controls. Of the 15 glycoproteins studied, 7 (transferrin, alpha$_2$-macroglobulin, GC-globulin, IgA, IgD, IgG, and IgM) remained invariable in all these patient categories. Three (alpha$_1$-acid glycoprotein, cerulosplasmin, alpha$_1$-antitrypsin) were elevated in both the cancer and the chronically ill group. Two (haptoglobin, hemopexin) were elevated in cancer patients but not in the nonmalignant controls. Three (alpha$_2$HS-glycoprotein, beta$_2$-glycoprotein I, prealbumin) were decreased in malignancy and unchanged in the pathological control group. Alterations in haptoglobin levels in the course of Hodgkin's and nonHodgkin's lymphomas were reported by several investigators.[54,59,80] Elevation of serum haptoglobin has been observed in association with acute and chronic infections, burns, trauma, surgery, collagen disease, certain neoplastic diseases, scurvy, and following the administration of sex hormones, glucocorticoids and parathormones, whereas decreased haptoglobin levels are found in conditions associated with hemolysis. The usefulness of haptoglobin determinations as an indicator of disease activity is very limited.

Wood et al.[61] studied C-reactive protein in 121 patients with Hodgkin's disease. A change in Hodgkin's disease with lymphocyte predominance to a disseminating form of Hodgkin's disease was mirrored in the appearance of C-reactive protein in blood. Following early aggressive treatment of localized Hodgkin's granuloma, C-reactive protein disappears from the blood. With recrudescence of disease activity, the C-reactive protein reappears and, with generalization of the disease, persists. C-reactive protein, however, did not provide an accurate, sensitive measure of response to therapeutic agents in patients with generalized Hodgkin's disease, and there was no consistent effect on C-reactive protein by the various therapeutic agents, despite clinical improvement. Yocum and Doerner[83] evaluated C-reactive protein in 729 patients with various pathological states, including malignant conditions. They found C-reactive protein to be more accurate and sensitive than the erythrocyte sedimentation rate. The C-reactive protein test was positive before the erythrocyte sedimentation rate was elevated and became negative before the erythrocyte sedimentation rate returned to normal levels. A similar observation was made by Gewurz et al.[84] who regarded C-reactive protein as an alternative to and in some cases superior to the erythrocyte sedimentation rate for following disease activity.

The correlation between serum copper levels, ceruloplasmin activity and C-reactive protein in 275 hospital patients (126 males and 149 nonpregnant females) was reported by Rice.[85] A strong positive linear correlation between total serum copper content and phenylendiamine (PPD)oxidase (ceruloplasmin) was found. The "perfect" correlation between serum copper level and C-reactive protein was observed only in 62.9% of the 275 sera studied, suggesting that either hypercupremia is a more sensitive index of disease activity than ceruloplasmin or that there are factors other than inflammation which affect serum copper levels.

The usefulness of erythrocyte sedimentation rate in the monitoring of lymphomatous disease activity has been controversial. While some authors have reported the erythrocyte sedimentation rate as the most reliable indicator of active disease,[7-10,29] others consider the sedimentation rate to be nonspecific and of limited value in the evaluation of Hodgkin's disease activity.[1,13,15,20] It has been known for several years that the sedimentation rate represents a complex interplay of factors, not least the hematocrit and the levels of acute phase reactants, particularly fibrinogen. Increased sedimentation rates over 30 mm/hr were found in 48% of untreated Hodgkin's disease patients in Kaplan's[1] series of 100 cases. Le Bouregeois and Tubiana,[10] in a series of 68 patients with clinical stages I and II Hodgkin's disease, found erythrocyte sedimentation rate values above 30 mm/hr in 80% of patients with relapsing disease. Conversely, relapses were detected in 90% of those cases in which the erythrocyte sedimentation rate and the detection of relapse was 4.5 months. In contrast to serum copper levels, initially elevated sedimentation rates usually fall slowly to normal levels and may take 1 year to normalize even in patients who exhibit an apparently complete remission in response to specific therapy, and may again become elevated prior to or during clinical relapse. Jensen et al.[86] found fluctuations in sedimentation rates in some patients in the course of Hodgkin's disease and no definite significance could be attached to small changes, whereas the copper levels showed significant variation which reflected more accurately the activity of disease. Jaffe et al.[30] found increased sedimentation rate in 56% of male and 62% of female pediatric patients with Hodgkin's disease in relapse and in 42% of boys and 50% of girls in remission. Ray et al.,[18] Asbjörnsen,[87] and Pizzolo et al.[88] suggested that the sedimentation rate and the serum copper level should be considered as complementary examinations in monitoring disease activity and response to therapy. Gobbi et al.[40] reported that determination of plasma iron and copper showed a clearly higher diagnostic value in evaluation of Hodgkin's disease activity than sedimentation rate, alpha$_2$-globulins and fibrinogen combined.

Based on the experience of Kaplan,[1] Jensen et al.,[86] Thorling and Thorling,[20] as well as our own experience, serial determinations of the serum copper level are particularly useful in the 6- to 12-month period immediately following completion of radiotherapy and/or combination chemotherapy, since the erythrocyte sedimentation rate tends to rise and to remain elevated for many months following treatment and is of little help during this interval as a prognostic indicator of disease activity.

Conversely, the serum copper characteristically falls to normal levels and remains normal in successfully treated patients whereas failure of the serum copper values to return to normal following treatment, or a transient fall followed by a renewed elevation during the first 12 months after treatment usually denotes impending relapse. The limited opportunities for quantitation, the recognized nonspecificity, and a rather formidable set of other factors which influence the erythrocyte sedimentation rate, in our opinion, have been the reason for lack of dependence on this test. The serum copper levels, in our hands, have been a much more useful indicator of disease activity in lymphoma patients.

In view of the nonspecific nature of the cuproprotein response, it was of particular interest to us to compare the serum copper levels with other clinical laboratory tests. This study was done on 6002 sera submitted for routine SMA-12/60 analysis (simultaneous multiple analysis) of 12 laboratory tests, representing an estimated 2000 individuals with a wide variety of clinical conditions and/or reasons for the requested procedures. It is to be noted that the hospital and clinic population from which these samples were taken is unusual. An estimate has been made that 85% of the admissions to the University of Texas M.D. Anderson Hospital and Tumor Institute are by reason of some form of neoplastic disease, and the remainder represent those requiring diagnostic procedures which, in effect, determine that they are free of neoplastic disease. No specific information regarding this sample group has been contemplated, the study being designated to note any relationship between serum copper levels and other values in the SMA-12/60, in the hospital population. Each laboratory day, all available serum samples were gathered for copper determination. Approximately 10% of the total number were not available for serum copper determinations due either to inadequate sample volume or requirement for other use. However, an examination of this group in detail has revealed no systemic exclusions or selection.

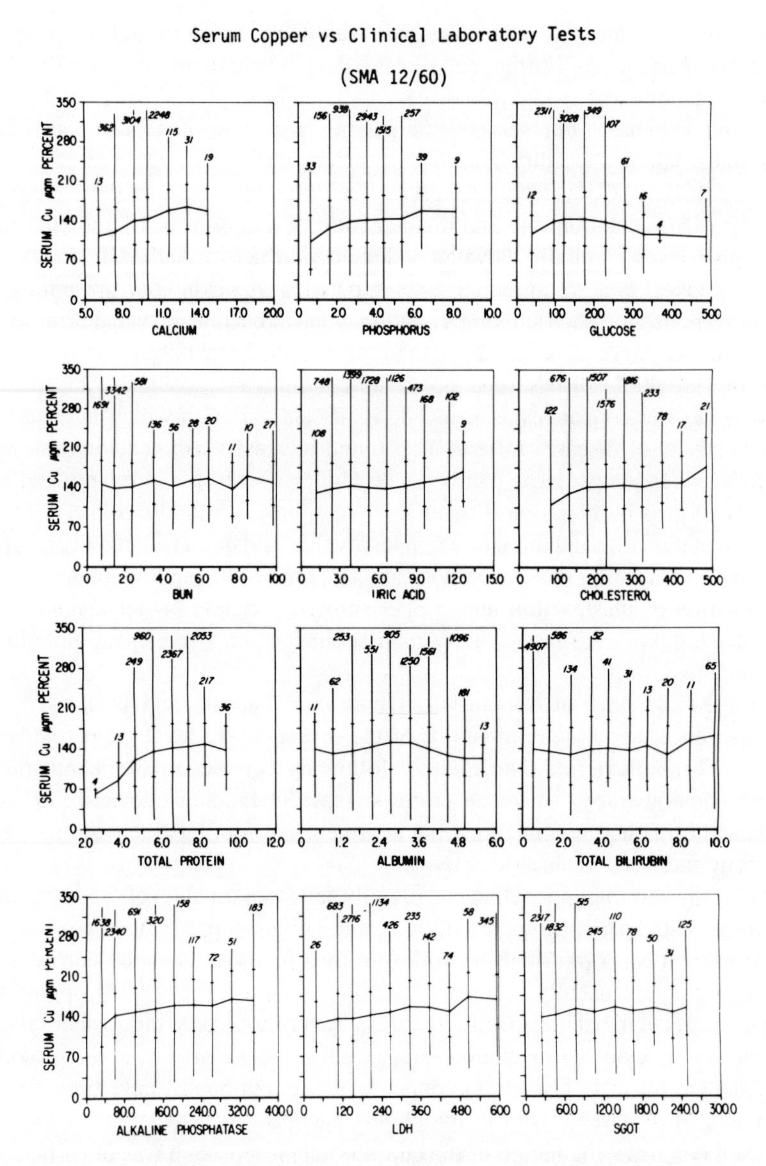

FIGURE 1. Relationships between the tests of the SMA-12/60 group and serum copper values determined on approximatly 6000 serum samples. The numeral at the top of the vertical line represents the number of values in the corresponding interval. Marks on vertical line indicate 1 SD from the mean, and the full vertical line the range of copper values. The limited correlations indicated in Tables 1 and 2 of the test may be seen graphically by degree and uniformity of slope.

To provide a uniform and a graphic presentation of the relation of serum copper level to the other determinations, the range of values of the other tests was divided into ten intervals. The mean, standard deviation, and range of the copper values in each interval were computed and plotted. The results are shown in Figure 1. The mean is indicated by a solid line. Tics are placed on the vertical lines at ±1 SD. The extent of the vertical line indicates the range of copper values for test values in the particular interval. The number of test values in each group is indicated by the numeral at the top of the vertical line.

The simple correlation coefficients of each test with serum copper were computed. These are presented in Table 1.[89]

Table 1
SIMPLE CORRELATION COEFFICIENTS OF SERUM COPPER VALUES WITH SMA 12/60 TEST VALUES

	SGOT	Calc.	Phos.	Gluc.	BUN	Uric acid	Chol.	T.P.	Alb.	Bili.	Alk. phos.	LDH	SCL
SGOT	1.000												
Calc.	−0.095	1.000											
Phos.	0.017	0.122	1.000										
Gluc.	0.032	−0.096	−0.163	1.000									
BUN	0.114	0.027	0.149	0.149	1.000								
Uric acid	0.054	0.212	0.149	0.010	0.369	1.000							
Chol.	−0.017	0.261	0.051	0.081	0.005	0.163	1.000						
T.P.	−0.162	0.507	0.172	−0.141	−0.079	0.155	0.330	1.000					
Alb.	−0.287	0.444	0.145	−0.202	−0.149	0.157	0.350	0.615	1.000				
Bili.	0.372	−0.205	−0.109	0.068	0.153	−0.033	0.024	0.248	−0.341	1.000			
Alk. phos.	0.435	−0.712	0.006	0.402	0.073	0.025	0.029	−0.190	−0.331	0.374	1.000		
LDH	0.436	−0.082	0.022	0.082	0.133	−0.023	−0.020	−0.202	0.282	0.217	0.370	1.000	
SCL	0.092	0.143	0.067	0.003	0.005	0.052	0.119	0.115	−0.127	0.101	0.242	0.198	1.000

From Hrgovcic, M., Tessmer, C. F., Brown, W. B., Wilbur, J. R., Mumford, D. M., Thomas, F. B., Shullenberger, C. C., and Taylor, G. H., in *Progress in Clinical Cancer*, Ariel, I. M., Ed., Grune & Stratton, New York, 1973, 121—153. With permission.

Table 2
MULTIPLE REGRESSION ANALYSIS[a]

Serum copper level	= 20.263135		Copper variance explained by variable
	+ 0.12164060	(Alk. Phos.)	
	+ 11.541515	(T.P.)	
	− 16.333263	(Albumin)	
	+ 6.4409416	(Calcium)	
	+ 0.053276883	(LDH)	
	+ 0.088860582	(Cholesterol)	
	− 3.0650002	(Bilirubin)	
	− 0.057581885	(SGOT)	
	− 0.21647651	(BUN)	
	+ 2.4322687	(Phos.)	17.4%

[a] Standard error of estimate = 39.510981. Noncontributory variables: glucose and uric aicd.

From Hrgovcic, M., Tessmer, C. F., Brown, B. W., Wilbur, J. R., Mumford, D. M., Thomas, F. B., Shullenberger, C. C., and Taylor, G. H., in *Progress in Clinical Cancer,* Ariel, I. M., Ed., Grune & Stratton, New York, 1973, 121—153. With permission.

An indication of the joint contribution of the SMA 12/60 test values to the variance of the serum copper level is provided in a stepwise multiple regression analysis. A summary of the results is given in Table 2.[89] Of the 12 tests, 10 contributed significantly ($p < 0.01$) to the prediction. When these ten were included, glucose and uric acid added less to the prediction than would a totally unrelated random variable. The multiple r, i.e., the correlation of the predictor based on the other tests with the observed copper values, was 0.42 — an indication that 17.4% of the variation in copper values could be accounted for by the other tests.

In the simple correlation coefficients (Table 1) it is apparent that a significantly positive but relatively low order of correlation exists for two tests, alkaline phosphatase (0.2418) and LDH (0.1975). These deserve some consideration individually, and perhaps in combination, since they might well be part of a disease pattern, for example, liver involvement. It has been the experience that the serum LDH activity is so frequently elevated in disease states of many types that it has limited specific value. Alkaline phosphatase similarly has multiple factors leading to its elevation. A finding of some interest, since liver involvement was one of the first conditions considered in the patterns of serum copper variation, it is the lack of correlation with SGOT.

The analysis of variance indicates in this instance that the relationships to other laboratory tests examined, while definite, is relatively small, and that by the same measure the ability of these other laboratory tests to predict the serum copper level is also relatively small. For example, the standard deviation of the 6002 serum copper values of approxiately 43 μg% would be reduced only to 39 μg% by the contribution of the other laboratory determinations.

The graphic presentation of copper values in relation to each of the SMA 12/60 tests illustrates the limited degree of correlation noted in LDH and alkaline phosphatase, and the virtual lack of correlation in the others (Figure 1).

Another aspect not indicated by the overall correlations above may be found in more detailed analysis of individual cases. The example of the reaction of an individual to synthetic estrogen, ethinyl estradiol (Figure 2), shows transient responses of alkaline phosphatase and SGOT, while the change in serum copper level is equally prompt, but sustained. This pattern is cited not as specifically related to estrogen, to serum copper, or other test responses, but as an indication that serum copper appears to act as an independent reacting factor. This

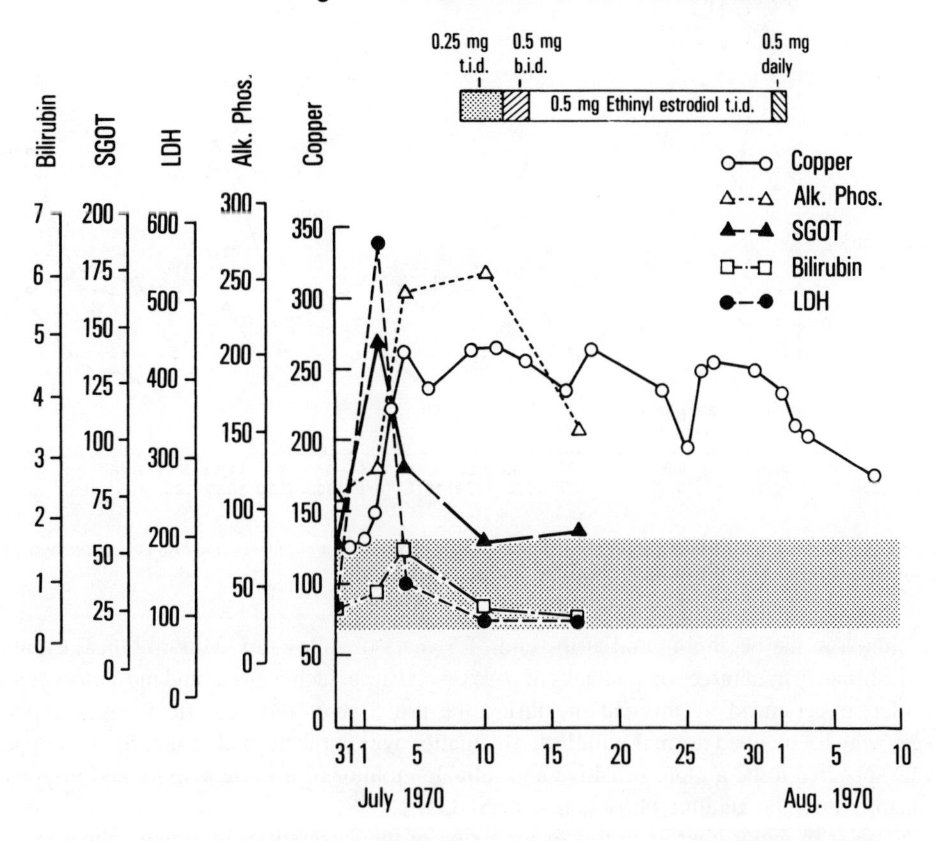

FIGURE 2. Sequential values of alkaline phosphatase, SGOT, LDH, and bilirubin for comparison with serum copper values. Ranges of each test have been set to place the normal portion approximately at the shaded level. Values represent response of one individual to ethinyl estradiol, a synthetic estrogen used in therapy of breast cancer. The graphic data suggest a basis for a degree of correlation, particularly in early response, but also suggest, by the sustained serum copper level, an essentially independent index.

type of response may also account for a degree of correlation, although as pointed out above, virtually none was identified for SGOT.

The second case illustrating the relationship between the sequential laboratory parameters and clinical course in a 34-year-old white female with stage (C.S., P.S.) IIA nodular sclerosing Hodgkin's disease is shown in Figure 3. Two weeks after excisional node biopsy, the patient underwent routine staging procedures. Elevated serum copper level, sedimentation rate, alkaline phosphatase, SGOT, and platelet count of 650,000 were found. Since no clear-cut evidence of Hodgkin's disease in the liver was detected, although it was clinically suspected, staging laparotomy was performed and no evidence of abdominal disease was found. Complete control of disease was achieved by irradiation therapy and the patient has been in complete remission since. Six weeks after completion of irradiation the initially elevated serum copper level and erythrocyte sedimentation rate fell to normal range, whereas alkaline phosphatase and SGOT remained elevated. During subsequent 5-year follow-up at regular 3- to 6-month intervals, serum copper levels, erythrocyte sedimentation rate (not regularly followed after initial normalization), and alkaline phosphatase remain within normal range, whereas SGOT showed occasional unexplained fluctuation. Sedimentation rate was

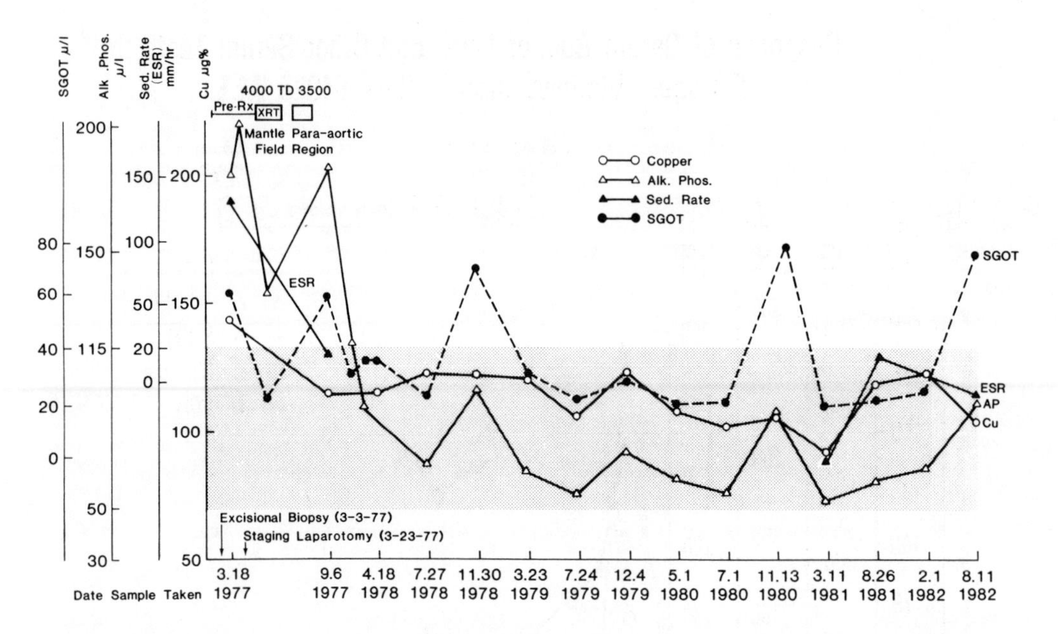

FIGURE 3. Correlation between laboratory indicators (SCL, ESR, AP, and SGOT) and clinical course in a 34-year-old female with stage IIA nodular sclerosing Hodgkin's disease.

determined in the beginning and at the end of 5 years of follow-up. Minor clinical events, i.e., respiratory tract infections, attacks of paroxysmal atrial tachycardia, and musculoskeletal disorder, experienced by this patient during the last 5 years did not affect serum copper levels, which remained normal with little fluctuation within the normal range. Serum copper levels appeared to be a more sensitive and reliable monitor of disease activity and response to therapy than the alkaline phosphatase or SGOT.

The point of major interest in this examination of the correlations of serum copper levels in a more or less random hospital and clinical population is that a limited number of relationships have been observed between the serum copper and the SMA-12/60 test groups, but the majority of interaction is not explained by individual tests per se, or any combinations.

An attempt at correlation also was made between serum copper levels and 630 serum protein electrophoreses performed on the University of Texas M.D. Anderson Hospital and Clinic patients. As can be seen from graphic presentations (Figures 4 and 5), no correlation was found between serum copper levels and the total protein or albumin fraction levels. Correlations of the serum copper level with globulin fractions (Figure 6) disclose some, although insignificant, correlation between alpha$_2$-globulins and serum copper levels, with a correlation coefficient of 0.4. This is not surprising, because ceruloplasmin, a major copper transfer protein, and haptoglobin are main components of the alpha$_2$-globulin fraction. These studies of correlations of the serum copper level, SMA-12/60, and serum protein electrophoresis suggest the relative independence of the serum copper level as a laboratory determination having increased value as an index of those changes with which it has been linked. This tends to clarify one portion of the complex pattern of the kinetics of serum copper, namely that it is not reflecting the activity of a state indicative of any of the other clinical laboratory tests examined.

Most recently, Swedish authors[90] reported that serum β_2-microglobulin correlates well with the activity and extent of malignant lymphoproliferative diseases, while other investigators found no correlation.[91,92]

In view of the nonspecificity of all these laboratory indicators, the total dependence on any one laboratory test is not advisable. It is hoped that continuing research concerning the

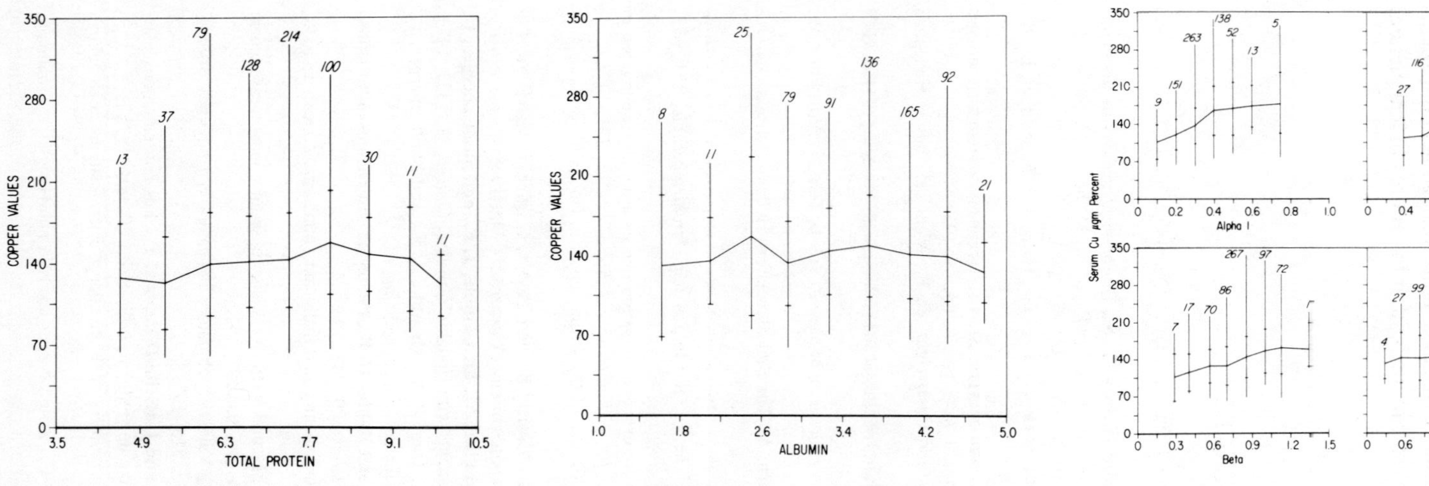

FIGURE 4. Correlation between SCL and total protein values in 623 serum samples from patients with malignant disease.

FIGURE 5. Correlation between SCL and albumin fraction (electrophoretic) in 628 hospital and clinical samples from patients with malignant diseases.

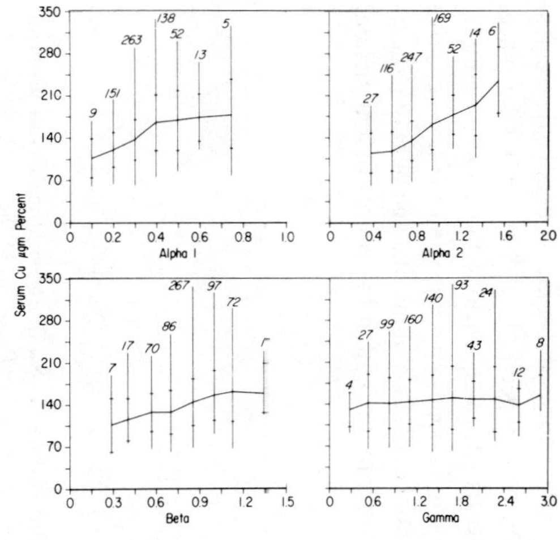

FIGURE 6. Correlation between SCL and globulin fractions in 631 hospital and clinical serum samples from patients with malignant diseases.

acute phase reacting proteins and metals (copper, zinc, and iron) will help us to better characterize and distinguish the clinical events in the course of disease — neoplastic (regression or progression) or nonneoplastic (infection or nonspecific), as they are taking place. At the present time, and in the absence of more specific and sensitive indicators, it appears that the determination of serum copper or the serum copper zinc ratio and the less sensitive erythrocyte sedimentation rate can be of substantial assistance in the management of patients with lymphomatous disease, if properly interpreted, in the light of the entire clinical picture.

REFERENCES

1. **Kaplan, H. S.,** *Hodgkin's Disease,* 2nd ed., Harvard University Press, Cambridge, Mass., 1980, 126 and 579.
2. **Child, J. A., Copper, E. H., Illingworth, S., and Worthy, T. S.,** Biochemical markers in Hodgkin's disease and non-Hodgkin's lymphoma, in *Recent Results in Cancer Research,* Mathe, G., Seligman, M., and Tubiana, M., Eds., Springer-Verlag, Berlin, 1978, 180.
3. **Reynoso, G.,** Biochemical tests in cancer diagnosis, in *Cancer Medicine,* Holland, J. F. and Frei, E., III, Eds., Lea & Febiger, Philadelphia, 1973, 335.
4. **Bethell, F. H., Andrews, G. A., Neligh, R. B., and Meyers, M. C.,** Treatment of Hodgkin's disease with roentgen irradiation and nitrogen mustards, *Am. J. Roentgenol.,* 64, 61, 1950.
5. **Levinson, B., Walter, B. A., Wintrobe, M. M., and Cartwright, G. E.,** A clinical study in Hodgkin's disease, *Arch. Intern. Med.,* 99, 519, 1957.
6. **Teillet, F., Boiron, M., and Bernard, J.** A reappraisal of clinical and biologic signs in staging of Hodgkin's disease, *Cancer Res.,* 31, 1723, 1971.
7. **Craver, L. F.,** Hodgkin's disease, in. *Tice-Harvey Practice of Medicine,* Vol. 6, Tice, F., Ed., Prior, W. F. Hagerstown, Md., 1964, 1017—1064.
8. **Ultmann, J. E., Cunningham, J. K., and Gellhorn, A.,** The clinical picture of Hodgkin's disease, *Cancer Res.* 26, 1047, 1966.
9. **Musshoff, K., Boutis, L., Laszlo, A. M., and Laszlo, I.,** Prognostische Krankheitssymptome und-zeichen bei Morbus Hodgkin und ihre Bedeutung für die Therapie der Erkrankung, *Strahlentherapie,* 131, 482, 1966.
10. **Le Bourgeois, J. P. and Tubiana, M.,** The erythrocyte sedimentation rate as a monitor for relapse in patients with previously treated Hodgkin's disease, *Int. J. Radiat. Oncol. Biol. Phys.,* 2, 241, 1977.
11. **Pagliardi, E. and Giangrandi, E.,** Clinical significance of the serum copper in Hodgkin's disease, *Acta Haematol.,* 24, 201, 1960.
12. **Hrgovcic, M., Tessmer, C. F., Minckler, T. M., Mosier, B., and Taylor, G.,** Serum copper levels in lymphoma and leukemia: special reference to Hodgkin's disease, *Cancer,* 21, 743, 1968.
13. **Hrgovcic, M., Tessmer, C. F., Thomas, F. B., Fuller, M. L., Gamble, J. F., and Shullenberger, C. C.,** Significance of serum copper levels in adult patients with Hodgkin's disease, *Cancer,* 31, 1337, 1973.
14. **Hrgovcic, M., Tessmer, C. F., Thomas, F. B., Ong, P. S., Gamble, J. F., and Shullenberger, C. C.,** Serum copper observations in patients with malignant lymphoma, *Cancer,* 32, 1512, 1973.
15. **Warren, L. R., Jelliffee, A. M., Watson, J. V., and Hobbs, C. B.,** Prolonged observations on variations in the serum copper in Hodgkin's disease, *Clin. Radiol.,* 20, 247, 1969.
16. **Illicin, G.,** Serum copper and magnesium levels in leukemia and malignant lymphoma, *Lancet,* 1, 1036, 1971.
17. **Mortazavi, S. H., Bani-Hashemi, A., Mozafari, M., and Raffi, A.,** Values of serum copper measurement in lymphomas and several other malignancies, *Cancer,* 29, 1193, 1972.
18. **Ray, G. R., Wolf, P. H., and Kaplan, H. S.,** Value of laboratory indicators in Hodgkin's disease: preliminary results, *Natl. Cancer Inst. Monogr.,* 36, 315, 1973.
19. **Cappelaere, P., Sulman, Ch., Chechan, Ch., and Gosselin-Delaquais, P.,** Les concentrations plasmatiques du cuivre et du zinc au cours de la maladie de Hodgkin, *Lillie Med.,* 20, 904, 1975.
20. **Thorling, E. B. and Thorling, K.,** The clinical usefulness of serum copper determinations in Hodgkin's disease, *Cancer,* 38, 255, 1976.
21. **Shah-Reddy, I., Khilanani, P., and Bishop, R. C.,** Serum copper levels in non-Hodgkin's lymphoma, *Cancer,* 45, 2156, 1980.

22. **Cohen, Y., Haim, N., and Zinder, O.,** Serum copper levels in non-Hodgkin's lymphoma (Abstr.), *Proc. Am. Assoc. Cancer Res.*, 22, 184, 1981.
23. **Davidioff, G. N., Votaw, M. L., Coon, W. W., Heicker, L. H., Richardson, R. J., Finkel, J. D., and Weitz, J.,** Elevations in serum copper and ceruloplasmin levels as an indicator of disease activity in lymphoma, *Clin. Pathol.*, 21, 359, 1968.
24. **Tura, S., Bernardi, L., Baccarani, M., Sanguinetti, F., and Branzi, A.,** La cupremia e la ceruloplasminemia nelle emolinfopatie. Nota 1. J. linfomi Maligni, *G. Clin. Med.*, 49, 1090, 1968.
25. **Pisi, E., Di Feliciantonio, R., Figus, E., and Ferri, S.,** Comportamento e significato prognostico della cerulopslamina sierica in relazione al quadro isto-patologico nella malattia di Hodgkin, *Minerva Med.*, 59, 944, 1968.
26. **Mosi, M., Vecchi, V., Vivarelli, F., and Paolucci, P.,** La ceruloplasminemia nella leucosi acuta linfoblastica e nel morbo di Hodgkin della infanzia, *Minerva Pediatr.*, 27, 1223, 1975.
27. **Sirsat, A. V.,** Serum ceruloplasmin levels in Hodgkin's disease and malignant lymphomas in correlation with serum copper levels, *Indian J. Cancer,* 16, 32, 1979.
28. **Foster, M., Pocklington, T., and Dawson, A. A.,** Ceruloplasmin and iron transferrin in human malignant disease, in *Metal ions in Biological Systems,* Vol. 10, Sigel, H., Ed., Marcel Dekker, New York, 1981, chap. 5.
29. **Westling, P.,** Studies on the prognosis in Hodgkin's disease, *Acta Radiol. (Stockholm),* Suppl. 245, 1, 1965.
30. **Jaffe, N. and Bishop, Y. M. M.,** The serum iron level, hematocrit, sedimentation rate, and leukocyte alkaline phosphatase level in pediatric patients with Hodgkin's disease, *Cancer,* 26, 332, 1970.
31. **Aisenberg, A. C.,** Lymphocytopenia in Hodgkin's disease, *Blood,* 25, 1037, 1965b.
32. **Brown, R. S., Haynes, H. A., Foley, H. T., Godwin, H. A., Berard, C. W., and Carbone, P. P.,** Hodgkin's disease. Immunologic, clinical and histologic features of 50 untreated patients, *Ann. Int. Med.*, 67, 291, 1967.
33. **Heilmeyer, L., Mossner, G., and Hunstein, W.,** Die Lymphogramulomatose, *Dtsch Med. Wochenschr.*, 82, 1046, 1957.
34. **Vogelgesang, K. H. and Tobben, A.,** Ein Beitrag Zur Prognose und Therapie der Lymphogranulomatose, *Sthrahlentherapie,* 101, 77, 1956.
35. **Auerbach, S.,** Zinc content of plasma, blood, and erythrocytes in normal subjects and in patients with Hodgkin's disease and various hematologic disorders, *J. Lab. Clin. Med.*, 65, 628, 1965.
36. **Bucher, W. C. and Jones, S. E.,** Serum copper-zinc ratio in patients with malignant lymphoma (Abstr.), *Am. J. Clin. Pathol.*, 68(1), 104, 1977.
37. **Giannopoulos, P. O. and Bergsagel, D. E.,** The mechanism of the anemia associated with Hodgkin's disease, *Blood,* 14, 856, 1959.
38. **Cline, M. J. and Berlin, N. I.,** Anemia in Hodgkin's disease, *Cancer,* 16, 526, 1963.
39. **Najean, Y., Dresch, C., and Ardaillou, N.,** Trouble de l' utilisation du fer hemoglobinique au cours des maladies de Hodgkin evolutives, *Nouv. Rev. Fr. Hematol.*, 7, 739, 1967.
40. **Gobbi, P. G., Scarpelini, M., Minoia, C., Pozzoli, L., and Perugini, S.,** Plasma iron and copper in Hodgkin's disease — a comparison with other laboratory indicators, *Hematologica,* 64, 416, 1979.
41. **Eilam, N., Johnson, P. K., Johnson, N. L., and Creger, W. P.,** Bradykininogen levels in Hodgkin's disease, *Cancer,* 22, 631, 1968.
42. **Lacher, M. J., Levy, A. B., and Pukite, A.,** The value of leukocyte alkaline phosphatase determinations in the malignant lymphomas, *Cancer,* 17, 402, 1964.
43. **Flury, R. and Wegmann, T.,** The behavior of leukoyctic alkaline phosphatase in Hodgkin's disease, *Schweiz. Med. Wochenschr.*, 94, 958, 1963.
44. **Lille-Szyszkowicz, I., Gabay, P., Saracino, R. T., and Bourdin, J. S.,** La phosphatase alcaline leucocytaire chez les malades atteints de tumeurs ou d'hemopathies malignes, *Nouv. Rev. Fr. Hematol.*, 6, 187, 1966.
45. **Bennett, J. M., Nathanson, L., and Rutenburg, A. M.,** Significance of leukocyte alkaline phosphatase in Hodgkin's disease, *Arch. Int. Med.*, 121, 338, 1968.
46. **Simmons, A. V., Spiers, A. S. D., and Fayers, P. M.,** Haematological and clinical parameters in assessing activity in Hodgkin's disease and other malignant lymphomas, *Q. J. Med.*, 42, 111, 1973.
47. **Levitan, R., Diamond, H. D., and Craver, L. F.,** Liver in Hodgkin's disease, *Gut.*, 2, 60, 1961.
48. **Aisenberg, A. C., Kaplan, M. M., Rieder, S. V., and Goldman, J. M.,** Serum alkaline phosphatase at the onset of Hodgkin's disease, *Cancer,* 26, 318, 1970.
49. **Rosenberg, S. A.,** Hodgkin's disease of the bone marrow, *Cancer Res.*, 31, 1733, 1971.
50. **Ferrante, W. A. and Maxfield, W. S.,** Comparison of the diagnostic accuracy of liver scans, liver function tests, and liver biopsies, *South. Med. J.*, 61, 1255, 1968.
51. **Glatstein, E., Guernsey, J. M., Rosenberger, A. S., and Kaplan, H. S.,** The value of laparotomy and splenectomy in the staging of Hodgkin's diseases, *Cancer,* 24, 709, 1969.

52. **Levine, P. H.,** Abnormal blood chemistry values in Hodgkin's disease. Lack of correlation with staging of disease, *J. Am. Med. Assoc.*, 220, 1734, 1972.
53. **Krauss, S., Schrott, M., and Sarcione, E. J.,** Haptoglobin metabolism in Hodgkin's disease, *Am. J. Med. Sci.*, 252, 184, 1966.
54. **Snyder, S. and Ashwell, G.,** Quantitation of specific serum glycoproteins in malignancy, *Clin. Chem. Acta*, 34, 449, 1971.
55. **Arends, T., Coonrad, E. V., and Rundles, R. W.,** Serum proteins in Hodgkin's disease and malignant lymphomas, *Am. J. Med.*, 16, 833, 1954.
56. **Neely, R. A. and Neill, D. W.,** Electrophoretic studies on the serum proteins in neoplastic disease involving the haemopoietic and reticuloendothelial systems, *Br. J. Haematol.*, 2, 32, 1956.
57. **Goulian, M. and Fahey, J. L.,** Abnormalities in serum proteins and protein-bound hexose in Hodgkin's disease, *J. Lab. Clin. Med.*, 57, 408, 1961.
58. **Malpas, J. S. and Fairley, G. H.,** Changes in serum alpha-2-globulins in reticuloses, *J. Clin. Pathol.*, 17, 651, 1964.
59. **Koj, A.,** Acute-phase reactants, in *The Structure and Function of Plasma Proteins*, Vol. 1, Allison, A. C., Ed., Plenum Press, New York, 1975, 73—132.
60. **LeRoy, E. C., Carbone, P. P., and Sjoerdsma, A.,** Elevated plasma levels of a hydroxyproline-containing protein in Hodgkin's disease and their relation to disease activity, *J. Lab. Clin. Med.*, 67, 891, 1966.
61. **Wood, H. F., Diamond, H. D., Craver, L. F., Pader, E., and Elster, S. K.,** Determination of C-reactive protein in the blood of patients with Hodgkin's disease, *Ann. Int. Med.*, 48, 823, 1958.
62. **Percori, V., Turrisi, E., Altucci, P., and Buonanno, G.,** Behavior of C-reactive protein in malignant blood diseases, with special regard to Hodgkin's lymphogranuloma, *Riforma Med.*, 73, 1, 1959.
63. **Spiers, A. S. D. and Malone, H. F.,** The significance of serum hexosamine levels in patients with cancer, *Br. J. Cancer*, 20, 485, 1966.
64. **Abdou, M. S., Salem, E., and el-Mahdy, H. M.,** Mucoprotein and protein bound hexose in reticulosis, *J. Egypt. Med. Assoc.*, 49, 619, 1966.
65. **Rachmilewitz, B. and Rachmilewitz, M.,** Serum transcobalamin-II levels in acute leukemia and lymphoma, *Isr. J. Med. Sci.*, 12, 583, 1976.
66. **Young, R. C., Corder, M. P., Haynes, H. A., and De Vita, V. T.,** Delayed hypersensitivity in Hodgkin's disease. A study of 103 untreated patients, *Am. J. Med.*, 52, 63, 1972.
67. **DeVita, V. T. and Hellman, S.,** Hodgkin's disease and the non–Hodgkin's lymphoma, in *Cancer — Principles and Practice of Oncology*, DeVita, V. T., Hellman, S., and Rosenberg, S. A., Eds., Lippincott, New York, 1982, 1357.
68. **McKenna, R. W., Blomfield, C. D., and Brunning, R. D.,** Nodular lymphoma: bone marrow and blood manifestations, *Cancer*, 36, 428, 1975.
69. **Alfrey, C. P., Jr., Lane, M., and Kariata, R. J.,** Modification of ferrokinetics in man by cancer chemotherapeutic agents, *Cancer*, 19, 428, 1966.
70. **Hrgovcic, M., Valkovic, V., Miljanic, D., Ong, P. S., and Phillips, G. C.,** Significance of Simultaneous Determinations of Serum Copper, Iron and Zinc in Non-Hodgkin's Lymphoma, unpublished data, 1974.
71. **Foster, M., Fell, L., Pocklington, T., Akinsete, F., Dawson, A., Hutchison, J. M. S., and Mallard, J. R.,** Electron spin resonance as a useful technique in the management of Hodgkin's disease, *Clin. Radiol.*, 28, 15, 1977a.
72. **Hughes, N. R.,** Serum transferrin and ceruloplasmin concentrations in patients with carcinoma, melanoma, sarcoma and cancers of hematopoetic tissues, *Aust. J. Exp. Biol. Med. Sci.*, 50, 97, 1972.
73. **Delves, H. T., Alexander, F. W., and Lay, H.,** Copper and zinc concentrations in the plasma of leukemic children, *Br. J. Haematol.*, 24, 525, 1973.
74. **Fisher, G. L., Byers, V. S., Shifrine, M., and Levin, A. S.,** Copper and zinc levels in serum from human patients with sarcomas, *Cancer*, 37, 356, 1976.
75. **Andrews, G. S.,** Studies of plasma zinc, copper, ceruloplasmin and growth hormone, *J. Clin. Pathol.*, 32, 325, 1979.
76. **Abdula, M., Björklund, A., Mathur, A., and Wallenius, K.,** Zinc and copper levels in whole blood and plasma from patients with squamous cell carcinoma of head and neck, *J. Surg. Oncol.*, 12, 107, 1979.
77. **Koch, H. J., Smith, E. R., and McNelly, J.,** Analysis of trace elements in human tissues. II. The lymphomatous diseases, *Cancer*, 10, 151, 1957.
78. **Hoffbrand, B. I.,** Hodgkin's disease, autoimmunity and thymus, *Br. Med. J.*, 1, 1592, 1965.
79. **Moore, D. F., Migliore, P. J., Shullenberger, C. C., and Alexanian, R.,** Monoclonal macroglobulinemia in malignant lymphoma, *Ann. Int. Med.*, 72, 43, 1970.
80. **Irunberry, J. and Colonna, P.,** Intérêt de l'immunoélectrophorèse des protéines seriques au cours de la maladie de Hodgkin, *Presse Med.*, 78, 187, 1970.
81. **Goodman, S. I., Rodgerson, D. O., and Kaufman, J.,** Hypercupremia in a patient with multiple myeloma, *J. Lab. Clin. Med.*, 70, 57, 1967.

82. **Lewis, R. A., Hultquist, D. E., Baker, B. L., Falls, H. F., Gershowitz, H., and Penner, J. A.,** Hypercupremia associated with a monoclonal immunogloublin, *J. Lab. Clin. Med.,* 88, 375, 1976.
83. **Yocum, R. C. and Doerner, A. A.,** A clinical evaluation of the C-reactive protein, *Arch. Int. Med.,* 99, 74, 1957.
84. **Gewurz, H., Mold, C., Siegel, J., and Fidel, B.,** C-reactive protein and the acute phase response, *Adv. Intern. Med.,* 27, 345, 1982.
85. **Rice, E. W.,** Correlation between serum copper, ceruloplasmin activity and C-reactive protein, *Clin. Chem. Acta,* 5, 632, 1960.
86. **Jensen, K. B., Thorling, E. B., and Andersen, C. J.** Serum copper in Hodgkin's disease, *Scand. J. Haematol.,* 1, 63, 1964.
87. **Asbjörnsen, G.,** Serum copper compared to erythrocyte sedimentation rate as indicator of disease activity in Hodgkin's disease, *Scand. J. Haematol.,* 22, 193, 1979.
88. **Pizzolo, G., Savarin, T., Molino, A. M., Ambrosetti, A., Todeschini, G., and Vettore, L.,** The diagnostic value of serum copper levels and other hematochemical parameters in malignancies, *Tumori,* 64, 55, 1978.
89. **Hrgovcic, M., Tessmer, C. F., Brown, B. W., Wilbur, J. R., Mumford, D. M., Thomas, F. B., Shullenberger, C. C., and Taylor, G.,** Serum copper studies in the lymphomas and acute leukemias, in *Progress in Clinical Cancer,* Ariel, I. M., Ed., Grune & Stratton, New York, 1973, 121—153.
90. **Hagberg, H., Killander, A., and Simonsson, B.,** Serum β_2-microglobulin in malignant lymphoma, *Cancer,* 51, 2220, 1983.
91. **Schuster, J., Gold, P., and Poulik, M. D.,** β_2-microglobulin levels in cancerous and other disease states, *Clin. Chem. Acta,* 67, 307, 1976.
92. **Amlot, P. L. and Adinolfi, M.,** Serum β_2-microglobulin and its prognostic value in lymphomas, *Eur. J. Cancer,* 15, 791, 1979.

Chapter 11

COPPER AND CERULOPLASMIN IN ANIMAL NEOPLASIA

I. INTRODUCTION

The relationship between serum copper and ceruloplasmin levels and neoplasia in experimental animals has been studied extensively in the past two decades. The alterations in copper and ceruloplasmin levels parallel each other and appear to reflect the same underlying pathophysiologic effect of tumor cell proliferation. Although it was suggested initially that the increase in ceruloplasmin level reflected a nonspecific acute phase reactant type of response, it is now felt that the change is a more specific and direct effect of tumor cell proliferation. This relationship has been studied in mice, rats, rabbits, dogs, and monkeys using spontaneous or induced tumors including leukemias, lymphomas, squamous cell carcinomas, carcinosarcomas, sarcomas, and hepatomas.

II. ANIMAL SYSTEMS STUDIED AT THE UNIVERSITY OF TEXAS M.D. ANDERSON HOSPITAL AND TUMOR INSTITUTE

In conjunction with the clinical studies of serum copper levels in lymphoreticular malignancies,[1-6] several animal models were evaluated at the University of Texas M.D. Anderson Hospital and Tumor Institute.[7-9] One of the animal systems evaluated initially was the SJL/J strain of mice.[7] Spontaneous development of a reticuloendothelial tumor closely resembling human Hodgkin's disease has been reported in this strain. No significant alteration in serum copper level was noted during periods of tumor growth. Studies of tumor tissue did not show significantly elevated copper levels. Subsequent studies reported later,[9] however, did show a direct relation between blood copper and the amount of tumor present in the liver. This relationship was not seen with tumor tissue present in other organs such as the spleen and the thymus. The correlation was noted to be higher in female mice. A study of *Herpes samiri* virus-induced reticulum cell sarcoma in marmosets also showed a rise in serum copper with full development of neoplastic changes.[7] Of all the systems evaluated, virus-induced murine leukemia and MSV (Moloney)rhabdomyosarcoma were found to be most suitable for study of alterations of serum copper levels associated with tumor proliferation.[8]

Rauscher leukemia virus (RLV) — The murine oncogenic virus RLV was injected into 100 Balb/c mice; 25 of the animals were selected at random and provided the data for splenomegaly and death rates. The remaining animals (in three groups) provided serum for copper values at 7, 28, and 51 days after injection. Figure 1 indicates the high serum copper values during and slightly preceding evidence of disease, continuing to death from leukemia.

MSV (Moloney) murine rhabdomyosarcoma — MSV-9, a Moloney strain of murine sarcoma virus induces a tumor (rhabdomyosarcoma) at the site of injection within a few days. In young mice the tumor may progress rapidly, resulting in death. In adult mice, however, the tumor usually regresses spontaneously and the animals survive. This regression is related to development of a specific immunologic response and can be reduced by exposure to X-ray or injection of cortisone acetate. A group of 100 female Yale mice were inoculated with MSV-9 and a similar group of control animals injected with diluent. Blood was collected at day 0, day 2 (infection without tumor), day 14 (peak of tumor incidence), day 21 (onset of regression), and day 35 (completion of regression). Figure 2 clearly shows a rise in the serum copper level before identification of tumor and a fall before regression. To make sure that the tumor state was responsible for increased serum copper, the experiment was repeated with a system that would not regress. This was accomplished by injecting 2.5 mg of cortisone

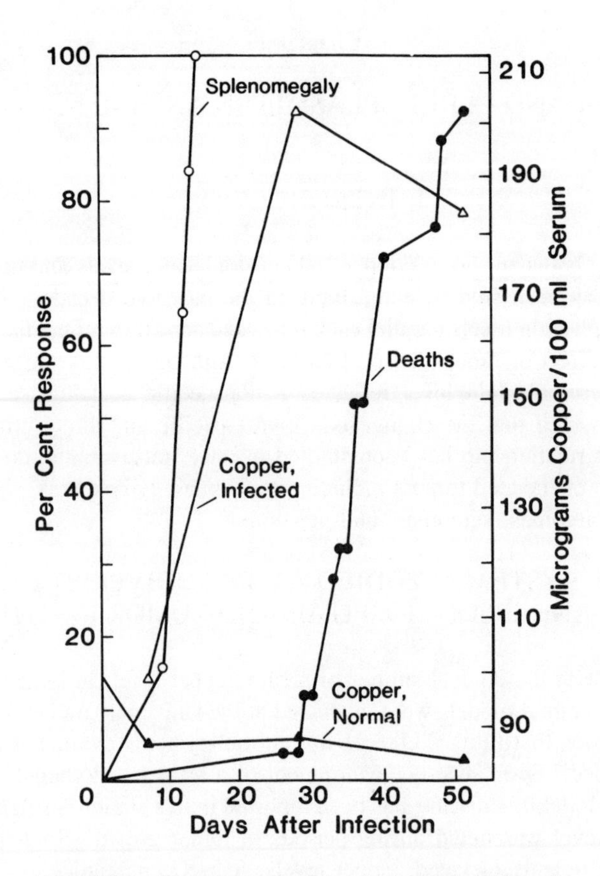

FIGURE 1. Serum copper levels in mice injected with Rauscher murine leukemia virus (RLV): correlation with splenomegaly and death.

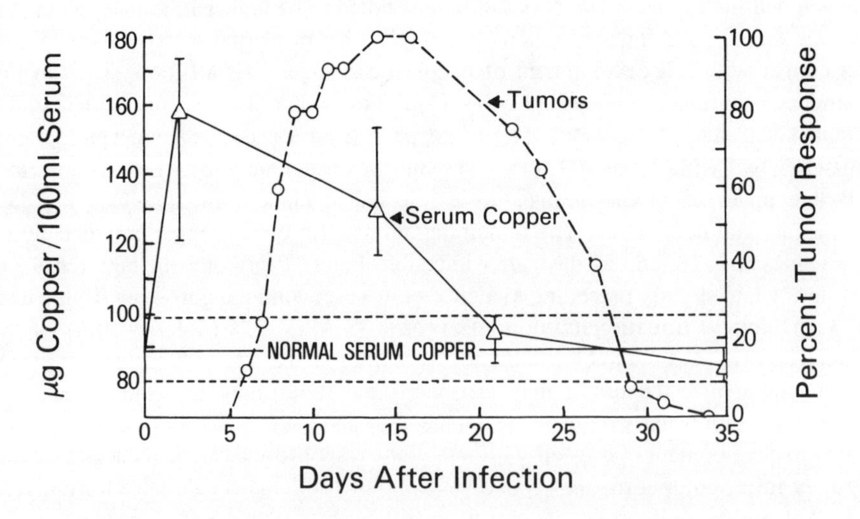

FIGURE 2. Serum copper levels in murine rhabdomyosarcoma (MSV-9): correlation with tumor development and spontaneous tumor regression. (From Hrgovcic, M., Tessmer, C. F., Brown, B. W., Wilbur, J. R., Mumford, D. M., Thomas, F. B., Shullenberger, C. C., and Taylor, G. H., in *Progress in Clinical Cancer,* Vol. 5, Ariel, I. M., Ed., Grune & Stratton, New York, 1973, 121—153. With permission.)

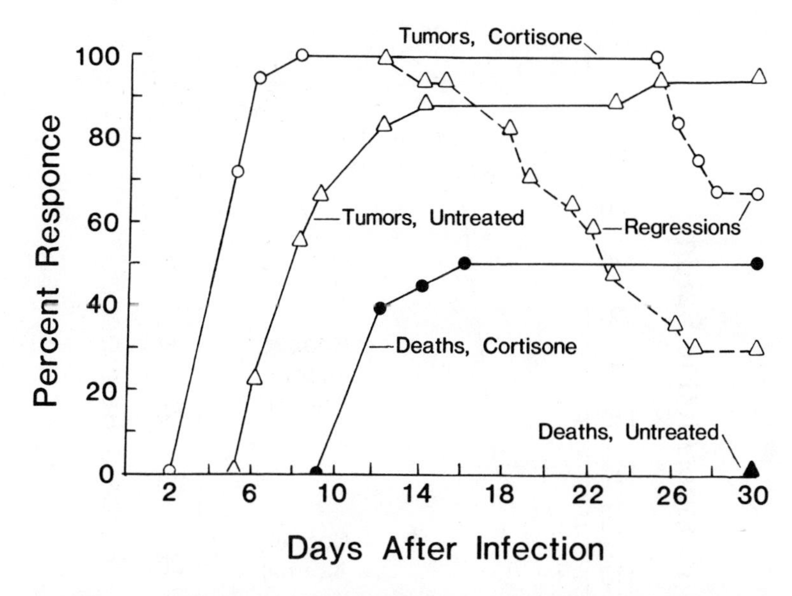

FIGURE 3. Murine rhabdomyosarcoma (MSV-9): effect of pretreatment with cortisone on tumor development and regression (see text for details).

acetate i.m. on day 1. Figure 3 demonstrates the effect of cortisone injection on tumor development after MSV-9 inoculation. The latency period was decreased, 100% of the mice developed tumor in 8 days compared to 94% in 30 days in the untreated group, and the number of regressions was also decreased. While 71 of all tumors had regressed by day 30 in the untreated group, only 22% of cortisone-treated animals showed tumor regression and 50% had died. Serum copper values in the group receiving cortisone at the time of inoculation are shown in Figure 4, demonstrating increasing levels peaking at time of death. Figure 4 also shows copper levels in cortisone-treated control animals demonstrating a rise related to the cortisone injection itself. This rise is clearly of a lesser magnitude and is transient, the levels returning to normal by day 13.

Both studies indicate that the rise in copper levels reflects tumor proliferation and, in the system with spontaneous regression, the copper levels also show a corresponding decline.

III. COPPER AND CERULOPLASMIN IN OTHER ANIMAL MODELS

Copper and/or ceruloplasmin levels have been studied in animal neoplasia using various experimental tumor models. The earliest studies were in mice with Ehrlich ascites tumor.[10,11] Hano and Akashi,[10] demonstrated that increased copper levels associated with tumor proliferation in mice were restored to normal with chemotherapy (8-azaguanine or 6-mercatopurine). Potopalski et al.[11] reported an increase in ceruloplasmin activity during active tumor growth in mice with Ehrlich ascites tumor and in rats with MTX sarcoma. At about the same time, Thomas and Constantinescu[12] reported on two studies on rats, one with Jensen carcinoma[12] and another with O-Ya ascitogen tumor.[13] Both sudies showed an increase in ceruloplasmin levels 2 days after tumor transplantation. This increase persisted for 15 days in the first study and for 6 days with the O-Ya tumor. They also demonstrated that this increase could be inhibited by administration of phosphate (HNa_2PO_4) to the rats 24 hr prior to obtaining a serum sample for ceruloplasmin estimation. The first American study of the relationship between ceruloplasmin and animal tumors was reported by Enneking et al.[14,15] While investigating the metabolic effects of intraosseous implantation of the VX-2 carcinoma in rabbits, they noted that the sera of the tumor-bearing animals were blue. This was determined (Sunderman et al.[16]) to be due to hyperceruloplasminemia. Further studies showed

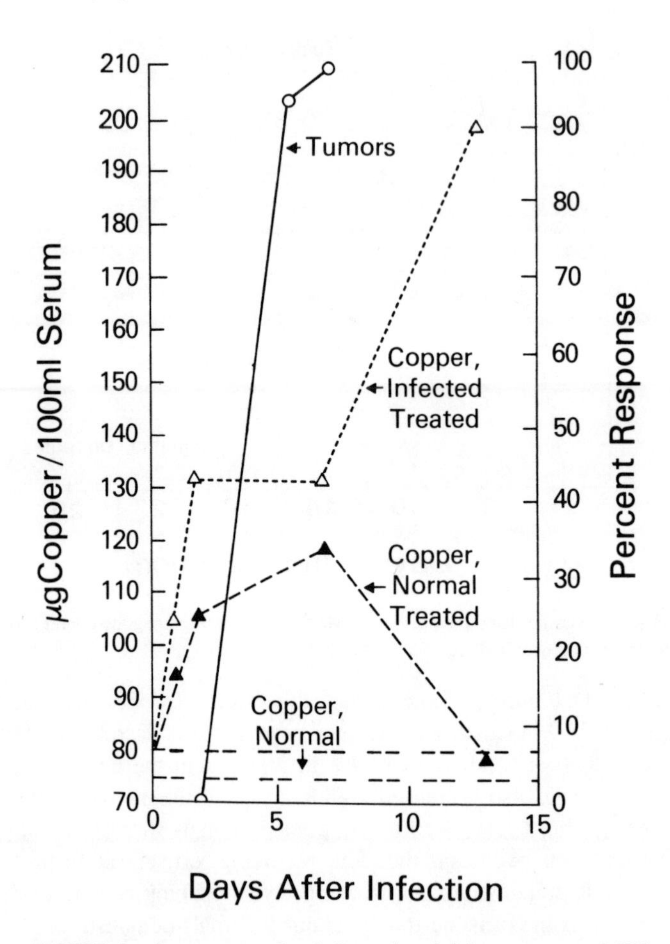

FIGURE 4. Serum copper levels in murine rhabdomyosarcoma (MSV-9) in animals pretreated with cortisone. (From Hrgovcic, M., Tessmer, C. F., Brown, B. W., Wilbur, J. R., Mumford, D. M., Thomas, F. B., Shullenberger, C. C., and Taylor, G. H., in Progress in Clinical Cancer, Vol. 5, Ariel, I. M., Ed., Grune & Stratton, New York, 1973, 121—153. With permission.)

progressive increase in ceruloplasmin commencing at 2 weeks after implantation and reaching a sixfold increase at 6 weeks. Studies with $^{64}Cu(II)$[15] showed greater uptake in the liver than in the tumor tissue, implying that the tumor might be stimulating the liver to increase ceruloplasmin synthesis.

VX-2 and VX-7 carcinomas in rabbits were also studied by Satoh et al.,[17] Ungar-Waron et al.,[18] and Voelkel et al.[19] These studies showed elevation of ceruloplasmin levels to between 4 and 20 times normal. In the study reported by Ungar-Waron et al.[18] VX-2 carcinoma was implanted in the gastrocnemius of rabbits, resulting in tumor growth which could be made to regress by a second tumor cell inoculation. Ceruloplasmin levels increased four- to eightfold normal during progression, often before tumors could be palpated. With tumor regression, apparently on an immunologic basis, ceruloplasmin levels came back to normal. The levels remained high when metastases developed. The other tumors studied in rabbits were Brown-Pearce carcinoma[20] and Shope papilloma.[21] The latter study is interesting since it implies association of ceruloplasmin alteration with malignancy. Papillomas were induced in neonatal domestic rabbits by intradermal injection of Shope papilloma virus and they converted into squamous cell carcinomas after a lapse of a year or more. Disc electrophoretic analysis of the serum showed one band of ceruloplasmin activity before, and two bands after, malignant conversion of the benign tumors. Ceruloplasmin level was also elevated, dropping to normal 1 month after total tumor resection. Malignant conversion was also associated with raised levels of serum copper, the increase being two to six times the level at 4 months of age.

Dogs with spontaneous osteosarcoma and beagles with radiation-induced (^{90}Sr and ^{226}Ra) osteosarcoma were studied by Fisher and Shifrine.[22] Serum copper and zinc levels were elevated with tumor growth in both groups, decreasing to near normal levels after amputation. Benign bone lesions were seen on biopsy in dogs clinically suspected to have spontaneous osteosarcoma but with normal or subnormal copper levels. They also examined serum copper levels in radiation-induced (^{90}Sr and ^{60}Co) hematopoietic neoplasia in beagles. One dog exposed to ^{60}Co developed blastic leukemia with infiltration of organs. Elevated copper levels similar to those seen in human leukemia were reported. Another dog exposed to ^{90}Sr died of chronic granulocytic leukemia. Serum copper was elevated from the time of diagnosis to death. One ^{90}Sr-exposed dog with lymphosarcoma also had markedly elevated copper levels.

Most of the studies of copper and ceruloplasmin in animal neoplasia have been on induced and transplantable tumors in rats. These include studies on Yoshida sarcoma,[23] Walker carcinosarcomas,[23-25] sarcomas,[16,26] Svec's erythroleukemia,[27] Morris hepatomas,[23,28,29] and DMBA mammary tumors.[28,29] They showed increased levels of copper and ceruloplasmin of between 1.5 and 2.9 times the normal level. The study reported by Sunderman et al.[16] is of interest. They measured ceruloplasmin concentration in 30 control Fischer rats, in 5 rats with primary sarcoma induced by intramuscular injection of nickel subsulfide, and in 12 rats at intervals of up to 6 weeks after subcutaneous transplantation of four nickel subsulfide-induced sarcomas. The primary induced sarcomas were slow growing and did not cause significant increase in serum ceruloplasmin levels. The rats with the transplanted sarcomas, however, showed a 1.6 $\pm$ 2-fold increase in ceruloplasmin level. This finding was attributed to the greatly enhanced growth rates of the transplanted neoplasms. This report[16] includes an excellent review of tumor-induced hyperceruloplasminemia in animals. Linder et al.[29] examined transplantable tumors of mammary, hepatic, and kidney origin in three different strains of rats. Ceruloplasmin oxidase activity in the plasma of most tumor-bearing rats was significantly elevated over that in controls, by 25 to 200%. Elevations were not seen in animals with slower growing tumors. The degree of elevation showed some correlation with tumor size, especially in the case of larger tumors. Plasma copper levels showed a similar increase in the initial phases of tumor growth. The ceruloplasmin activity continued to increase as the tumors grew much larger, whereas the copper levels did not. The incorporation of ^{3}H-leucine into all plasma protein classes was depressed in the tumor-bearing animls, but the incorporation into ceruloplasmin was increased appreciably. In the study reported by Owen,[25] subcutaneous injection of Walker 256 carcinosarcoma cells into rats was followed by a pronounced increase in plasma ceruloplasmin activity and serum copper levels starting within 3 days and peaking at 1 week after injection. The increased levels started falling subsequently and fell to normal at 3 weeks despite continued tumor growth. Studies with ^{67}Cu seemed to indicate increased production of ceruloplasmin by the liver. Injection of turpentine also raised plasma ceruloplasmin, but to a significantly lesser extent, and normal values were reached within 2 weeks. The activity of serum ceruloplasmin oxidase was studied by Abreu and Abreu[26] during growth of sarcoma II in rats. The activity was increased in rats with tumor and remained high during all phases of tumor evolution. Administration of cyclophosphamide to the sarcoma-bearing rats produced an initial decrease in ceruloplasmin activity with subsequent return to abnormally high levels.

It is clear from these studies that the elevation of levels of copper and ceruloplasmin is related to rapidly growing malignant tumor proliferation demonstrable in numerous animal tumor models. This has been shown in spontaneous tumors and in tumors induced by oncogenic viruses, chemical carcinogens, and radioisotopes and in transplanted tumors. The elevations parallel growth rates of tumors and spontaneous or induced regressions result in the fall of the levels to normal. Excision of tumor results in a fall in the increased levels unless metastatic disease is present. These observations indicate that serum copper and ceruloplasmin levels can be used as markers for tumor progression or response to therapy in animals with rapidly growing tumors. The mechanism responsible for these alterations in copper and ceruloplasmin is not clear. Several studies[15,25,29] have shown that ceruloplasmin synthesis is increased in livers of tumor-bearing animals. Cohen et al.[28] demonstrated that, in comparison with controls, rats with transplantable mammary tumors showed a threefold increase in absorption of copper from the gut following intragastric intubation with $^{64}Cu(NO_3)_2$. The distribution of copper in the body was also altered in tumor-bearing rats. It is possible that the elevation in serum ceruloplasmin may be at least partly related to production of an isoenzyme, as suggested by studies of Seto et al.[21] The increased synthesis of ceruloplasmin in the liver may be mediated by prostaglandin E_2 or its metabolites, such as PGE-M_2. This was suggested by studies of Wolfe et al.[30] and Voelkel et al.[19]

The studies cited above have clearly demonstrated the relationship between copper and ceruloplasmin levels and tumor proliferation in various animal models. Further studies may elucidate the pathophysiologic mechanism involved in the alterations that have been observed.

REFERENCES

1. **Hrgovcic, M., Tessmer, C. F., Minckler, T. M., Mosier, B., and Taylor, G. H.,** Serum copper levels in lymphoma and leukemia, *Cancer,* 21, 743—755, 1968.
2. **Hrgovcic, M., Tessmer, C. F., Thomas, F. B., Ong, P. S., Gamble, J. F., and Shullenberger, C. C.,** Serum copper observations in patients with malignant lymphoma, *Cancer,* 32, 1512—1524, 1973.
3. **Hrgovcic, M., Tessmer, C. F., Thomas, F. B., Fuller, L. M., Gamble, J. F., and Shullenberger, C. C.,** Significance of serum copper levels in adult patients with Hodgkins' disease, *Cancer,* 31, 1337—1345, 1973.
4. **Tessmer, C. F., Hrgovcic, M., Thomas, F. B., Wilbur, J. R., and Mumford, D. M.,** Long-term serum copper studies in acute leukemia in children, *Cancer,* 30, 358—365, 1972.
5. **Tessmer, C. F., Hrgovcic, M., and Wilbur, J. R.,** Serum copper in Hodgkin's disease in children, *Cancer,* 31, 303—315, 1973.
6. **Tessmer, C. F., Hrgovcic, M., Thomas, F. B., Fuller, L. M., and Castro, J. R.,** Serum copper as an index of tumor response to radiotherapy, *Radiology,* 106, 635—639, 1973.
7. **Hrgovcic, M., Tessmer, C. F., Brown, B. W., Wilbur, J. R., Mumford, D. M., Thomas, F. B., Shullenberger, C. C., and Taylor, G. H.,** Serum copper studies in the lymphomas and acute leukemias, in *Progress in Clinical Cancer,* Vol. 5, Ariel, I. M., Ed., Grune & Stratton, New York, 1973, 121—153.
8. **Pienta, R. J., Tessmer, C. F., and Thomas, F. B.,** Effect of murine oncogenic viruses on serum copper levels (abstr.), *Proc. Am. Assoc. Cancer Res.,* 10, 69, March 1969.
9. **Tessmer, C. F. and Lichtiger, B.,** Blood copper in SJL/J mice in relation to spontaneous tumor development (abstr.), *Am. J. Pathol.,* 59, 919, 1970.
10. **Hano, K. and Akashi, A.,** Influences of anticancer agents on the metabolism of δ-amino levulinic acid in normal and tumor bearing mice, *Gann.,* 55, 25—40, 1964.
11. **Potopalski, A. I., Neiko, E. M. W., Busdartschuk, W. K., and Fedortschuk, A. M.,** Changes in activity of ceruloplasmin during tumor growth, *Vopr. Eksp. Onkol.,* 1, 169-172, 1965.
12. **Thomas, E. and Constantinescu, R.,** Ceruloplasmin modification by HNa_2PO_4 administration to rats bearing Jensen carcinoma, *Clin. Chim. Acta,* 13, 708—710, 1966.
13. **Thomas, E., Olinescu, R., and Constantinescu, R.,** Ceruloplasmin modifications by HNa_2PO_4 administration to O-Ya ascitogen tumor-bearing rats, *Clin. Chim. Acta,* 13, 711—712, 1966.
14. **Enneking, W. F., Flynn, L., and Vogel, S.,** VX-2 carcinoma in bone, *J. Bone Jt. Surg.,* 49A, 795, 1967.
15. **Enneking, W. F. and Flynn, L.,** Effects of VX-2 carcinoma implanted in bone in rabbits, *Cancer Res.,* 28, 1007—1013, 1968.

16. **Sunderman, F. W., Jr., Trudeau, E. A., Jr., Horak, E., Mitchell, J. M., and Allpass, P. R.,** Serum ceruloplasmin concentrations in rats with primary and transplanted sarcomas induced by nickel subsulfide, *Ann. Clin. Lab. Sci.*, 9, 60—67, 1979.
17. **Satoh, T., Nakagawa, N., Sato, A., Ishikura, M., Sakurai, S., and Saito, T.,** Increase in copper containing protein in serums from VX-7 and VX-2 carcinoma bearig rabbits, *Igaku to Seitbutsugaku,* 93, 29—33, 1976.
18. **Ungar-Waron, H., Gluckman, A., Spira, E., Waron, M., and Trainin, Z.,** Ceruloplasmin as a marker of neoplastic activity in rabbits bearing the VX-2 carcinoma, *Cancer Res.*, 38, 1296—1299, 1978.
19. **Voelkel, E. F., Levine, L., Alper, C. A., and Tashjian, A. H., Jr.,** Acute phase reactants, ceruloplasmin and haptoglobin, and their relationship to plasma prostaglandins in rabbits bearing the VX-2 carinoma, *J. Exp. Med.*, 147, 1078—1088, 1978.
20. **Neporadny, D. D. and Goinatskii, M. N.,** Change in ceruloplasmin activity in blood serum of rabbits during development of Brown-Pearce carcinoma, *Mikroelem. Med.*, 3, 190—193, 1972.
21. **Seto, A., Tokuda, H., and Ito, Y.,** Malignant conversion of Shope papillomas and associated changes of serum ceruloplasmin in domestic rabbits, *Proc. Soc. Exp. Biol. Med.*, 157, 694—696, 1978.
22. **Fisher, G. L. and Shifrine, M.,** Serum copper and serum zinc levels in dogs and humans with neoplasia, in *Biologic Implications of Metals in the Environment (ERDA Symp. Ser.),* Vol. 42, U.S. Department of Energy Research and Development Administration, Washington, D.C., 1977, 507—522.
23. **Wesch, H., Zimmerer, J., Wayss, K., and Volm, M.,** Einfluss des Tumorwachstums auf den Plasmak-upferund Caeruloplasmin-spiegel der Ratte, *Fortschr. Geb. Roentgenstr. Nuklearmed.*, Suppl., 182, 1972.
24. **Volm, M., Wayss, K., Wesch, H., and Zimmerer, J.,** Correlation between copper and ceruloplasmin concentrations in plasma and ^{3}H-thymidine incorporation into Walker carcinosarcoma in rats, *Arch. Geschwulstforsch.*, 40, 248—258, 1972.
25. **Owen, C. A., Jr.,** The response of serum ceruloplasmin to injections of Walker 256 tumor cells or turpentine into rats, *Biol. Trace Element. Res.*, 3, 217—224, 1981.
26. **Abreu, L. A. and Abreu, R. R.,** Effect of cyclophosphamide on serum ceruloplasmin oxidase activity in sarcoma bearing rats, *Arch. Geschwulstforsch.*, 51(5), 394—397, 1981.
27. **Mazepa, I. V.,** Copper metabolism in experimental leukemia, *Mikroelem. Med.*, 3, 129—132, 1972.
28. **Cohen, D. I., Illowski, B., and Linder, M. C.,** Uptake and distribution of copper in rats with transplantable tumors, *Proc. Am. Assoc. Cancer Res.*, 17, 200, 1976.
29. **Linder, M. C., Bryant, R. R., Lim, S., Scott, L. E., and Moore, J. E.,** Ceruloplasmin elevation and synthesis in rats with transplantable tumors, *Enzyme*, 24(2), 85—95, 1979.
30. **Wolfe, H. J., Bitman, W. R., Voelkel, E. F., Griffiths, H. J., and Tashjian, A. H., Jr.,** Systemic effects of the VX-2 carcinoma on the osseous skeleton, *Lab. Invest.*, 38, 208—215, 1978.

Chapter 12

DISCUSSION

Copper metabolism in patients with malignant diseases has been the subject of many investigations. A literature survey revealed over 150 reports, most of them summarized in Chapter 6. All references known to the authors concerning serum copper or ceruloplasmin variations in neoplastic diseases are included in this review, regardless of positive or negative correlations. The majority of investigators have studied serum copper, although some have examined ceruloplasmin concentrations in the course of malignant diseases.

In 1962, it was reported that increases in serum copper were associated with increased ceruloplasmin concentrations.[1] It appears that alterations in serum copper levels in patients with malignant diseases, in general, are reflections of changes in ceruloplasmin concentrations.[2-10] In two studies ceruloplasmin concentrations correlated better with tumor activity in carcinomas of the lung than did serum copper levels.[5,11]

There exist a number of general references to the elevation of serum copper in lymphomas, leukemias, and solid tumors. In some reports, single types of tumors have been studied; in others, several types of malignant diseases have been investigated and reported upon.

On the average, increased serum copper levels from 50 to 250% over normal have been observed. The highest serum copper levels of 3350 and 1250 μg/100 mℓ were found in two cases of multiple myeloma.[12,13] Hypocupremia associated with malignant diseases has not been reported. Most of the studies in humans (58 reports) have been done on serum copper levels or ceruloplasmin in adult patients with Hodgkin's disease, considerably less (about a dozen) in non-Hodgkin's lymphoma, followed by studies in acute leukemia (8 reports) and 58 studies concerned with serum copper and/or ceruloplasmin concentrations in various types of solid tumors have been reported. Tissue copper concentrations in patients with malignant diseases have been investigated much less (about a dozen reports), and experimental animal-related studies are noted in approximately 30 reports.

Based on a literature review including clinical data (Chapter 6) and experimental data (Chapter 11), and the authors' own reports (Chapter 9), the serum copper and ceruloplasmin levels have been shown to correlate reasonably well with activity and extent ("volume") of malignant disease. This is particularly true in patients with malignant lymphoreticular and myeloproliferative diseases. A sampling of a much larger segment of investigators' data concerning serum copper levels in Hodgkin's disease, according to activity and extent of disease, is summarized in Table 1. High serum copper levels are found in active disease, and normal levels in remission. It is also obvious that patients with advanced stages of Hodgkin's disease have higher serum copper levels as compared to those with early stages of disease. Increased serum copper levels in active phases of disease, decreasing with reduction of disease activity as a response to therapy, and normal levels in the remission phase, have been reported by all investigators (with two exceptions, both in studies in children)[18,19] who studied serum copper levels throughout the course of disease.

In patients responding to therapy which results in complete remission, the serum copper level decreases more or less rapidly towards normal values, whereas in those who respond poorly and complete remission is not achieved, the fall is slight or absent. It has been also observed that the decrease in serum copper level is often the first sign of response to treatment, and that an increase in serum copper is one of the earliest indicators of impending relapse.[2-4,19,20-22,24-30] Most studies have been made on adults, but similar findings have been reported in children.[20-22]

Literature concerning serum copper studies in children is scanty.[18-22] Perhaps this is because of an age-related serum copper pattern and difficulties in interpretations which require use

Table 1
SERUM COPPER LEVELS IN PATIENTS WITH HODGKIN'S DISEASE ACCORDING TO ACTIVITY AND EXTENT OF DISEASE

Authors	Type of lymphoma	No. of patients	Mean SCL (μg/100 mℓ) active/remission	Mean SCL (μg/100 mℓ) no. of controls	Ref.
Hrgovcic et al.	Hodgkin's disease	191	165/123	101/117	14
	Stage I	11	133		
	Stage II	20	164		
	Stage III	39	167		
	Stage IV	10	190		
Cappelaere et al.	Hodgkin's disease	92	174/137	109/103	15
	Stage I and II	13	167		
	Stage III and IV	14	196		
Pizzolo et al.	Hodgkin's disease	74	187/131		16
	Stage I and II		154		
	Stage III and IV		201		
Gobbi et al.	Hodgkin's disease	55	171/113	97/60	17
Tura et al.	Hodgkin's disease	25	218/169 (partial and complete)		8

of an age correction factor.[23] Only five reports concerning serum copper in Hodgkin's disease in children have been noted.[18-22] Three studies have demonstrated good correlation between serum copper/or ceruloplasmin levels and disease activity,[20-22] whereas in two other reports[18,19] elevated serum copper levels were found in some patients in remission. In a first conflicting report,[18] only seven of ten patients in relapse had elevated serum copper levels, while nine of ten patients at relapse had increased serum copper levels over the preceding value. Of patients in remission 19% also had increased serum copper levels over the preceding values. It is uncertain what "degree of remission" these patients had achieved or whether they were about to have relapse. If a patient had active disease and was entering into remission phase, it may require some time for the serum copper levels to return to normal. Since a number of patients in this study group consisted of adolescent girls, it is possible that oral contraceptives constituted a factor. Furthermore, it is possible that some of inconsistent serum copper patterns in remission were caused by occult nonspecific infections frequently seen in children. Another conflicting report concerning serum copper in children was found in abstract form only, and due to its lack of detailed information it is not possible to comment on this particular study.[19] No conflicting report in the literature concerning the usefulness of serum copper levels in monitoring disease activity in adult patients with Hodgkin's disease has been noted.

In addition to reflection of disease activity in serum copper levels, data from the literature, including the authors', point to a relationship between serum copper levels and extent and/or volume of disease, being higher in more advanced stages as compared to early stages.[3,4,14-17] Also, all reports seem to agree that patients with systemic or constitutional symptoms have higher serum copper levels than those without.[3,4,14-17]

With regard to histologic types of Hodgkin's disease, it appears that serum copper levels associated with the lymphocytic predominance pattern are lower than those seen with the mixed cellularity and lymphocytic depletion types.[3,4,14,24] In one report, the lymphocytic predominance combined with mixed cellularity types showed higher serum copper levels than the nodular sclerosing and lymphocytic depletion types combined.[17] The lumping together of data and differences arising from criteria of tissue classification in Hodgkin's disease may contribute to a lack of correlation. In the authors' studies,[14,20] serum copper showed a significant relation to disease activity and therapy in the nodular sclerosing, mixed cellularity, and lymphocytic depletion types.

The ability to monitor lymphomatous disease activity is of great importance to clinicians. It is also of importance that a given test will show significant and consistent differences when used to follow an individual patient's course. Most reports have indicated that the changes in serum copper levels in Hodgkin's disease can meet these requirements.[2-4,7-10,14-22,24-28] Also of importance is the timing of the change. The serum copper level appears to increase or decrease before other clinical changes appear in Hodgkin's disease patients about to enter a period of increased or decreased disease activity.[2,3,9,14,15,20,26-28]

In the majority of reports, including the authors', inconsistent serum copper patterns (normal in active phase and elevated in remission) have been observed in approximately 15 to 20% of cases. While not proposing that all inconsistencies could be explained, there are considerations which suggest that each case be evaluated by general criteria and by individual characteristics. Normal serum copper levels in active localized disease seemingly inconsistent could be explained by a low volume of diseased tissue. Within the range of two standard deviations of ''normality'' commonly used as a laboratory guide, a Hodgkin's disease patient's serum copper changing from upper portion of normal range to a low normal, or vice versa, may be equally revealing the change of disease activity, as we observed in a few patients with localized disease. As mentioned previously, a relative increase or decrease in serum copper during a patient's clinical course is more important in the evaluation of disease activity than the absolute quantity of this trace element. The predictive or anticipatory role of serum copper levels preceding the verified clinical relapse by a very significant period of time or superposition of nononcological events (infections, hormonal imbalances) may be responsible for elevated serum copper levels in remission. Speculations concerning the observations of normal serum copper levels in relapse, particularly in the preterminal phases of disease, include many systems known to fail, especially liver function, protein metabolism, and breakdown of already comprised immune mechanisms. From these observations, it is obvious how important it is to know the clinical details of each patient under study in order to properly interpret serum copper level variations associated with the course of malignant disease or nononcological event.

Several comparative studies have been done comparing serum copper and/or ceruloplasmin levels, serum copper and iron/transferring levels, Cu:Zn ratio, with erythrocyte sedimentation rate, leukocyte alkaline phosphatase, alpha-2-globulin, C-reactive protein, and fibrinogen.[2,4,7-10,15-17,27,29-32] The best correlation was found between serum copper levels and the erythrocyte sedimentation rate and it has been suggested that these two tests are complementary and both can have a role in the management of Hodgkin's disease.[4,27,31] The reliability, sensitivity, advantages, and pitfalls of each test have already been discussed.

All groups of investigators (with the exception of two reports on serum copper levels in children already discussed), particularly those carrying out large scale studies,[2,4,7-10,14-17,20-35] have demonstrated a clinically useful relationship between serum copper levels and the activity of Hodgkin's disease. Since the return to normal copper levels with complete remission is a constant feature of this disease, it has been suggested that a normal serum copper level be included among the criteria for complete remission.[3]

The usefulness of serum copper levels in following the clinical course of malignant lymphoreticular diseases has recently received fairly wide acceptance. Several current texts comment on this fact.[36-38] ''Elevated serum copper levels have been demonstrated in active Hodgkin's disease, and this parameter has been found useful in monitoring disease activity and in detecting recurrence before it is demonstrable by other means.''[36] It can be recommended, therefore, that determinations of serum copper should be a part of patient management in Hodgkin's disease. Whether or not this test should be supported either by measurement of the erythrocyte sedimentation rate, or of zinc or iron (transferrin) would be dependent upon available resources, clinicians' experience, and familiarity with the tests.

Copper metabolism has been less well studied in patients with non-Hodgkin's lym-

Table 2
SERUM COPPER LEVELS IN PATIENTS WITH NON-HODGKIN'S LYMPHOMA ACCORDING TO ACTIVITY AND STAGE OF DISEASE

Authors	Type of lymphoma	No. of patients	Mean SCL (μg/100 mℓ) active/remission	Mean SCL (μg/100 mℓ) no. of controls	Ref.
Hrgovcic et al.	Non-Hodgkin's	236	158/117	101/117	45
	Stage I and II	17	133		
	Stage III and IV	71	165		
Shah-Reddy et al.	Non-Hodgkin's	34	191/115	113/16	43
Pizzolo et al.	Non-Hodgkin's	36	174/139		16
Ravat and Vijayvargiya	Malignant lymphoma				33
	Stage I and II	10	330/161	119/10	
	Stage III and IV	40	370/160		
Ilicin	Malignant lymphoma	15	298/168	102/	39
Sirsat	Hodgkin's and non-Hodgkin's	54	272/?	105/	10
Mortazari et al.	Hodgkin's and non-Hodgkin's	42	170/	100/	28
Davidoff et al.	Hodgkin's and non-Hodgkin's	22	197/143	99/40	7

phoma.[7,8,10,16,28,30,33,39-46] Perhaps a reason for this is the heterogeneous nature of these malignancies, which makes analysis of results more difficult. Also, the non-Hodgkin's lymphomas are frequently found in older people who tend to have more intercurrent diseases which mask any clear relationship of serum copper levels to the neoplastic process. In some studies all data pertaining to lymphoma have been mixed together, even to the extent of not separating Hodgkin's and non-Hodgkin's lymphoma. The changes in serum copper levels in non-Hodgkin's lymphoma appear somewhat less striking and, perhaps, less consistent than in Hodgkin's disease. It has been reported by the authors, as well as by other study groups, that non-Hodgkin's lymphoma disease activity is indeed reflected in serum copper levels.[7,8,10,16,28,30,33,39-46] In patients who responded well to therapy, elevated serum copper levels returned to the normal range preceding, or coinciding with, clinical signs of regression of disease. Nonresponders had a persistently increased serum copper level. Relapse was associated with an elevation of serum copper concentration, which generally preceded clinical signs of disease recurrence. A relationship between serum copper level and extent of disease was also found in non-Hodgkin's lymphoma patients, being higher in generalized disease as contrasted to localized disease. Again, patients with Hodgkin's disease showed serum copper levels consistently higher than patients with non-Hodgkin's lymphoma.[45] A sampling of several authors' findings is summarized in Table 2, elevated serum copper levels occurring in active phases of disease and decreased or normal levels in patients achieving partial or complete remission. This relationship of serum copper levels to disease activity is noted in all histologic groups and types of malignant lymphoma. In two reports[44,46] no difference in serum copper levels was found as between localized or generalized disease, or nodular vs. diffuse types of lymphoma, although there was a correlation related to disease activity. In two other studies serum copper levels were slightly higher in undifferentiated as compared to other non-Hodgkin's lymphomas.[40-45] Further studies are needed to verify the predictive role of copper in recurrent disease. At this point, in the absence of more specific and sensitive tests, serum copper levels, when related to the total clinical picture, appear to represent a reasonably sensitive biochemical index for evaluating disease activity and efficacy of therapy in adult patients with non-Hodgkin's lymphoma.

Table 3
SERUM COPPER LEVELS IN PATIENTS WITH ACUTE LEUKEMIA ACCORDING TO ACTIVITY OF DISEASE

Authors	Type of leukemia	No. of patients	Mean SCL (μg/100 mℓ) active/remission	Mean SCL (μg/100 mℓ) no. of controls	Ref.
Legutko	Acute lymphocytic (children)	57	261/129		50
Delves et al.	Acute lymphocytic (children)	21	256/106	123/13	47
Hrgovcic et al.	Acute lymphocytic (children)	20	130/110	Age related	51
Pizzolo et al.	Acute leukemia (adults)	38	180/116		16
Cherry et al.	Acute leukemia (adults)	8	273/decreased		47
Rechenberger	Acute leukemia (adults)	28	204		55

Copper metabolism is also altered in patients with myeloproliferative disorders, particularly in acute leukemia.[16,47-54] Increased serum copper levels are found in active disease and normal levels in remission. A sampling of literature data is shown in Table 3. Regression of disease activity in response to therapy is preceded by normalization of the serum copper level, before bone marrow remission is confirmed.[51,52] Early workers found a direct correlation between increased serum copper levels and leukocytosis.[47] Subsequent studies measuring serum copper and zinc suggested that the best correlation was with an index based on the total leukemic cell mass (using white cell count and liver and spleen enlargement) rather than with leukocyte count alone. They also confirmed the authors' findings that serum copper showed a rise with increasing percentage of blast cells in the bone marrow.[48]

In our pilot[51] as well as subsequent studies,[52,53] the correlation between serum copper levels and the percentage of blast cells in the bone marrow in untreated acute childhood leukemia was striking. Based on these observations, it was considered that the serum copper level might serve as a partial substitute for unpleasant bone marrow aspiration procedures needed in the management of leukemic disease. However, in long-term serum copper studies in childhood acute leukemia, positive (100%) correlation between serum copper levels and bone marrow blast cells in the initial (30 days) treatment period fell to 57% over extended periods and was generally lost in terminal phase of disease.[53] Therefore, there is little clinical value in serum copper measurements for monitoring leukemic disease activity except in evaluation of initial treatment response and as a possible index of extramedullary leukemic disease, as has also been suggested by others.[50]

Some elevation of serum copper or ceruloplasmin levels have also been reported in chronic leukemia,[16,55-57] and activity of chronic leukemic disease has also been reported to be reflected in serum copper levels.[55-57]

With the exception of two patients with multiple myeloma[12,13] associated with hypercupremia, in which a cause-effect relationship was not clear, it is of interest to note that serum copper levels, in general, are not significantly altered in this type of malignant disease. In eight of nine untreated multiple myeloma patients studied the serum copper levels were normal, ranging from 82.7 to 123.4 μg%, with a mean of 105 μg/100 mℓ. Only one patient had increased serum copper levels (203.6 μg%) in the pretreatment period which decreased (164.6 μg%) following treatment and remission induction. In another study, 11 of 12 IgG myeloma patients studied showed a mean serum copper level of 137 μg/100 mℓ.[13] An exception to the above findings was noted in one report in which three of four multiple

Table 4
SERUM COPPER LEVELS IN PATIENTS WITH SOLID TUMORS, BEFORE AND AFTER THERAPY

Authors	Type of cancer	No. of patients	Mean SCL (μg/100 mℓ) pre-Rx/post-Rx	Ref.
Fisher et al.	Osteosarcoma	11	180/decreased	58
	Localized	5	162/normal	
	Generalized	6	195/ ed	
Breiter et al.,	Osteosarcoma	18	173/no change	59,
Fisher et al.,	Melanoma	37	164/107	60,
O'Leary, and				79
Feldman	Cervix			
	Stage I		162/122	
	Stage II		184/132	
	Stage III		202/122	
	Stage IV		219/	
	Ovary		162/129	
	Endometrium		171/129	
	Vagina		178/145	
	Vulva		172/140	
Pizzolo et al.,	Breast	48	186/132	16
DeJorge et al.	Breast	23	229/	69
Scanni et al.	Lung	20	188/	68
Kolaric et al.	Lung	34	197/normal	61
Scanni et al.	Stomach	35	195/146	77
	Colon	22	197/138	
Graff	Colon	43		76
	Stage I		154	
	Stage III		184	
Albert et al.	Bladder	56		87
	Stage I		119/	
	Stage II		144/	
	Stage III		175/	
	Stage IV		217/	

myeloma patients studied had serum copper levels ranging from 159 to 200 μg/100 mℓ.[28] In another study done on 13 patients with myeloma, serum copper levels were found to be 167 ± 79 μg/100 mℓ.[16]

Copper metabolism has also been studied in a variety of solid tumors. In some studies patients with all types of cancer have been studied as a group but this, from available evidence, would appear to be too much of generalization. Many investigators studying patients with various types of cancer have reported elevation in either serum copper or ceruloplasmin concentration. Increased copper or ceruloplasmin concentrations have been reported in osteogenic sarcoma,[58,59] melanoma,[60] carcinoma of the lung,[5,61-68] and malignancies of head and neck,[69-71] breast,[16,61,72-74] gastrointestinal tract,[60,75-77] reproductive organs,[72-82] prostate,[83-86] bladder,[87] and central nervous system.[88] Several reports of data concerning serum copper levels in pre- and posttreatment phases are shown in Table 4. As can be seen, elevated serum copper levels in pretreatment periods decreased to the normal range following effective therapy. In some types of cancer studied, a good correlation with the extent (stage) of disease was found. Conflicting reports concerning serum copper levels and their significance in following solid tumor disease activity have been published. For example, the degree of elevation of serum copper levels in osteogenic sarcoma was reported to correlate with the extent and activity of this disease.[58] Other investigators confirm increased serum copper levels in osteogenic sarcoma but did not find serum copper and zinc levels to correlate with extent of disease or response to therapy.[59] Several workers have reported increased serum copper and ceruloplasmin levels in patients with bronchial carcinoma,[5,61-68] others have confirmed this, but also reported some increase with conditions such as pulmonary infection or chronic obstructive lung disease.[66] Regression of lung malignancy as a result of therapy was associated with normalization of serum copper or ceruloplasmin levels.[5,61,62] With regard to diagnosis and follow-up of carcinoma of the lung, it has been suggested that periodic checks of serum ceruloplasmin oxidase activity be included among the tests used to monitor the health of smokers and perhaps other groups at risk of developing or redeveloping lung cancer.[5] At this time, the authors do not believe that measurement of this metaloenzyme has a place as an overall "test for cancer or lymphoma" screening. Activity of breast carcinoma disease has been reported to be reflected in serum copper levels,[16,89] whereas other investigators found no correlation.[73] Several investigators found a good correlation between serum copper levels and the activity and extent of disease in patients with gastrointestinal malignancy,[60,75-77] while others could not confirm significant changes.[61] Data in the literature concerning serum copper level and ceruloplasmin concentrations in other solid tumors,[78-87] with the possible exception of reproductive organ malignant diseases,[78-82] are also controversial and sometimes conflicting, and they are summarized in Chapter 6.

The subject of tissue copper content in malignant diseases has also been controversial. There is much to be done to properly investigate this issue. Uniform tissue classification and analytical techniques for each type of malignant disease, and comparing copper content of benign and malignant tissues obtained at laparotomy are indicated for better understanding of the role of trace elements in cancer.

Based on the literature review, one can conclude that only in certain types of solid tumors and in clearly defined situations, serial serum copper and/or ceruloplasmin determinations to identify trends may be helpful in following disease activity, efficacy of therapy, and prognosis. An encouraging report concerning the clinical usefulness of ceruloplasmin measurements in monitoring response and relapse in advanced solid tumor malignancies has appeared recently.[89]

The authors have obtained serum copper levels twice weekly in their control group, consisting of 45 patients with carcinoma of the prostate undergoing irradiation. No significant change in serum copper levels was noted before, during, or after irradiation in this particular group of solid tumor patients, with the exception of one patient on estrogen therapy.

It is apparent that certain trends emerge from studying copper metabolism in patients with malignant diseases, although the results are not conclusive in many forms of neoplasia. It appears that the serum copper level as a significant parameter is best established in patients with malignant lymphoreticular diseases.

Experimental animal data confirmed clinical observations that changing serum copper levels reflected, but preceded, the development and regression of tumor.

An interesting aspect of copper-neoplasia relationship is the possible role of this trace element in cancer chemotherapy. Several recent reports reveal that certain chemotherapeutic agents, such as bleomycin and metronidazole, have been shown to bind strongly to copper.[90,91] Forming a complex with copper may reduce the antitumor properties of bleomycin, since the copper-bound bleomycin is known to be inactive against DNA.[92] Based on the observations that the serum copper level in some patients with malignant disease is high and the copper chelating agents have been shown to inhibit tumor growth,[93] it was thought that bleomycin combined with D-penicillamine might enhance the potency of the former agent.[94,95] The chelating effect alone, or incombination with bleomycin, resulted in decrease of total serum and ceruloplasmin level without improvement in the patient's condition.[94-96] The exact mechanism of serum copper and/or ceruloplasmin elevation in the tumor-bearing host is not known.

The general behavior pattern of ceruloplasmin levels has led various workers to suggest that it acts as an acute phase reactant.[97,98] Various stimuli such as cellular injury, inflammation, tumor activity, and even pregnancy evoke production of a heterogeneous group of proteins, the so-called acute phase reactants, functionally integrated with the effectors of the immunologic and inflammatory responses. They are involved in a spectrum of biologic activities, some of them not yet clearly defined. Biomedical research attention to acute phase reactant proteins has waxed, waned, and recently waxed again. In the past, serum copper and ceruloplasmin have sometimes been written off as indicators for monitoring lymphoreticular malignant diseases on the basis that the metaloenzyme is an ''acute phase reactant'' that responds to many physiologic signals, some originating in disease states and some not. Several groups of observers have believed differently for a variety of reasons.[5,57] First, the degree of correlation found between serum copper and acute phase reactants (alpha-2-globulins, C-reactive protein, haptoglobin, and fibrinogen) was limited. Second, except in certain well-defined conditions such as estrogen administration, severe inflammatory conditions, obstructive hepatobiliary disease, and myocardial infarction the degree of increase in copper-ceruloplasmin concentrations is much less than that occurring in certain types of malignant diseases. Although the effects of some nonmalignant conditions known to disturb copper metabolism cannot be dismissed and must be considered (if this metaloenzyme is used for monitoring neoplastic disease), they do not prevent the limited use of copper/ceruloplasmin assays in the following of disease activity in certain types of malignancies. An overview of literature suggested that ceruloplasmin levels are rather constant, except under specific unusual conditions, mentioned above.[5] Furthermore, in some disease states in which there is physiological stress, if ceruloplasmin was acting as an acute phase reactant only, one would expect an increase, which is not found.[57]

Ceruloplasmin does indeed seem to be a multifunctional protein, as reported by many investigators. There is evidence to suggest that the whole protein is absorbed by cells, especially cells rich in respiratory enzymes.[99] The various functions of ceruloplasmin may explain its elevation in the last trimester of pregnancy when large amounts of iron and copper are transferred to the fetus.[100] The exact mechanism and process by which estrogen therapy induces copper and ceruloplasmin elevation is not clear. It is possible that the increase of ceruloplasmin in neoplastic disease is due to the dependence of tumor cell proliferation on the synthesis of respiratory enzymes and the availability of iron.[101] The increase in serum copper concentrations found in the malignant diseases studied has special significance since it has recently been recognized that many, if not all, tumor cells have decreased copper-dependent superoxide dismutase enzymatic activity.[102] The elevation of serum copper/ceruloplasmin following release from the liver stores may induce superoxide dismutase activity or other enzymatic activity in cancer cells and play a role in the facilitation of remission and the subsequent return to normal copper levels. As a result of the observed decreased superoxide dismutase activity in tumor cells and antitumor activity of copper complexes in rodents, resulting in the inhibition of DNA synthesis and tumor growth, it was suggested that some copper-dependent process is required for remission or anticancer response.[102] Copper complexes have been shown to be effective against sarcoma 180 cells and Ehrlich carcinoma cells, increasing survival time of the tumor-bearing mice and inhibiting tumor growth.[103,104]

Changes in ceruloplasmin in neoplastic states must also be viewed in the context of overall changes which occur in copper metabolism. It was found that absorption of inorganic copper from the intestinal tract is increased 60 to 100% in rats with transplantable tumors.[105] This phenomenon is believed to be due to a changed capacity of the intestine, as shown in vitro, and is evident when tumors are still very small. In addition, at least in the rat, tumors contain concentrations of copper in the range of 3 to 4 μg/g,[5] a concentration second only to that found in the liver.[105]

A direct parallelism between high dietary copper intake and increased incidence of certain carcinomas and leukemias has been reported.[106-108] There was also a significant correlation between dietary copper intake and mortality rates in the same group. Since the high copper intake appears to induce selenium deficiency it has been suggested that copper, like zinc, interferes with the anticarcinogenic properties of selenium.[109] On the other hand, some investigations have observed that increasing the amount of copper in diet of rats fed 4-dimethylaminoazobenzene lengthened hepatic tumor induction time.[110,111]

The basic questions requiring answers are first, what is the basic control mechanism that produces a consistent serum copper and/or ceruloplasmin alteration reflecting the activity of lymphoma disease, and second, do the observed changes in this metaloenzyme have a clearly defined, purposeful role as a host defensive mechanism or alternatively, do the variations merely represent secondary pathophysiological consequences of a malignant condition?

It is believed that the acute phase proteins are synthesized mainly in the liver as a response to various types of disease and injury, incuding tumor activity. There is evidence that these proteins can be synthesized or at least localized in lymphocytes and tumors.[112-114] Some investigations suggest that serum protein levels, or peptides (especially those related to the alphaglobulin fraction of serum) can alter cellular immunity.[115,116]

Other investigators reviewed additional evidence implicating normal serum proteins in immunosuppression and proposed that glycoproteins synthesized in the liver have immunosuppressive properties believed to be linked with the antigen-screening function of the liver.[117,125] Elevations in one or more of the glycoprotein-associated carbohydrates in the sera of cancer patients have been reported by a number of investigators.[118-121] For example, increases in the protein-bound carbohydrate fucose have been shown in patients with breast cancer,[122,123] gynecologic malignancies,[120] and cancer patients in general.[116,124] There are glycoproteins in serum that are found in very low concentrations and that also increase during malignancy, including hormone-related substances such as human chorionic gonadotropin.[124] The speculations concerning the possible source of the increased protein-bound carbohydrates occurring in cancer patients include increased carbohydrate content of normal proteins, glycoprotein production by the tumor itself, or increased synthesis of glycoprotein by the liver and/or lymphoreticular tissue.[120] Some investigators have suggested that increased levels of chemically measured carbohydrate in the serum of rats with tumors is elaborated by the tumor.[126,127] Studies using labeled glycoprotein precursors indicate that although the liver is the major source of synthesized serum protein, transplantable tumors in hepatectomized rats can also synthesize serum glycoproteins.[128,129] The elevation in alpha$_1$-antitrypsin, haptoglobin, alpha$_1$-acid glycoprotein, ceruloplasmin, and hemopexin have been reported in patients with various types of cancer.[118,120,121] Based on these observations, it was concluded that most of the increased protein-bound carbohydrate is from liver-produced normal serum proteins, especially the acute-phase proteins of the alpha globulin fraction of the serum.[120]

Recently, improvement in clinical status of some patients after plasmapheresis has been reported and it was suggested that the removal of blocking activity[130,131] possibly related to elevated levels of circulating acute-phase proteins might be responsible for the responses observed. On the other hand, human ceruloplasmin was found to have a neutralizing effect against the toxohormone activities of the basic protein isolated from Ehrlich carcinoma cells. It was also shown that ceruloplasmin has antitumor activity against sarcome-180 cells implanted in ICR mice.[132] A recent speculation suggested that some increased ceruloplasmin levels may be associated with a change in catabolism. Ceruloplasmin contains about nine sialic acid residues per molecule and these may be involved in the catabolism of the protein, being cleaved by neuraminidase to form asialoceruloplasmin, which can be removed from the circulation by the liver. Elevated serum sialic acid concentrations have been reported in Hodgkin's disease and other malignant conditions, and malignant cells have higher sialyl transferase activity than normal cells. This led some investigators to suggest that in malignant

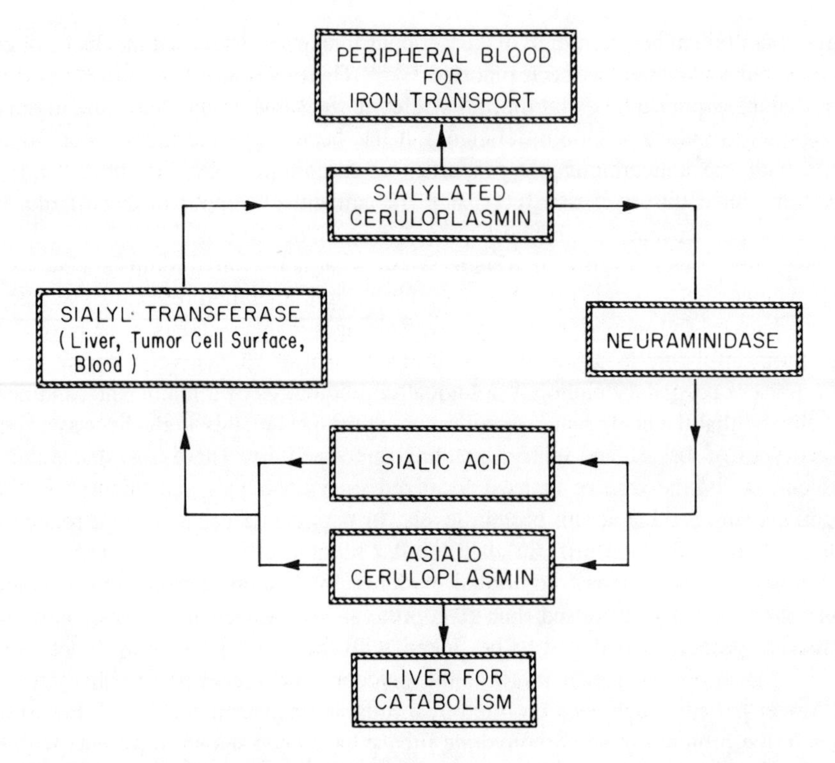

FIGURE 1. Proposed possible mechanism of decreased catabolism of ceruloplasmin. (From Fisher, G. L. and Shifrine, M., *Oncology,* 35; 22, 1979. With permission of S. Karger AG, Basel, Switzerland.)

conditions either the desialylation of ceruloplasmin is less effective or that it is resialylated by the malignant cells so that catabolism of ceruloplasmin is decreased[133] (Figure 1). As yet no experimental evidence (e.g., reduced ceruloplasmin turnover in patients with malignant diseases) have been offered to support this possible mechanism.

According to another hypothesis thymic hormone functions as a catalyst in combining copper and an organic component to form the enzyme found in normal functioning lymphocytes. If the thymic hormone is lacking, the combining does not occur and the serum copper level is elevated, as is seen in the acute phase of Hodgkin's disease and malignant lymphomas. Excessive serum copper found in thymic dysfunction combines with an organic component in epithelial cells if antigens are present. The result is an abnormal metaplasia. However, the new cell formed is not capable of destroying the antigen, which normally would have been destroyed by the lymphocytes. The copper and organic component found in the epithelial cell interfere with normal mitosis, resulting in carcinoma.[134] Other hypotheses have been offered but none has, as yet, been substantiated.

Until a great deal more is known about copper metabolism and the functions of ceruloplasmin, the reasons for many of their pathophysiologic changes will remain obscure.

CONCLUSION

Based on the literature review and clinical and experimental animal data, the serum copper level appears to represent a reasonably sensitive although nonspecific test for monitoring the activity of Hodgkin's disease (nodal or extranodal) and the effectiveness of therapy.

Serum copper levels are also helpful in following disease activity and response to therapy in non-Hodgkin's lymphoma. The general activity of lymphoreticular malignant diseases, is reflected in serum copper levels — rising levels indicate increasing activity of disease, declining levels indicate decreasing activity. Elevation of the serum copper level in active disease decreases as disease activity is reduced in response to therapy. Normal serum copper levels exist during remission and increased serum copper levels in relapse, often preceding clinical signs of increased or decreased disease activity. Until proven otherwise, sequential increases in a patient's serum copper level can be considered an indication of exacerbation of disease activity and the need for further investigations.

Higher serum copper levels are found in more advanced stages of lymphomatous disease, in contrast to localized disease, and in patients with systemic symptoms. Also, more aggressive and rapidly progressing types of lymphomatous disease are associated with higher serum copper levels.

Serum copper levels reflect bone marrow activity and response to treatment in acute leukemic individuals, particularly in the first treatment period. These relationships seems to be reduced in later stages and are generally lost in the terminal phase, and therefore, are of doubtful value in daily oncological practice.

In an examination of the relationship between the serum copper levels and other clinical laboratory determinations (SMA 12/60, and serum protein electrophoresis) studied in a large number of patients with neoplastic diseases, a limited degree of correlation was found, suggesting that serum copper is a substantially independent test providing reasonably specific information for clinical use.

In order to clarify a true relation of copper/ceruloplasmin to the malignant disease process, not apparent as yet, a better understanding of copper metabolism in normal and disease states is needed.[135]

It seems probable that a more specific link exists between cuproproteins and certain of the neoplastic diseases than has been demonstrated heretofore. Although the specific biologic significance of an increase in the serum copper or ceruloplasmin levels is still obscure, their variations in Hodgkin's disease and non-Hodgkin's lymphoma are of value for following disease activity and response to therapy.

Therefore, despite the possible influences of interfering substances and conditions and the nonspecificity of the test, measurement of serum copper level is still one of the most sensitive assays available for monitoring the clinical course of lymphomatous disease activity. With proper understanding of the factors controlling interpretation, serum copper patterns can increase the physician's perception of the clinical events — neoplastic and nonneoplastic — taking place.

REFERENCES

1. **Pineda, E. P., Ravin, A. H., and Rutenburg, A. M.,** Serum ceruloplasmin: observations in patients with cancer, obstructive jaundice and other diseases, *Gastroenterology,* 43, 266, 1962.
2. **Warren, L. R., Jelliffee, A. M.,Watson, J. V., and Hobbs, C. B.,** Prolonged observations of variations in the serum copper in Hodgkin's disease, *Clin. Radiol.,* 20, 247, 1969.
3. **Thorling, E. B. and Thorling, K.,** The clinical usefulness of serum copper determinations in Hodgkin's disease, *Cancer,* 38, 255, 1976.
4. **Ray, G. R., Wolf, P. H., and Kaplan, H. S.,** Value of laboratory indicators in Hodgkin's disease: preliminary results, *Natl. Cancer Inst. Monogr.,* 36, 315, 1973.
5. **Linder, M. C., Moor, J. R., and Wright, K.,** Ceruloplasmin assays in diagnosis and treatment of human lung, breast and gastrointestinal cancers, *J. Natl. Cancer Inst.,* 67, 263, 1981.
6. **Shifrine, M. and Fisher, G. L.,** Ceruloplasmin levels in sera from human patients with osteosarcoma, *Cancer,* 38, 244, 1976.

7. **Davidoff, G. N., Wotaw, M. L., Coon, W. W., Hultquist, D. E., Filter, B. J., and Wexler, S. A.,** Elevations in serum copper, erythrocyte copper and ceruloplasmin concentrations in smokers, *Am. J. Clin. Pathol.*, 70, 790, 1978.
8. **Tura, S., Bernardi, L., Baccarani, M., Sanguinetti, F., and Branzi, A.,** La cupremia e la ceruloplasminemia nelle emolinfopatie. Nota I. I linfomi Maligni, *G. Clin. Med.*, 49, 1090, 1968.
9. **Hobbs, C. B., Warren, R. L., and Jeliffe, A. M.,** The flame spectrometric measurement of serum copper and its use in the assessment of patients with malignant lymphoma, *Proc. Assoc. Clin. Biochem.*, 197, 1967.
10. **Sirsat, A. V.,** Serum ceruloplasmin levels in Hodgkin's disease and malignant lymphomas in correlation with serum copper levels, *Indian J. Cancer*, 16, 32, 1979.
11. **Thorling, E. B., Jensen, B. K., Andersen, C. J., and Jensen, Y.,** Cobre serico y ceruloplasmina en la enfermedad de Hodgkin, *Folia Clin. Int. (Barcelona)*, 15, 2, 1965.
12. **Goodman, S. J., Rodgerson, D. O., and Kauffman, J.,** Hypercupremia in a patient with multiple myeloma, *J. Lab. Clin. Med.*, 70, 57, 1967.
13. **Lewis, R. A., Hultquist, D. E., Baker, B. L., Falks, H. F., Gershowitz, H., and Penner, J. A.,** Hypercupremia associated with a monoclonal immunoglobulin, *J. Lab. Clin. Med.*, 88, 375, 1976.
14. **Hrgovcic, M., Tessmer, C. F., Thomas, F. B., Fuller, L. M., Gamble, J. F., and Shullenberger, C. C.,** Significance of serum copper levels in adult patients with Hodgkin's disease, *Cancer*, 31, 1337, 1973.
15. **Cappelaere, P., Sulman, Ch., Chechan, Ch., and Gosselin-Delaquais, P.,** Les concentrations plasmatiques du cuivre et du zinc au cours de la maladie de Hodgkin, *Lille Med.*, 20, 904, 1975.
16. **Pizzolo, G., Savarin, T., Molino, A. M., Ambroseti, A., Todeschini, G., and Vettore, L.,** The diagnostic value of serum copper levels and other hematochemical parameters in malignancies, *Tumori*, 641, 55, 1978.
17. **Gobbi, P. G., Scampellini, Minoia, C., Pozzoli, L., and Ferugini, S.,** Plasma iron and copper in Hodgkin's disease a comparison with other laboratory indicators, *Haematologica (Pavia)*, 64(4), 416, 1979.
18. **Williams, J., Thompson, E., and Smith, K. L.,** Value of serum copper levels and erythrocyte sedimentation rates as indicators of disease activity in children with Hodgkin's disease, *Cancer*, 42, 1929, 1978.
19. **Mita, S. K. and Tan, C.,** Serum copper levels in children with Hodgkin's disease, in Proc. AACR and ASCO, Waverly Press, Baltimore, 1979, C-377(abstract).
20. **Tessmer, C. F., Hrgovcic, M., and Wilbur, J.,** Serum copper in Hodgin's disease in children, *Cancer*, 31, 303, 1973.
21. **Masi, M., Vecchi, V., Vivarelli, F., and Paolucci, P.,** La ceruloplasminemia nella leucosi acuta linfoblastica e nel morbo di Hodgkin della infanzia, *Minerva Pediatr.*, 27, 1223, 1975.
22. **Sullivan, M. P., Hrgovcic, M., Wilbur, J., and Tessmer, C. F.,** Progress in the treatment of Hodgkin's disease in children. Lecture, *Forum of Original Res. (Abstr. 4)*, Texas Medical Association, Houston, 1971.
23. **Tessmer, C., Krohn, W., Johnston, D., Thomas, F., Hrgovcic, M., and Brown, B.,** Serum copper in children (6—12 years old): an age-correction factor, *Am. J. Clin. Pathol.*, 60, 870, 1973.
24. **Pisi, E., Di Feliciantonio, R., Figus, E., and Ferri, S.,** Comportamento e significato prognostico della ceruloplasmina sierica in relazione al quadron iso-patologico nella malattia di Hodgkin, *Minerva Med.* 59, 944, 1968.
25. **Pagliardi, E. and Giangrandi, E.,** Clinical significance of the blood copper in Hodgkin's disease, *Acta Haematol.*, 24, 201, 1960.
26. **Jelliffe, A. M.,** Invited discussion: value of fluctuations in the serum copper level in the control of patients with Hodgkin's disease, *Natl. Cancer Inst. Monogr.*, 36, 325, 1973.
27. **Foster, M., Fell, L., Pocklington, T., Akinsete, A. D., Dawson, A., Hutchinson, M. S., and Mallard, J. R.,** Electron spin resonance as a useful technique in the management of Hodgkin's disease, *Clin. Radiol.*, 28, 15, 1977.
28. **Mortazavi, S. H., Bani-Hashemi, A., Mozafari, M., and Raffi, E.,** Values of serum copper measurement in lymphomas and several other malignancies, *Cancer*, 29, 1193, 1972.
29. **Jensen, K. B., Thorling, E. G., and Anderson, C. F.,** Serum copper in Hodgkin's disease, *Scand. J. Haematol.*, 1, 62, 1964.
30. **Bucher, W. C. and Jones, S. E.,** Serum copper-zinc ratio in patients with malignant lymphoma (abstr.), *Am. J. Clin. Pathol.*, 68, (1), 104, 1977.
31. **Asbjörnsen, G.,** Serum copper compared to erythrocyte sedimentation rate as indicator of disease activity in Hodgkin's disease, *Scand. J. Haematol.*, 22, 193, 1979.
32. **Wolf, P. L., Ray, G., and Kaplan, H.,** Evaluation of copper oxidase (ceruloplasmin) and related tests in Hodgkin's disease, *Clin. Biochem.*, 12, 202, 1979.
33. **Rawat, M. and Vijayvargiya, R.,** Serum copper estimation and lymphomas, *Indiana J. Med. Res.*, 66, 815, 1977.
34. **Kessava Rao, K. V., Shetty, P. A., Bapat, C. V., and Jussawalla, D. T.,** Serum copper assay as a biochemical marker to assess the response to therapy in Hodgkin's disease, *Indian J. Cancer*, 14(4), 320, 1977.

35. **Cravario, A., Brusa, L., De Filippi, P. G., and Giangrandi E.,** La frazioni sieriche del rame nel decorso del morbo di Hodgkin, *Cancro,* 19, 399, 1966.
36. **Sutow, W. W., Vietti, T. J., and Feunbach, D. J.,** *Clinical Pediatric Oncology,* C. V. Mosby, St. Louis, 1973, 277.
37. **Holland, G. F. and Frei, E.,** *Cancer Medicine,* Lea & Febiger, Philadelphia, 1973, 338.
38. **Kaplan, H. S.,** *Hodgkin's Disease,* Harvard University Press, Cambridge, Mass., 1980, 131.
39. **Illicin, G.,** Serum copper and magnesium levels in leukemia and malignant lymphoma, *Lancet,* 1, 1036, 1971.
40. **Foster, M., Dawson, A., Pocklington, T., and Fell, L.,** Electron spin resonance measurements of blood ceruloplasmin and iron transferrin levels in patients with non-Hodgkin's lymphoma, *Clin. Radiol.,* 28, 23, 1977.
41. **Subrahmaniyam, K., Rao, B. N., Shukla, P. K., Rastogi, B. L., Roy, S. K., Sanyal, B., Pant, G. C., and Khanna, N. N.,** Serum copper in patients in non-Hodgkin's lymphomas — A preliminary appraisal of its significance in management, *Indian J. Radiol.,* 33(2), 101, 1979.
42. **De Bellis, R., Boulard, M. R., Kasdorf, H., Rodgriguez, I., Ferrando, R., Di Landro, J., Ferrari, M., Sanguinett, C. M., and Tanzer, J.,** Metabolic changes in red blood cells in malignant lymphomas, *Br. J. Haematol.,* 42, 35, 1979.
43. **Shah-Reddy, I., Khilanani, P., and Bishop, R. C.,** Serum copper levels in non-Hodgkin's lymphoma, *Cancer,* 45, 215, 1980.
44. **Cohen, Y., Haim, N., and Zinder, O.,** Serum copper levels in non-Hodgkin's lymphoma (abstr.), *Proc. Am. Assoc. Cancer Res.,* 22, 184, 1981.
45. **Hrgovcic, M., Tessmer, C. F., Thomas, F. B., Ong, P. S., Gamble, J. F., and Shullenberger, C. C.,** Serum copper observations in patients with malignant lymphoma, *Cancer,* 32, 1512, 1973.
46. **Bhardwaj, D. N., Chander, J., Singh, R. P., Thapar, S. P., and Kaur, K.,** Clinical evaluation of malignant process by serum copper estimation, *J. Indian Med. Assoc.,* 75, 119, 1980.
47. **Cherry, N. H., Kalas, J. P., and Zarafonetis, C. J. D.,** Study of plasma copper levels in patients with acute leukemia, *J. Einstein Med. Cent.,* 9, 24, 1961.
48. **Delves, H. T., Alexander, F. W., and Lay, H.,** Copper and zinc concentrations in the plasma of leukemic children, *Br. J. Haematol.,* 24, 525, 1973.
49. **El-Haddad, S., Mahfouz, M., Magahed, Y., Mahmoud, F., Kamel, M., and Ali, M. A.,** Value of serum copper measurement in acute leukaemia of childhood, *Gaz. Egypt. Paediatr. Assoc.,* 26, 67, 1977.
50. **Legutko, L.,** Serum copper investigations in children with acute lymphoblastic leukemia, *Folia Hematol. (Leipz),* 105, 248, 1978.
51. **Hrgovcic, M., Tessmer, C. F., Minckler, T. M., Mosier, B., and Taylor, G. H.,** Serum copper levels in lymphoma and leukemia, *Cancer,* 21, 43, 1968.
52. **Tessmer, C. F., Hrgovcic, M., Brown, G. W., Wilbur, J., and Thomas, F. B.,** Serum copper correlations with bone marrow, *Cancer,* 29, 173, 1972.
53. **Tessmer, C. F., Hrgovcic, M., Thomas, F. B., Wilbur, J., and Mumford, D. M.,** Long-term serum copper studies in acute leukemia in children, *Cancer,* 30, 358, 1972.
54. **Andronikashvili, E. L. and Mosuhishvili, L. M.,** Human leukemia and trace elements, in *Metal Ions in Biological Systems,* Vol. 10, Sigel, H., Ed., Marcel Dekker, New York, 1980, chap. 6.
55. **Rechenberger, J.,** Serumweisen und Serumkupfer bei akuten and chronischen Leukamien sowie bei Morbus Hodgkin, *Dtsch. Z. Verdau. Stoffwechselkr.,* 17, 79, 1957.
56. **Green, J. A., Pocklington, T., Dawson, A. A., and Foster, M.,** Electron spin resonance studies on caeruloplasmin and iron transferrin in patients with chronic lymphocytic leukaemia, *Br. J. Cancer,* 41(3), 356, 1980.
57. **Foster, M., Pocklington, T., and Dawson, A. A.,** Ceruloplasmin and iron transferrin in human malignant disease, in *Metal Ions in Biological Systems,* Vol. 10, Sigel, H., Ed., Marcel Dekker, New York, 1981, chap. 5.
58. **Fisher, G. L., Byers, V. S., Shifrine, M., and Levin, A. S.,** Copper and zinc levels in serum from human patients with sarcomas, *Cancer,* 37, 356, 1976.
59. **Breiter, D. N., Diasio, R. B., Neifeld, J. P., Roush, M. L., and Rosenberg, S. A.,** Serum copper and zinc measurements in patients with osteogenic sarcoma, *Cancer,* 42, 598, 1978.
60. **Fisher, G. L., Spitler, L. E., McNeill, B. S., and Rosenblatt, L. S.,** Serum copper and zinc levels in melanoma patients, *Cancer,* 47, 1838, 1981.
61. **Kolaric, K., Roguljic, A., and Fuss, V.,** Serum copper levels in patients with solid tumors, *Tumori,* 61, 173, 1975.
62. **Huhti, E., Pokkula, A., and Uksila, E.,** Serum copper levels in patients with lung cancer, *Respiration,* 40, 112, 1980.
63. **Bariety, M. and Gajdos, A.,** Etude de la sidérémie et de la cuprémie au cours des cancers, particulierement des cancers bronchiques, *Presse Med.,* 12, 3259, 1964.

64. **Tani, P. and Kokkola, K.,** Serum iron, copper and iron-binding capacity in bronchogenic pulmonary carcinoma, *Scand. J. Respir. Dis.,* Suppl. 80, 121, 1972.
65. **Andrews, G. S.,** Studies on plasma zinc, copper, ceruloplasmin and growth hormone, *J. Clin. Pathol.,* 32, 325, 1979.
66. **Wiljasalo, M. A. and Haikonen, M.,** Blood copper in pulmonary cancer and in chronic pulmonary infections, *Ann. Med. Intern. Fenn.,* 55, 107, 1966.
67. **Martin Mateo, M. C., Bustamante, J., and Arellano, I. F.** Serum copper, ceruloplasmin, lactic dehydrogenase and α_2 globulin in lung cancer, *Biomedicine,* 31, 66, 1979.
68. **Scanni, A., Licciardello, L., Trovato, M., Tomirotti, M., and Biraghi, M.,** Serum copper and ceruloplasmin levels in patients with neoplasias localized in the stomach, large intestine or lung, *Tumori,* 63(2), 175, 1977.
69. **DeJorge, F. B., Paiva, L., Mion, D., and Da Nova, R.,** Biochemical studies on copper, copper oxidase, magnesium, sulfur, calcium, phosphorus in cancer of the larynx, *Acta Otolaryngol.,* 61, 454, 1966.
70. **Abdulla, M., Biorklund, A., Mathur, A., and Walleniusm, K.,** Zinc and copper levels in whole blood and plasma from patients with squamous cell carcinomas of head and neck, *J. Surg. Oncol.,* 12, 107, 1979.
71. **Garofalo, J. A., Erlandson, E., Strong, E. W., Lesser, M., Garold, F., Spiro, R., Schwartz, M., and Good, R. A.,** Serum zinc, serum copper, and the Cu/Zn ratio in patients with epidermoid cancers of the head and neck, *J. Surg. Oncol.,* 15, 381, 1980.
72. **De Jorge, F. B., Sampaio Goes, Jr., Greedes, J. L., and De Ulhoa Cintra, A. B.,** Biochemical studies on copper, copper oxidase, magnesium, phosphorus in cancer of the breast, *Clin. Chim. Acta,* 12, 403, 1965.
73. **Garofalo, J. A., Ashikari, H., Lesser, M. L., Menendez-Bodet, C., Cunningham-Rundles, S., Schwartz, M. K., and Good, R. A.,** Serum zinc, copper and the Cu/Zn ratio in patients with benign and malignant breast lesions, *Cancer,* 46, 2682, 1980.
74. **LeBorgne de Kaouel, C., Aubert, C., and Juret, P.,** Etude de la cuprémie et de la sidérémie chez les femmes atteintes d'affections mammaires benignes ou malignes, *Pathol. Biol.,* 16, 85, 1968.
75. **Inutsuka, S. and Araki, S.,** Plasma copper and zinc levels in patients with malignant tumors of digestive organs, *Cancer,* 42, 626, 1978.
76. **Graf, H.,** Ceruloplasmin bei Killumkarzinomen verschiedener Stadien, *Med. Welt.,* 19, 1059, 1965.
77. **Scanni, A., Tomirotti, M., Licciardello, L., Annibali, E., Biraghi, M., Trovato, M., Fittipaldi, M., Adamoli, P., and Curtarelli, G.,** Variations in serum copper and ceruloplasmin levels in advanced gastrointestinal cancer treated with polychemotherapy, *Tumori,* 65, 331, 1979.
78. **Zamello, J.,** Copper and iron levels in patients with cancer of the reproductive organs, *Chem. Abstr.,* 65, 1165b, 1966.
79. **O'Leary, J. A. and Feldman, M.,** Serum copper alterations in genital cancer, *Surg. Forum,* 21, 411, 1970.
80. **Rummel, A. C.,** Evaluation of iron and copper levels in normal and sick women. A contribution to the significance of heavy metals in abdominal cancer of women, *Medizinische,* 22, 1062—1067, 1959.
81. **Piskazeck, K., Billek, K., and Rothe, K.,** Copper in carcinoma of female genitalia in relation to the localization and the form of therapy, *Arch. Gynaekol.,* 196, 447, 1962.
82. **Vescavo, R. and Lorenzoni, L.,** Serum copper in patients with uterine new growths, with special reference to cervical localization, *Excerpta Med.,* 18, 361, 1965.
83. **Wiederaniders, R. E. and Evans, G. W.,** The copper concentration of hyperplastic and cancerous prostates, *Invest. Urol.,* 6, 531, 1969.
84. **Belanger, L.,** Evaluation du taux de céruloplasmine serique au cours des maladies de la prestata, *Union Med. Can.,* 100, 1554, 1971.
85. **Jafa, A., Mahendra, A. R., Chowdhury, A. R., and Kamboj, V. P.,** Trace elements in prostatic tissue and plasma in prostatic disease of man, *Indian J. Cancer,* 17, 34, 1980.
86. **Habib, F. K., Dembinski, T. C., and Stich, S. R.,** The zinc and copper of blood leukocytes and plasma from patients with benign and malignant prostases, *Clin. Chim. Acta,* 104, 329, 1980.
87. **Albert, L., Hienzsch, E., Arndt, J., and Kriester, A.,** Bedeutung und Veranderungen des Serum-Kupferspiegels während und nach der Bestrahlung von Harnblasenkarzinomen, *J. Urol.,* 8, 561, 1972.
88. **Manami, M., Retamal, C., Galvez, S., and Cordero, I.,** Normal values of ceruloplasmin concentration in the serum and in cerebrospinal fluid. Variations in some neurological diseases and in neoplastic affections, *Neurocirugia,* 34(1/2), 5, 1976.
89. **Shapira, D. V. and Shapira, M.,** Ceruloplasmin levels to monitor response and relapse on adjuvant therapy, Proc. ASCO, 1982 (C_2).
90. **Nunn, A. D.,** Interactions between Bleomycins and metals, *Jpn. J. Antibiot.,* 29, 1102, 1976.
91. **Chien, Y. W., Lambert, H. W., and Sandvordeker, D. R.,** Interaction of Metronidazole with metallic ions of biological importance, *J. Pharm. Sci.,* 64, 957, 1975.

92. **Shirakawa, I., Azegami, M., Ishi, S., and Umezava, H.,** Reaction of Bleomycin with DNA, Strand scission of DNA in the absence of sulfhydryl or peroxide compounds, *J. Antibiot. (Tokyo),* 24, 761, 1971.
93. **Takamiya, K.,** Anti-tumour activities of copper chelates, *Nature (London),* 185, 190, 1960.
94. **Light, A. P., Preece, W. A., and Evans, A. P.,** Effects of 3, 3-dimethylcysteine (D-pencillamine) on serum copper in subjects with malignant disease, *Biochem. Soc. Trans.,* 5, 1460, 1977.
95. **Preece, A. W., Light, P. A., Evans, P. A., and Nunn, A. D.,** Serum-copper, penicillamine and cytoxic therapy, *Lancet,* 1, 953, 1977.
96. **Pagliardi, E., Cravario, A., Brusa, L., Cantino, D., Giangrandi, E.** Indagine sul metabolismo del rame nel morbo di Hodgkin, *Hematologica,* 48, 209, 1963.
97. **Rice, W. W.,** Correlation between serum copper, ceruloplasmine activity and C-reactive protein, *Clin. Chim. Acta,* 5, 652, 1960.
98. **Beisel, W. R.,** Trace elements in infectious processes, *Med. Clin. North Am.,* 60, 831, 1976.
99. **Linder, M. C. and Moor, J. R.,** Plasma ceruloplasmin: Evidence for its presence in and uptake by heart and other organs of the rat, *Biochim. BIophys. Acta,* 499, 329, 1977.
100. **Linder, M. C. and Munro, H. N.,** Iron and copper metabolism during development, *Enzyme,* 15, 111, 1973.
101. **Robbins, E. and Pederson, T.** Iron: its intracellular localization and possible role in cell division, *Proc. Natl. Acad. Sci. U.S.A.,* 66, 1244, 1970.
102. **Oberley, L. W. and Buettner, G. R.,** Role of superoxide dismutase in cancer patient: a review, *Cancer Res.,* 39, 1141, 1979.
103. **Sorenson, J. R. J.,** The anti-inflammatory activities of copper complexes, in *Metal Ions in Biological Systems,* Vol. 14, Sigel, H., Ed., Marcel Dekker, New York, 1982, chap. 4.
104. **Leuthauser, S. W. C., Oberley, L. W., Oberley, T. D., and Sorenson, J. R. J.,** Antitumor effect of a copper coordination compound with superoxide dismutase-like activity, *Natl. Cancer Inst.,* 66, 1077, 1981.
105. **Cohen, D. I., Illowsky, B., and Linder, M. C.,** Altered copper absorption in tumor-bearing and estrogen treated rats, *Am. J. Physiol.,* 236, E209, 1979.
106. **Schrauzer, G. N.,** Cancer mortality correlation studies. II. Regional association of mortalities with the consumptions of food and other commodities, *Medic. Hypoth.,* 2, 39, 1976.
107. **Schrauzer, G. N., White, D. A., and Schneider, C. J.,** Inhibition of the genesis of spontaneous mammary tumors in C3H mice: effects of Selenium and Selenium-antagonistic elements and their possible role in human breast cancer, *Bioinorg. Chem.* 6, 265, 1976.
108. **Schrauzer, G. N., White, D. A., and Schneider, C. J.,** Cancer mortality correlation studies. IV. Associations with dietary intakes and blood levels of certain elements, notably Seantagonists, *Bioinorg. Chem.,* 7, 35, 1977.
109. **Jensen, K.,** Modification of a Selenium toxicity in chicks by dietary silver and copper, *J. Nutr.,* 105, 769, 1975.
110. **Sharples, G. R.,** The effects of copper on liver tumor induction by *p*-dimethyl-aminoazobenzene, *Fed. Proc.,* 5, 239, 1946.
111. **Brada, Z. and Altman, N. H.,** The inhibition effect of copper on ethionine carcinogenesis, in *Inorganic and Nutritional Aspects of Cancer,* Schrauzer, G. N., Ed., Plenum Press, New York, 1978, chap. 14.
112. **Yoshimura, S., Tamaoki, N., Veyama, Y., and Hata, J. I.,** Plasma protein production by human tumors xenotransplanted in nude mice, *Cancer Res.,* 38, 3474, 1978.
113. **Hauptman, S. P. and Kansu, E.,** T-cell origin of human macromolecular insoluble cold globulin, *Nature (London),* 276, 393, 1978.
114. **Gahmberg, C. G. and Andersson, L. C.,** Leukocyte surface origin of human a_1-acid glycoprotein (orosomucoid), *J. Exp. Med.,* 148, 507, 1978.
115. **Larsen, B., Heron, I., and Thorling, B.,** Elevated serum Cu in Hodgkin's disease and inhibitory effects of ceruloplasmin or lymphocyte response in vitro, *Eur. J. Cancer,* 16, 415, 1980.
116. **Israel, L., Edelstein, R., and Samak, R.,** In vitro depression of lymphocyte response to PHA by acute phase reactants, *Proc. Am. Assoc. Cancer Res.* 19, 10, 1978.
117. **Apffel, C. A. and Peters, J. H.,** Tumors and serum glycoproteins. The "symbodies", *Prog. Exp. Tumor Res.,* 12, 1, 1969.
118. **Snyder, S. and Ashwell, G.,** Quantitation of specific serum glycoproteins in malignancy, *Clin. Chim. Acta,* 34, 449, 1971.
119. **Child, J. A., Cooper, E. H., Illingworth, S., and Worthy, T. S.,** Biochemical markers in Hodgkin's disease and non-Hodgin's lymphoma, in *Recent Results in Cancer Research,* Vol. 64, Mathé, G., Seligman, M., and Tubiana, M., Eds., Springer-Verlag, Berlin, 1978, 180.
120. **Bradley, W. P., Blasco, A. P., Weiss, J. F., Alexander, J. C., Jr., Silverman, N. A., and Chretien, P. B.,** Correlations among serum protein-bound carbohydrates, serum glycoproteins, lymphocyte reactivity, and tumor burden in cancer patients, *Cancer,* 40, 2264, 1977.

121. **Baskies, A. M., Chretien, P. B., Weiss, J. F., Makuch, R. W., Beveridge, R. A., Catalona, W. J., and Spiegel, H. E.,** Serum glycoproteins in cancer patients: First report of correlations with in vitro and in vivo paratmeters of cellular immunity, *Cancer,* 45, 3050, 1980.
122. **Johannsen, R., Carlsson, A. B., Haupt, H., and Heide, K.,** Human blutproteine mit hemmwirkung auf die lymphozyten-transformation in vitro, *Behring Inst. Mitt.,* 54, 33, 1974.
123. **Lee, Y.-T. N.,** Quantitative change of serum protein and immunoglobulin in patients with solid cancers, *J. Surg. Oncol.,* 9, 179, 1977.
124. **Tormey, D. C., Waalkes, T. P., Ahmann, D., Gehrke, C. W., Zumwatt, R. W., Snyder, J., and Hansen, H.,** Biological markers in breast carcinoma. 1. Incidence of abnormalities of CEA, HCG, three polyamines, and three minor nucleosides, *Cancer,* 35, 1095, 1975.
125. **Sample, W. F., Gertner, H. R., Jr., and Chretien, P. B.** Inhibition of phytohemagglutinin-induced in vitro lymphocyte transformation by serum from patients with carcinoma, *J. Natl. Cancer Inst.,* 46, 1291, 1971.
126. **Watkins, E., Jr., Gray, B. N., Anderson, L. L., Baralt, O. L., Nebril, L. R., Waters, M. F., and Connery, C. K.,** Neuroaminidase-mediated augmentation of in vitro immune response of patients with solid tumors, *Int. J. Cancer,* 14, 799, 1974.
127. **Kim, U., Baumler, A., Carruthers, C., and Bielat, K.,** Immunological escape mechanism in spontaneously metastasizing mammary tumors, *Proc. Natl. Acad. Sci. U.S.A.,* 72, 1012, 1975.
128. **Bekesi, J. G., Macbeth, R. A. L., and Bice, S.,** The metabolism of plasma glycoproteins. II. Studies on the rate of incorporation of glucosamine-1-^{14}C into protein-bound hexosamine in the rat bearing Walker 256 carcinoma, *Cancer Res.,* 26, 2307, 1966.
129. **Macbeth, R. A. L., Boorman, M. G., and Gellatly, J.,** Serum glycoprotein synthesis in intact and hepatectomized Walker 256 tumor-bearing rats following glucosamine-1-^{14}C administration, *Can. J. Physiol. Pharmacol.,* 51, 437, 1973.
130. **Israel, L. and Edelstein, R.,** In vivo and in vitro studies on nonspecific blocking factors of host origin in cancer patients. Role of plasma exchange as an immunotherapeutic modality, *Isr. J. Med. Sci.,* 14, 105, 1978.
131. **Israel, L., Edelstein, R., Mannoni, P., Radot, E., and Greenspan, E. M.,** Plasmapheresis in patients with changes in serum protein. The concept of "non-specific blocking factors", *Cancer,* 40, 3146, 1977.
132. **Itoh, O., Torikai, T., Satoh, M., Okumura, O., and Osawa, T.,** Antitumor and toxohormone-neutralizing activities of human ceruloplasmin, *Gann.,* 72, 370, 1981.
133. **Fisher, G. L. and Shifrine, M.,** Hypothesis for the mechanism of elevated serum copper in cancer patients, *Oncology,* 35, 22, 1978.
134. **Bedrick, A. E.** The regression of malignancies, *Am. Lab.* 182, February 1982.
135. **Deur, C. J., Stone, M. J., and Frenkel, E. P.,** Trace metals in hematopoiesis, *Am. J. Hematol.,* 11, 309, 1981.

INDEX

A

B

C

D

E

F

K

L

M

N

O

P

Q

R

S

T

U

V

W

X

Y

Z